Drug Delivery Systems Based on Polysaccharides

Drug Delivery Systems Based on Polysaccharides

Editors

Marcel Popa
Leonard Ionut Atanase

MDPI • Basel • Beijing • Wuhan • Barcelona • Belgrade • Manchester • Tokyo • Cluj • Tianjin

Editors

Marcel Popa
"Gheorghe Asachi" Technical University
Romania

Leonard Ionut Atanase
"Apollonia" University of Iasi
Romania

Editorial Office
MDPI
St. Alban-Anlage 66
4052 Basel, Switzerland

This is a reprint of articles from the Special Issue published online in the open access journal *Molecules* (ISSN 1420-3049) (available at: https://www.mdpi.com/journal/molecules/special_issues/Drug_Delivery_Polysaccharides).

For citation purposes, cite each article independently as indicated on the article page online and as indicated below:

LastName, A.A.; LastName, B.B.; LastName, C.C. Article Title. *Journal Name* **Year**, *Volume Number*, Page Range.

ISBN 978-3-0365-1676-9 (Hbk)
ISBN 978-3-0365-1675-2 (PDF)

Contents

About the Editors

Marcel Popa graduated in Chemical Engineering in 1972 and obtained his Ph.D. degree in Macromolecular Chemistry at the Polytechnic Institute of Iasi, Romania. Dr. Popa has authored 440 scientific papers, 9 books, 29 book chapters, has presented 530 papers at scientific meetings and has 20 patents in the fields of polymer mechanochemistry, polysaccharides and biologically active polymer chemistry; he is the editor of two books. He is the recipient of the following awards: the Gheorghe Spacu Romanian Academy prize for his pioneering work in the field of bioactive polymers, the Opera Omnia prize awarded by the Romanian Education Ministry for lifelong scientific achievements, the Henri Coanda gold medal granted by the Romanian Inventor Society, the Excellence in Research award of the Gheorghe Asachi Technical University of Iasi, Romania, etc.; he is Doctor Honoris Causa of Pau et Pays de l'Adour University and of Haute Alsace University from France and a member of the Romanian Academy of Scientists.

Leonard Ionut Atanase studied macromolecular chemistry at Gheorghe Asachi Technical University in Iasi. Under the guidance of Prof. Riess he continued his postgraduate studies in France, and in 2010 received a Ph.D. in Materials Chemistry from the University of Haute Alsace, Mulhouse. Until 2014, he continued his research, at the same university in France, as a university researcher in the field of polymers and emulsions biocompatible with applications in the biomedical field. From 2014 to 2015 he worked in the field of nanoemulsions with cosmetic applications at the University of Pau, France in collaboration with Louis Vuitton Moët Hennessy (LVMH), the Dior branch. At the beginning of 2016, he returned in Romania, to the Apollonia University of Iasi and obtained the title of full professor in 2017.

Currently, he is the Dean of the Faculty of Medical Dentistry at the Apollonia University of Iasi and the coordinator of the Biomaterials Laboratory.

Preface to "Drug Delivery Systems Based on Polysaccharides"

We live in a constantly changing society in which life expectancy has continuously increased, and therefore important health issues need to be solved. The development of nanotechnology with applications in the medical field—nanomedicine—has been proven to have strong therapeutic potential, especially by combining drugs with natural polymers, polysaccharides being most commonly used in the development of sustained/controlled release systems of biologically active principles. Polymeric nanoparticles loaded with drugs can actively target various diseases, being able to penetrate cells more effectively or succeed in overcoming some physiological barriers such as the blood–brain barrier. Drug-loaded hydrogels are used to treat dermal and dental conditions, and can act as scaffolds for the development of cell cultures with applications in tissue engineering. The recent literature abounds with articles discussing aspects of obtaining new polymer drug systems and their use in various biomedical applications.

The editors of this special edition of the journal *Molecules*, entitled *Drug Delivery Systems Based on Polysaccharides* are researchers with decades of experience in this field, and they consider justified and useful these several articles which report recent results of drug delivery systems based on polysaccharides and derivatives, respectively, and their biomedical applications.

The authors of the articles are experts in the field, and the editors express their gratitude for the kindness and promptness with which they responded to the call to contribute the recently obtained results of their research to this specific edition of the journal *Molecules*.

Marcel Popa, Leonard Ionut Atanase
Editors

molecules

MDPI

Article

Immobilization and Release Studies of Triazole Derivatives from Grafted Copolymer Based on Gellan-Carrying Betaine Units

Nicolae Baranov [1,2], Stefania Racovita [3], Silvia Vasiliu [3,*], Ana Maria Macsim [3], Catalina Lionte [4], Valeriu Sunel [2], Marcel Popa [1,5,*], Jacques Desbrieres [6] and Corina Cheptea [7]

1 Department of Natural and Synthetic Polymers, Faculty of Chemical Engineering and Environmental Protection, Gheorghe Asachi Technical University of Iasi, Prof. Dr. Docent Dimitrie Mangeron Street, No. 73, 700050 Iasi, Romania; Baranov_nicolae@yahoo.com

2 Faculty of Chemistry, Al. I. Cuza University, Carol I Bvd., No. 11, 700506 Iasi, Romania; vsunel@uaic.ro

3 Department of "Mihai Dima" Functional Polymers, Petru Poni Institute of Macromolecular Chemistry, Grigore Ghica Voda Alley, No. 41A, 700487 Iasi, Romania; stefania.racovita@icmpp.ro (S.R.); macsim.ana@icmpp.ro (A.M.M.)

4 Faculty of Medicine, Gr. T. Popa University of Medicine and Pharmacy, Universitatii Street, No. 16, 700115 Iasi, Romania; clionte@yahoo.com

5 Academy of Romanian Scientists, Splaiul Independentei Street No. 54, 050085 Bucuresti, Romania

6 Institut des Sciences Analytiques et de Physico-Chimie pour l'Environnement et les Materiaux (IPREM), Pau and Pays de l'Adour University (UPPA), UMR CNRS 5254, Helioparc Pau Pyrenees, 2 av. President Angot, CEDEX 09, 64053 Pau, France; jacques.desbrieres@univ-pau.fr

7 Department of Biomedical Sciences, Faculty of Biomedical Bioengineering, Gr. T. Popa University of Medicine and Pharmacy, Kogalniceanu Street No. 9–13, 700454 Iasi, Romania; coricheptea@yahoo.com

* Correspondence: silvia.vasiliu@icmpp.ro (S.V.); marpopa2001@yahoo.fr (M.P.)

Citation: Baranov, N.; Racovita, S.; Vasiliu, S.; Macsim, A.M.; Lionte, C.; Sunel, V.; Popa, M.; Desbrieres, J.; Cheptea, C. Immobilization and Release Studies of Triazole Derivatives from Grafted Copolymer Based on Gellan-Carrying Betaine Units. *Molecules* **2021**, *26*, 3330. https://doi.org/10.3390/molecules26113330

Academic Editor: Roberta Cassano

Received: 29 April 2021
Accepted: 29 May 2021
Published: 1 June 2021

Abstract: New polymer-bioactive compound systems were obtained by immobilization of triazole derivatives onto grafted copolymers and grafted copolymers carrying betaine units based on gellan and N-vinylimidazole. For preparation of bioactive compound, two new types of heterocyclic thio-derivatives with different substituents were combined in a single molecule to increase the selectivity of the biological action. The 5-aryl-amino-1,3,4 thiadiazole and 5-mercapto-1,2,4-triazole derivatives, each containing 2-mercapto-benzoxazole nucleus, were prepared by an intramolecular cyclization of thiosemicarbazides-1,4 disubstituted in acidic and basic medium. The structures of the new bioactive compounds were confirmed by elemental and spectral analysis (FT-IR and ^{1}H-NMR). The antimicrobial activity of 1,3,4 thiadiazoles and 1,2,4 triazoles was tested on gram-positive and gram-negative bacteria. The triazole compound was chosen to be immobilized onto polymeric particles by adsorption. The Langmuir, Freundlich, and Dubinin–Radushkevich adsorption isotherm were used to describe the adsorption equilibrium. Also, the pseudo-first and pseudo-second models were used to elucidate the adsorption mechanism of triazole onto grafted copolymer based on N-vinylimidazole and gellan (PG copolymer) and grafted copolymers carrying betaine units (PGB1 copolymer). In vitro release studies have shown that the release mechanism of triazole from PG and PGB1 copolymers is characteristic of an anomalous transport mechanism.

Keywords: grafted copolymer; triazole derivative; adsorption studies

1. Introduction

The design of new bioactive compounds that can be immobilized onto a polymeric support to achieve a controlled/sustained release and to find a way to treat some diseases represents an important goal in medical and pharmaceutical research [1,2].

Depending on the radicals present on the heterocycle, the organic compounds that contain in their molecule 1,3,4-thiadiazole and 1,2,4-triazole heterocycles have gained a remarkable interest due to their different biological properties, such as: antibacterial [3–5], antifungal [6–9], tuberculostatic [10,11], analgesic [12,13], anti-inflammatory [14,15], hypnotic and sedative [16], cardioprotective [17], anticancer [18–22], anti-ulcerative [23], anticonvulsant [24], antidiabetic agents [25], allosteric modulators [26], and cathepsin B and tubulin inhibitors [27,28].

At the same time, the addition of a benzoxazole structure can be useful for the improvement of the biological activity. An argument in favor of this statement is that the compounds with benzoxazole heterocycle in their structure have practically the same biological properties like thiadiazole and triazole derivatives, such as antimicrobial, antifungal, antiviral, anti-inflammatory, antipyretic, analgesic, anticonvulsant, tuberculostatic, and antitumor agents [29–36]. Benzoxazole and triazole heterocyclic rings played an important role in the improving of antimicrobial and anticancer activity of the bioactive compounds [37].

These facts served as an argument in the synthesis of new heterocyclic compounds containing 1,3,4-thiadiazole and 1,2,4-triazole nucleus and the rest of 2-mercaptobenzoxazole, considering their mutual influence as well as the biological effect that can be generated throughout the molecule.

Although the researches on compounds with a benzoxazole structure are multiple and varied [38–41], some aspects of the synthesis and studies of these types of compounds in which both thiadiazole and triazole structures are grafted on the aromatic heterocycle, 2-mercaptobenzoxazole, have not been studied. Based on this premise, we expanded our research to the synthesis of new derivatives of 1,3,4-thiadiazole and 1,2,4-triazole classes of compounds using 1,4-disubstituted thiosemicarbazides as intermediates.

Also, the choice of macromolecular support for immobilization of bioactive compounds is very important. Encouraged by the preliminary results obtained in our previous study [42], we chose to use for the immobilization of these new bioactive compounds two types of macromolecular supports: grafted copolymers and grafted copolymers carrying betaine units based on gellan and N-vinylimidazole (PG and PGB1, respectively). The structure of PG and PGB1 copolymers are presented in Figure 1.

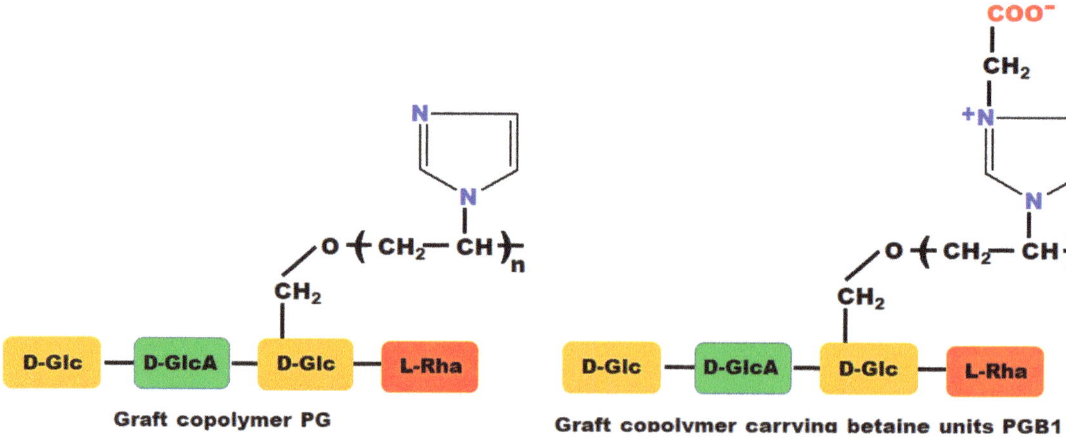

Figure 1. Chemical structure of PG and PGB1 graft porous copolymers.

In this paper, our investigations were directed to the following directions:

- Synthesis of new bioactive compounds in which thiadiazole and triazole structures are grafted onto the aromatic heterocycle, 2-mercaptobenzoxazole, followed by the choice of the compound with the best biological properties in order to be immobilized on a polymeric support;
- Synthesis of new polymer-bioactive compound systems; and
- Detailed studies on the immobilization of the bioactive compound by adsorption on grafted copolymers and those containing the betaine structure.

2. Results and Discussion

2.1. Synthesis of Thiadiazole and Triazole Derivatives

Among the heterocyclic combinations, the 1,3,4-thiadiazole and 1,2,4-triazole derivatives present a remarkable application interest. The synthesis of 1,3,4-thiadiazole and 1,2,4-triazole derivatives were performed in several steps as follows:

Step 1. Synthesis of 1-(benzoxazole-2′-yl-mercapto-acetyl)-4-aryl-thiosemicarbazides (II–VII) was performed in methanol solution by heating a mixture of benzoxazolyl-2-mercapto-acetic acid hydrazide (I) [43] with various types of isothiocyanate derivatives: phenyl, p-tolyl, p-methoxyphenyl, p-bromophenyl, p-chlorophenyl, and p-iodofenil, according to the procedures described in our previous papers for other types of thiosemicarbazides (Figure 2) [4,44,45].

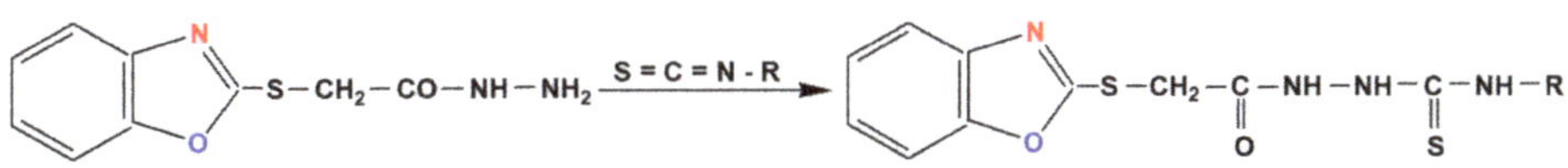

II - VII

II, R = - C_6H_5; III, R = - $C_6H_4 - CH_3$ (p); IV, R = - $C_6H_4 - OCH_3$ (p); V, R = - $C_6H_4 - Br$ (p); VI, R = - $C_6H_4 - Cl$ (p); VII, R = - $C_6H_4 - I$ (p).

Figure 2. Synthesis of 1-(benzoxazole-2′-yl-mercapto-acetyl)-4-aryl-thiosemicarbazides (II-VII).

The structure elucidation of the compounds II–VII was carried out by elemental analysis as well as by spectroscopic methods (FT-IR and ^{1}H-NMR).

The results obtained from the elemental analysis as well as the chemical formula and the yield of reaction of all bioactive compounds synthesized in this paper (compounds II-XIX) are presented in Table 1.

Table 1. Some characteristics of compounds II–XIX.

Samples	Elemental Analysis, Calc (Found) (%)								Chemical Formula	Yield (%)
	C	H	N	S	O	Br	Cl	I		
1,4-disubstituted thiosemicarbazides (compounds II–VII)										
II	53.6 (53.9)	3.9 (4.2)	15.6 (16)	17.9 (18.2)	8.9 (7.7)	-	-	-	$C_{16}H_{14}N_4O_2S_2$	65
III	54.8 (55.1)	4.3 (4.5)	15.1 (15.4)	17.2 (17.6)	8.6 (7.4)	-	-	-	$C_{17}H_{16}N_4O_2S_2$	75
IV	52.6 (52.9)	4.1 (4.3)	14.4 (14.8)	16.4 (16,9)	12.5 (11.1)	-	-	-	$C_{17}H_{16}N_4O_3S_2$	85
V	43.9 (44.1)	2.9 (3.2)	12.8 (12.9)	14.6 (15)	7.4 (6.3)	18.3 (18.6)	-	-	$C_{16}H_{13}N_4O_2S_2Br$	80
VI	48.9 (49)	3.3 (3.6)	14.3 (14.6)	16.3 (16.6)	8.2 (6.8)	-	9 (9.5)	-	$C_{16}H_{13}N_4O_2S_2Cl$	78
VII	39.7 (39.9)	2.7 (2.9)	11.6 (11.9)	13.2 (13.6)	6.6 (5.2)	-	-	26.2 (26.6)	$C_{16}H_{13}N_4O_2S_2I$	77
1,3,4-thiadiazoles (compounds VIII–XIII)										
VIII	56.5 (56.6)	3.5 (3.8)	16.5 (16.8)	18.8 (19.2)	4.7 (3.6)	-	-	-	$C_{16}H_{12}N_4OS_2$	89
IX	57.6 (58)	4 (4.1)	15.8 (16.1)	18.1 (18.5)	4.5 (3.3)	-	-	-	$C_{17}H_{14}N_4OS_2$	83
X	52.1 (52.4)	3.8 (4)	15.1 (15.4)	17.3 (17.6)	11.7 (10.6)	-	-	-	$C_{17}H_{14}N_4O_2S_2$	81
XI	45.8 (46.2)	2.6 (2.8)	13.4 (13.6)	15.3 (15.7)	22.9 (21.7)	19 (19.4)	-	-	$C_{16}H_{11}N_4OS_2Br$	79
XII	51.3 (51.4)	2.9 (3.1)	15 (15.3)	17.1 (17.4)	13.7 (12.8)	-	9.5 (9.7)	-	$C_{16}H_{11}N_4OS_2Cl$	74
XIII	41.2 (41.5)	2.4 (2.6)	12 (12.3)	13.7 (14.1)	30.7 (29.2)	-	-	27.3 (27.5)	$C_{16}H_{11}N_4OS_2I$	71

Table 1. *Cont.*

Samples	Elemental Analysis, Calc (Found) (%)								Chemical Formula	Yield (%)
	C	H	N	S	O	Br	Cl	I		
1,2,4-triazoles (compounds XIV–XIX)										
XIV	56.5 (56.7)	3.5 (3.9)	16.5 (16.9)	18.8 (19.2)	4.7 (3.3)	-	-	-	$C_{16}H_{12}N_4OS_2$	66
XV	57.6 (57.8)	4 (4.3)	15.8 (16.2)	18.1 (18.3)	4.5 (3.4)	-	-	-	$C_{17}H_{14}N_4OS_2$	74
XVI	55.1 (55.3)	3.8 (4)	15.1 (15.5)	17.3 (17.7)	8.7 (7.5)	-	-	-	$C_{17}H_{14}N_4O_2S_2$	88
XVII	45.8 (46)	2.6 (3)	13.4 (13.7)	15.3 (15.7)	22.9 (21.6)	19 (19.3)	-	-	$C_{16}H_{11}N_4OS_2Br$	79
XVIII	51.3 (51.4)	2.9 (3.3)	15 (15.2)	17.1 (17.4)	13.7 (12.7)	-	9.5 (9.9)	-	$C_{16}H_{11}N_4OS_2Cl$	77
XIX	41.2 (41.4)	2.4 (2.5)	12 (12.4)	13.7 (14)	30.7 (29.7)	-	-	27.3 (27.6)	$C_{16}H_{11}N_4OS_2I$	74

In the FT-IR spectra (Figure not shown), the following characteristic absorption bands can be observed: 700–749 cm^{-1} assigned to the C-S stretching vibration; 1400–1498 cm^{-1} attributed to the CH_2 bending vibration; 1620–1704 cm^{-1} assigned to the C=O stretching vibrations; the C=S stretching vibration was observed at 1120–1150 cm^{-1}; NH stretching vibration appears at 2880–3150 cm^{-1}; the absorption bands of the mono and disubstituted aromatic rings were identified in the region of 680–789 cm^{-1}; and the C-Br, C-Cl, and C-I stretching vibrations appear at 629 cm^{-1}, 725 cm^{-1}, and 630 cm^{-1}, respectively.

In the ^{1}H-NMR spectra, the two methylene protons generate a singlet at δ = 4.25–4.60 ppm and at δ = 8.67–8.89 ppm, while the protons from the -NH group appear at 10.06–10.85 ppm. The aromatic protons in phenyl and benzoxazole appear within the range 6.50–7.92 ppm. The protons belonging to the methyl groups are identified at δ = 2.11 ppm for the compound III and at δ = 3.38 ppm for the thiosemicarbazide IV (Figure 3).

Step 2. Synthesis of new 1,3,4-thiadiazoles and 1,2,4-triazoles starting from 1,4-disubstituted-thiosemicarbazides (compounds II–VII). The 5-aryl-amino-2-substituted-1,3,4-thiadiazoles (compounds VIII–XIII) were obtained by intramolecular cyclization in concentrated sulfuric acid medium, while the 5-mercapto-3,4-disubstituted -1,2,4-triazoles (compounds XIV–XIX) were prepared under the catalytic action of hydroxyl ions, according to our previously work described in literature (Figure 4) [4,44–47].

The idea of synthesis of grafted 1,3,4-thiadiazoles on the 2-mercapto-benzoxazole molecule is an extremely interesting possibility to find new derivatives, which can be active against some microbial strains.

The mechanism of the synthesis reaction of these types of heterocyclic compounds is presented in Figure 5 and contains the following steps:

In case of 1,2,4-triazoles, the mechanism of cyclization in the presence of hydroxyl ions (Figure 6) begins with the hydrogen extraction from the nitrogen atom situated at position 4 of 1,4-disubstituted thiosemicarbazide, which has an increased mobility due to the p-π conjugation between the lone pair electrons belonging to the nitrogen atom and the π electrons of thionic bond or of benzene nucleus, respectively. As a result of this conjugation, a negative charge appears on the nitrogen atom from position 4 that can be in equilibrium with another anion formed at the sulfur atom. In this mesomer ion, the negatively charged nitrogen atom (more stable) attacks the carbon belonging to the carbonyl group simultaneously with π electron delocalization, leading to the formation of the five-membered heterocycle, the negative charge being located at the oxygen atom outside the ring. This structure is stabilized by eliminating the hydroxyl group in the reaction medium leading to the formation of 5-mercapto-1,2,4-triazole structure. Depending on the preparation conditions and the reactant nature, as well as due to the presence of thioamide group in triazole molecule, 5-mercapto-3,4 disubstituted-1,2,4-triazoles show the phenomenon of double reactivity, reacting as if they had a thionic structure (A) or a thiol structure (B). This behavior explains the presence of a reaction center at the sulfur atom from 5-position of five-membered heterocycle [48].

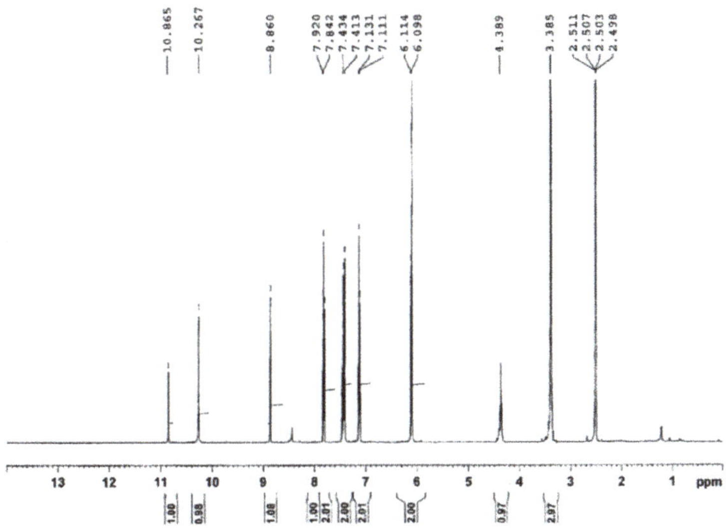

Figure 3. ^{1}H-NMR spectra of 1-(benzoxazole-2′-yl-mercapto-acetyl)-4-(p-methoxyphenyl)-thiosemicarbazide (compound IV).

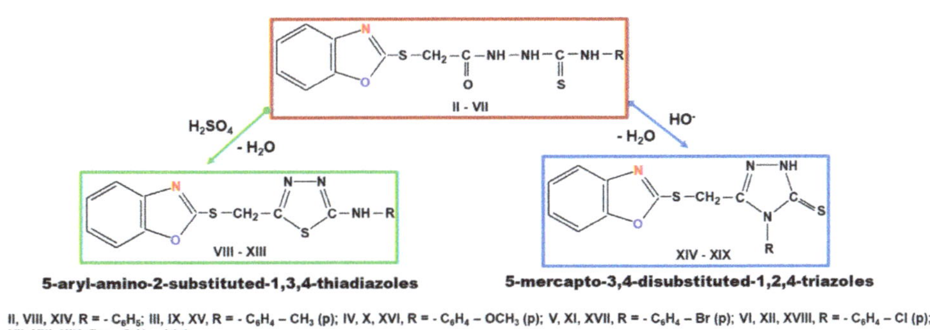

Figure 4. Synthesis of 1,3,4-thiadiazoles (compounds VIII–XIII) and 1,2,4-triazoles (compounds XIV–XIX).

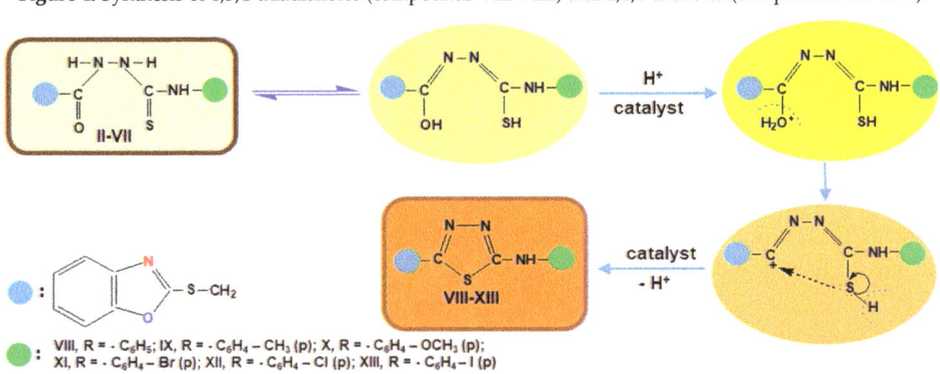

Figure 5. Intramolecular cyclization mechanism of reaction synthesis of 5-aryl-amino-2-substituted-1,3,4-thiadiazoles.

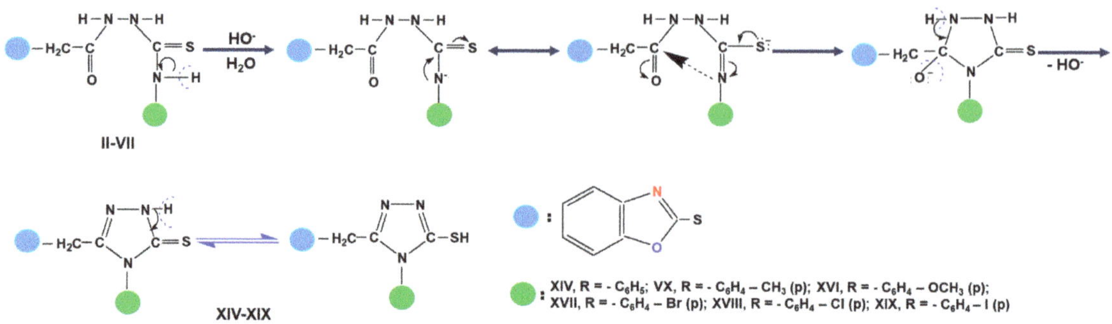

Figure 6. Cyclization mechanism of reaction synthesis of 5-mercapto-3,4-disubstituted-1,2,4-triazoles.

In order to obtain the information about the structure of 1,3,4-thiadiazoles (compounds VIII–XIII) and 1,2,4-triazoles (compounds XIV–XIX), the elemental analysis (Table 1) as well as the FT-IR (Figure 7) and ^{1}H-NMR spectra (Figure 8) were performed.

Generally, in FT-IR spectra of the 1,3,4-thiadiazole (compounds VIII–XIII), the following absorption bands can be observed: the absorption bands at 3318–2880 cm^{-1} corresponding to the NH stretching vibrations; the absorption band at 1438–1462 cm^{-1} can be attributed to the stretching vibration of -N=C-S group; the bending vibration of the CH_2 group has been assigned to the absorption bands situated between 1401–1496 cm^{-1}; characteristic CH bands from mono and disubstituted benzene nucleus are assigned to the absorption band situated in the range of 680–789 cm^{-1}; and the C-Br, C-Cl, and C-I stretching vibrations appear at 629, 725, and 630 cm^{-1}, respectively.

In the case of 1,2,4-triazole derivatives, the FT-IR spectra of the compounds XIV–XIX present a wide band at 3553 cm^{-1} due to the NH stretching vibrations; 1613–1628 cm^{-1} corresponding to the C=N stretching vibration; the C=S group appears at 1490–1495 cm^{-1} while the bands at 837–942 cm^{-1} can be attributed to the para-disubstituted benzene nucleus; the S-CH_2 stretching vibration has been assigned to an adsorption band at 765–785 cm^{-1}; and the C-Br, C-Cl, and C-I stretching vibrations show maximum adsorption bands at 753, 746, and 748 cm^{-1}, respectively.

For example, in Figure 7, the spectra of compound VIII (a) and compound XVI (b), respectively, are presented.

^{1}H-NMR spectra bring additional arguments regarding the structure of the obtained organic compounds. In case of 1,3,4-thiadiazole derivatives, the protons of the CH_2 group bound to the sulfur atom situated at position 2′ of the benzoxazole nucleus appear as a singlet in the region of 4.25–4.53 ppm. The proton bound to the nitrogen could be detected as a singlet in the region 10.27–10.98 ppm. For all types of 1,3,4-thiadiazoles, aromatic protons appear at 6.91–7.56 ppm and 8.21–8.57 ppm, respectively. The protons of the methyl group in 1,3,4-thiadiazoles (compounds IX and X) are found as a triplet at 2.98 and 3.38 ppm, respectively.

The ^{1}H-NMR spectra of 1,3,4-thiadiazoles (compound VIII) and 1,2,4-triazoles (compound XVI) are presented in Figure 8.

The ^{1}H-NMR spectra of the 1,2,4-triazole derivatives (compounds XIV–XIX) present a singlet due to the protons of the CH_2 group at 4.17–4.18 ppm, while at 14.04–14.33 ppm is observed a singlet specific to the proton belonging to the NH group. For all 1,2,4-triazoles, the aromatic protons appear at 6.99–7.85 ppm, and for compounds XV and XVI, the protons of the methyl group appear at 2.43 and 3.43 ppm, respectively.

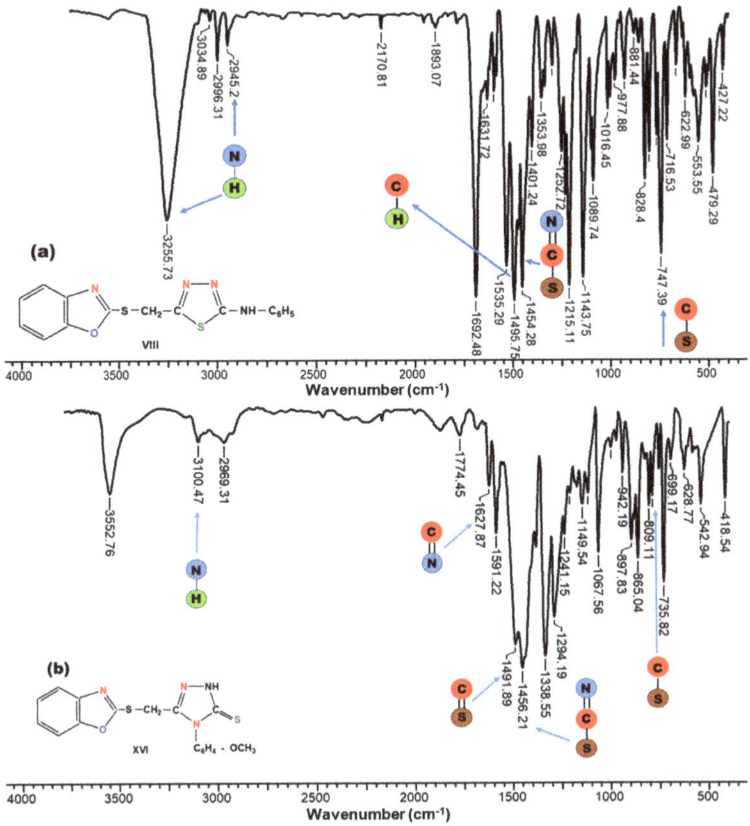

Figure 7. FT-IR spectrum of (**a**) 1,3,4-thiadiazoles (compound VIII) and (**b**) 1,2,4-triazole (compound XVI).

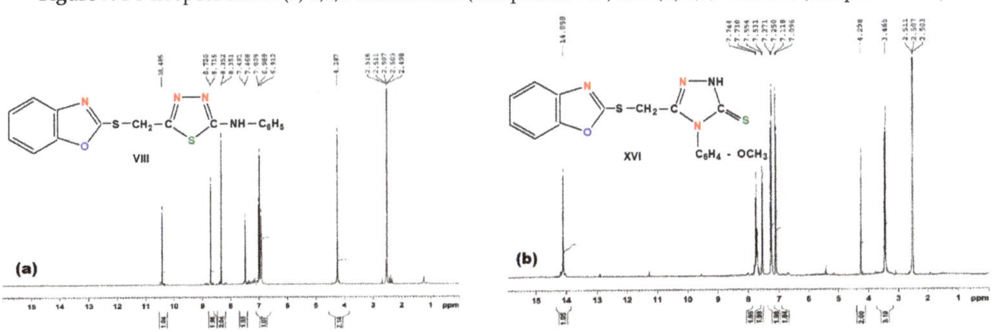

Figure 8. ^{1}H-NMR spectra of (**a**) 1,3,4-thiadiazoles (compound VIII) and (**b**) 1,2,4-triazoles (compound XVI).

2.2. Antibacterial Activities of Bioactive Compounds

The antimicrobial activity of 1,3,4 thiadiazoles and 1,2,4 triazoles was tested according to the National Committee for Clinical Laboratory Standards Institute NCCLS Approval Standard Document M2-A5, Vilanova, PA, USA (2000), on the following microbial strains: gram-positive bacteria (*Staphylococcus aureus* (ATCC-2593), *Bacillus subtilis* (ATCC-6638), *Bacillus cereus* (ATCC-10876)) and gram-negative bacteria (*Salmonella enteritidis* (P-1131) and *Escherichia coli* (ATCC-25922)).

The results regarding the antimicrobial activity of the bioactive compounds synthesized can be seen in Table 2.

Table 2. Antibacterial spectra of 1,3,4-thiadiazoles and 1,2,4- triazoles.

Sample Code	Time for Thermostating (h)	*Staphylococcus aureus*			*Bacillus subtilis*			*Bacillus cereus*			*Salmonella enteritidis*			*Escherichia coli*		
		mg/mL			mg/mL			mg/mL			mg/mL			mg/mL		
		1	0.5	0.25	1	0.5	0.25	1	0.5	0.25	1	0.5	0.25	1	0.5	0.25
VIII	24	-	+	•	-	-	•	+	+	+	+	+	+	-	-	•
	48	-	-	-	-	-	•	-	+	+	+	+	+	-	•	•
IX	24	•	•	-	•	•	+	+	+	+	+	+	+	-	•	•
	48	•	•	-	•	•	+	+	+	+	+	+	+	•	•	•
X	24	•	•	•	•	•	-	+	-	-	+	+	+	-	-	•
	48	•	•	•	•	+	-	+	+	+	+	+	-	•	•	•
XI	24	+	•	•	•	•	•	+	+	+	+	+	+	•	•	•
	48	+	•	+	•	•	+	+	+	+	+	+	+	•	•	•
XII	24	•	•	•	•	•	+	+	+	+	+	+	+	•	•	•
	48	•	+	•	•	•	+	+	+	+	+	+	+	•	•	•
XIII	24	•	+	•	•	•	+	+	+	+	+	+	+	•	•	•
	48	•	+	•	•	•	+	+	+	+	+	+	+	•	•	•
XIV	24	-	•	•	-	-	-	-	•	•	-	-	-	-	-	-
	48	•	+	+	•	+	+	+	+	+	-	-	-	-	-	-
XV	24	-	-	-	-	-	-	•	•	•	-	-	-	-	-	-
	48	-	•	•	-	-	-	+	+	+	-	-	-	-	-	-
XVI	24	-	-	-	-	-	-	•	•	•	-	-	-	-	-	-
	48	-	-	•	-	-	-	+	+	+	-	-	-	-	-	-
XVII	24	•	•	•	•	•	•	•	•	•	•	•	•	•	•	•
	48	+	+	+	+	+	+	+	+	+	+	+	+	+	+	+
XVIII	24	•	•	•	•	•	•	•	•	•	•	•	•	•	•	•
	48	+	+	+	+	+	+	+	+	+	+	+	+	+	+	+
XIX	24	•	•	•	•	•	•	•	•	•	•	•	•	-	•	•
	48	+	+	+	+	+	+	+	+	+	+	+	+	+	+	+
Kanamycine	24	-	-	-	-	-	-	-	-	-	-	-	+	-	-	-
	48	-	-	-	-	-	-	-	-	•	-	-	+	-	-	+

- no growth; • moderate growth; + normal growth.

Experimental data shows that 1,3,4-thiadiazoles (compounds VIII–XIII) are sensitive to *Staphyloccocus aureus* and present a moderate effect on *Bacillus subtilis* and *Escherichia coli*, while the culture of *Bacillus cereus* and *Salmonella enteritidis* are resistant to the action of the 1,3,4-thiadiazole test.

The 1,2,4-triazoles (compounds XIV–XVI) have been shown to be strong inhibitors on the growth of the bacteria *Escherichia coli*, *Salmonella enteritidis*, and *Bacillus subtilis* in any concentration and regardless of the contact time, while against *Staphylococcus aureus* and *Bacillus cereus* germs, the 1,2,4-triazoles show a moderate action. In contrast, the 1,2,4-triazoles (compounds XVII–XIX) show a moderate action on all tested germs only in a short period of contact time and at all concentrations. It can be stated that the tested 1,3,4-thiadiazoles and 1,2,4-triazoles show appreciable antimicrobial activity against microbial strains similar to that of the model drug, kanamycin.

2.3. Immobilization Studies of Triazoles

Among the new synthesized compounds, the 1,2,4-triazole derivative obtained with the highest yield (88%) and the best antimicrobial activity, namely 3-(benzoxazole-2′-yl-mercapto-methyl)-4-(p-methoxyphenyl)-5-mercapto-1,2,4-triazole (compound XVI), which was selected for immobilization by sorption onto grafted copolymers.

The interaction between PG and PGB1 copolymers and compound XVI was highlighted by FT-IR spectroscopy (Figure 9).

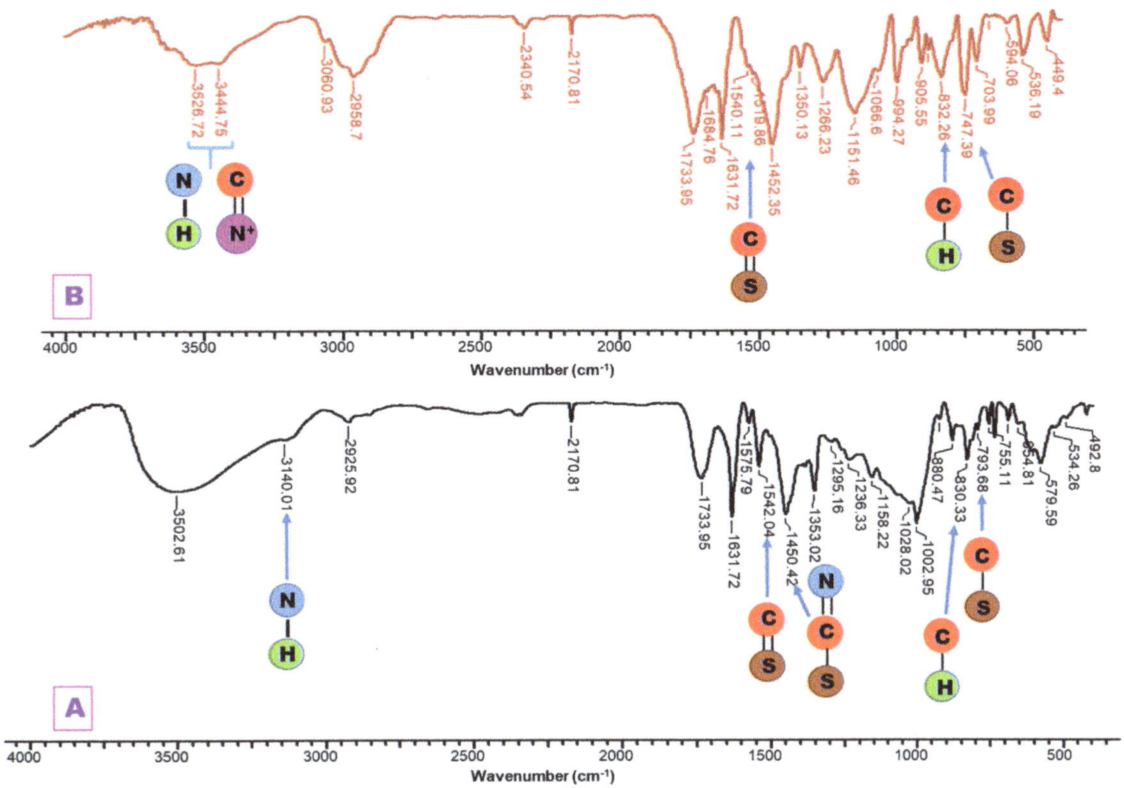

Figure 9. FT-IR spectra for (**A**) PG-T system; (**B**) PGB1-T system.

- For PG—1,2,4-triazole systems (PG-T), it can be observed some new absorption bands as follows: the absorption band at 3140 cm^{-1} due to the stretching vibration of -NH group; 1542 cm^{-1} assigned to the stretching vibration of C=S group; 1450 cm^{-1} corresponding to the stretching vibration of -N=C-S group; and 793 cm^{-1} assigned to the stretching vibration of S-CH_2 group.
- For PGB1—1,2,4-triazole systems (PGB1-T), the changes that occur in the FTIR spectrum are the following: the band vibration of 3439 cm^{-1} from PGB1 copolymer is shifted in PGB1-T spectra to 3445 cm^{-1} due to the overlap of the absorption band of -C=N^+ group from PGB1 copolymer, with absorption band corresponding to the -NH group belonging to the 1,2,4-triazole; absorption band at 832 cm^{-1} is assigned to the stretching vibration of C-H out of plane bending.

Also, XRD and SEM studies were performed for a better characterization of PG-T and PGB1-T systems. The XRD patterns of 1,2,4-triazole, PG-T, and PGB1-T are presented in Figure 10.

The XRD analysis of PG-T and PGB1-T systems showed the combined signals of both PG/PGB1 copolymers and 1,2,4-triazole, leading to the conclusion that the 1,2,4-triazole was successfully immobilized onto polymeric supports.

Visualization of surface morphology of PG-T and PGB1-T systems was achieved through SEM microscopy with a magnification of 1000, and the images are presented in Figure 11.

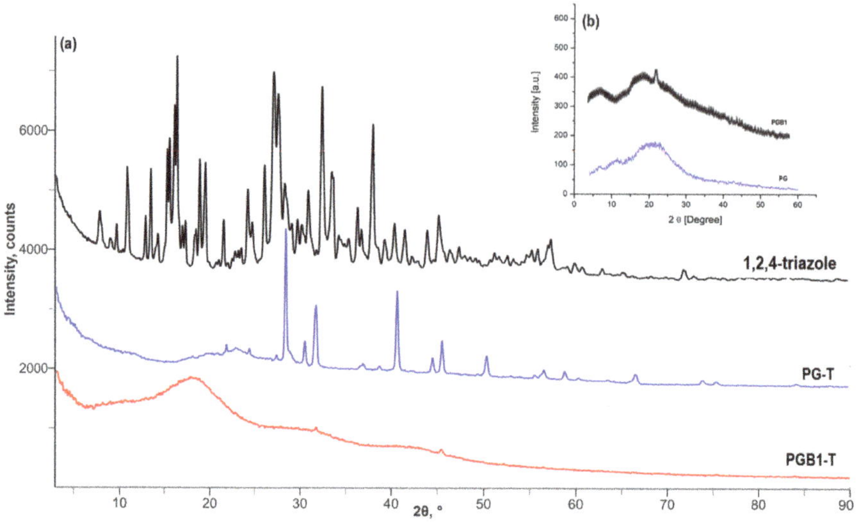

Figure 10. XRD patterns of 1,2,4-triazole, PG-T, and PGB1-T systems (**a**) and of PG and PGB1 copolymers (**b**).

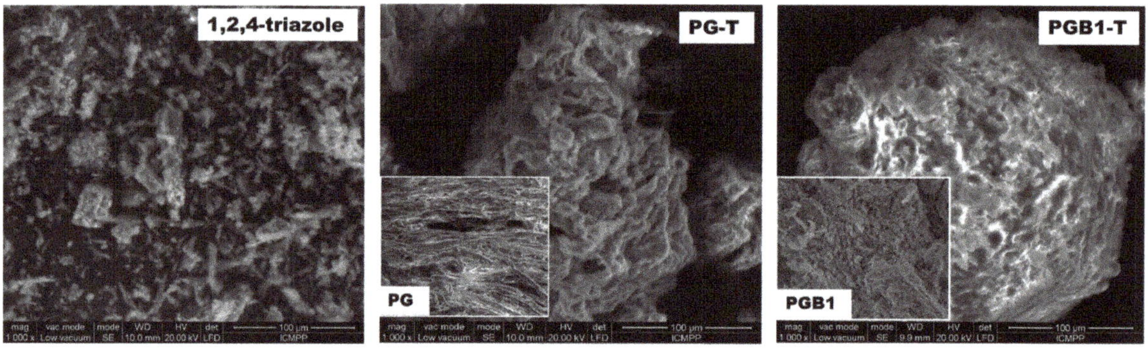

Figure 11. SEM images for 1,2,4-triazole, PG-T, and PGB1-T systems.

From SEM images, it can be observed that the PG-T and PGB1-T systems appear to have porous structures different from those of PG and PGB1 copolymers. The changes in the surface morphologies of PG-T and PGB1-T systems indicate that the sorption of 1,2,4-triazole onto polymeric support occurred.

Generally, in order to evaluate a sorption process of bioactive compound, it is important to consider two physico-chemical aspects, namely the equilibrium and the kinetics of sorption.

2.3.1. Sorption Isotherms

To describe the interactions between sorbate and sorbent that occurs in sorption processes, three model isotherms were used, namely: Langmuir, Freundlich, and Dubinin–Radushkevich.

- Langmuir isotherm

The Langmuir isotherm describes the sorption in a homogeneous system [49] and is given by the equation:

$$q_e = \frac{q_m \cdot K_L \cdot C_e}{1 + K_L \cdot C_e} \quad (1)$$

where: q_e represents the amount of 1,2,4-triazole sorbed at equilibrium (mg/g); q_m represents the maximum amount sorbed (mg/g); C_e is the concentration of 1,2,4-triazole solution at equilibrium (mg/L); and K_L is the Langmuir constant that reflects the affinity between sorbate and sorbent.

To determine whether the sorption system used is favorable or unfavorable to the sorption, the equilibrium parameter R_L, given by the equation [50], was also calculated:

$$R_L = \frac{1}{1 + K_L.C_i} \tag{2}$$

where C_i represents the initial concentration. According to the literature [51], if $R_L < 1$, then the sorption is unfavorable, for $R_L = 1$, the sorption is linear; if $0 < R_L < 1$, the sorption is favorable or irreversible if $R_L = 0$.

- Freundlich isotherm

Another model used in sorption studies is the Freudlich isotherm. This isotherm model is applied in the case of multilayer sorption of sorbate on a heterogeneous surface [52]. The Freundlich isotherm is described by the following equation:

$$q_e = K_F \cdot C_e^{1/n_f} \tag{3}$$

where K_F is the Freundlich constant which represents the amount of 1,2,4-triazole sorbed per gram of sorbent when the equilibrium concentration is equal to unity (L/g); $1/n_f$ indicates the type of isotherm as follows: favorable $0 < 1/n_f < 1$ or unfavorable $1/n_f > 1$.

- Dubinin–Radushkevich isotherm

The Dubinin–Radushkevich isotherm is an empirical model designed to estimate the apparent free energy of sorption as well as to differentiate between the physical and chemical sorption process [53]. The Dubinin–Radushkevich model equation is given by the following mathematical relation:

$$q_e = q_{DR} \cdot exp\left\{ -\beta \left[RTln\left(1 + \frac{1}{C_e} \right) \right]^2 \right\} \tag{4}$$

where q_{DR} represents the maximum amount sorbed (mg/g); β is the Dubinin–Radushkevich constant (mol^2/kJ^2); R represents the ideal gas constant (R = 8.314 kJ/mol·K); and T is the temperature (K).

The Dubinin–Radushkevich isotherm constant, β, is associated with the average free sorption energy, E (kJ/mol), calculated using the following equation:

$$E = 1(2\beta)^{\frac{1}{2}} \tag{5}$$

The value of E is used to obtain information about the nature of the sorption process. The sorption process can be physical when the E values are between 1 and 8 kJ/mol, ion exchange for values of E between 8 and 16 kJ/mol, and chemical nature for values of E greater that 16 kJ/mol [54].

The parameters of the studied isotherms as well as the values of the error functions R^2 and χ^2 are presented in Table 3.

Table 3. Parameter values corresponding to Langmuir, Freundlich, and Dubinin–Radushkevich isotherms calculated in case of 1,2,4-triazole sorption on PG and PGB1 copolymers at different temperatures.

	PG			PGB1		
	298	**303**	**308**	**298**	**303**	**308**
			Langmuir model			
q_m (mg/g)	478	533	636	537	616	685
K_L (L/mg)	0.068	0.086	0.109	0.098	0.126	0.161
R_L	0.01–0.29	0.01–0.25	0.10–0.20	0.01–0.22	0.01–0.18	0.004–0.15
χ^2	3.324	2.613	3.429	2.700	2.010	2.150
R^2	0.994	0.993	0.997	0.993	0.992	0.996
			Freundlich model			
K_F (L/g)	0.457	0.512	0.617	0.528	0.602	0.673
$1/n_f$	0.939	0.738	0.729	0.686	0.540	0.383
χ^2	33.462	23.958	38.332	25.213	19.219	18.782
R^2	0.911	0.920	0.915	0.913	0.921	0.914
			Dubinin–Radushkevich model			
q_{DR} (mg/g)	463	521	629	532	612	679
E (kJ/mol)	1.099	1.561	2.171	1.357	2.608	3.714
χ^2	0.402	0.359	0.413	0.371	0.226	0.277
R^2	0.998	0.997	0.998	0.998	0.999	0.998

From data presented in Table 3, the following conclusions can be drawn:

- The theoretical values obtained for the maximum sorption capacity (q_m) calculated on the basis of the Langmuir isotherm are close to the experimental values q_c (457, 512 and 618 mg 1,2,4-triazole /g PG copolymer and 529, 601 and 672 mg of 1,2,4-triazole /g PGB1 copolymer);
- The values of the equilibrium parameter, R_L, were in the range between 0 and 1, thus confirming that the PG and PGB1 copolymers are favorable supports for the sorption of 1,2,4-triazoleat at the three temperatures studied. It is also observed that the K_L values are higher in the case of the PGB1 copolymer than in the case of the PG copolymer, which indicates a higher affinity of the PGB1 copolymer for 1,2,4-triazole; this is in agreement with the highest sorption capacity obtained in the case of the PGB1 copolymer;
- The values for R^2 and χ^2 are in the range of 0992–0.997 and 2.010–3.429, respectively, indicating that the Langmuir isotherm describes well the experimental data;
- Although the values of $1/n_f$ are in the range of 0–1, which indicates that the Freundlich isotherm is favorable in the case of 1,2,4-triazolesorption onto PG and PGB1 copolymers, the small values of R^2 (0.911 to 0.921) associated with high values of χ^2 (18.782–38.332) shows that the Freundlich isotherm does not describe well the experimental data;
- The analysis of the parameter values obtained by applying the Dubinin–Radushkevich isotherm shows that the q_{DR} values are very close to the experimental values, which indicates that this isotherm describes well the experimental data. This is also supported by the fact that high values for R^2 (0.997–0.999) and low values for χ^2 (0.226–0.413) were obtained, confirming that the Dubinin–Radushkevich isotherm describes very well the sorption of 1,2,4-triazole onto PG and PGB1 copolymers;
- The calculated values of the average free energy of 1,2,4-triazole sorption onto PG and PGB1 copolymers are in the range of 1.099–3.714 kJ/mol, which indicates that the sorption process studied is physical in nature.

2.3.2. Thermodynamic Parameters

By means of thermodynamic parameters, such as Gibbs free energy change (ΔG), enthalpy change (ΔH), and entropy change (ΔS), the mechanism and the type of sorption process can be determined. Thus, according to the literature [55], it is known that in the case of physical sorption, the values of ΔH are less than 40 kJ/mol, while in the case of chemical sorption, the values of ΔH are in the range of 40–120 kJ/mol. The value of ΔH and ΔS can be determined by means of the Langmuir constant using the Van 't Hoff equation [56]:

$$lnK_L = \frac{\Delta S}{R} - \frac{\Delta H}{RT} \tag{6}$$

From the linear representation ln K_L versus $1/T$ (Figure 12), the values of ΔS and ΔH were obtained from the intercept and slope, respectively, and the results are presented in Table 4. The values of ΔG were obtained using the following equation:

$$\Delta G = \Delta H - T\Delta S \tag{7}$$

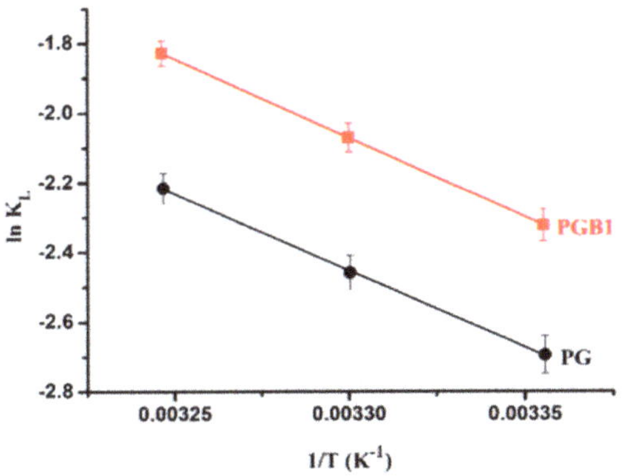

Figure 12. Plots of lnK_L versus $1/T$ for sorption of 1,2,4-triazole derivative onto PG and PGB1 copolymers.

Table 4. Thermodynamic parameters of sorption process.

Sample Code	ΔH (kJ/mol)	ΔS (J/mol·K)	R^2	ΔG (kJ/mol)		
				298 K	303 K	308 K
PG	36.49	100.05	0.995	−29.78	−30.278	−30.78
PGB1	37.67	107.14	0.998	−31.89	−32.425	−32.96

As it can be seen from Table 4, the negative values of ΔG show that the 1,2,4-triazole sorption onto PG and PGB1 copolymers is spontaneous and favorable. Also, the values of this thermodynamic parameter decrease with increasing of the temperature, indicating that there is a higher efficiency of the sorption process at higher temperatures. The values of ΔH < 40 kJ/mol indicate that the interactions between the PG or PGB1 copolymers and 1,2,4-triazole are physical in nature. On the other hand, the positive value of the ΔH shows that the sorption process is endothermic and the increase of the temperature leads to an increase in the amount of 1,2,4-triazole sorbed. The positive value of ΔS indicates the affinity of PG and PGB1 copolymers for 1,2,4-triazole as well as the increase randomness at the solid-solution interface during the sorption process [57]. Also, the positive value of ΔS can indicate an increase in the degree of freedom of 1,2,4-triazole. Moreover, the positive values of ΔH and ΔS lead to the conclusion that the sorption process occurs spontaneously at all temperatures.

2.3.3. Sorption Kinetic Study

The study of sorption kinetic describes the rate of sorption and is very important because it gives us information about the mechanism of sorption. Two mathematical models, such as the Lagergren model (pseudo-first order kinetic model) and the Ho model (pseudo-second order kinetic model), were used to elucidate the mechanism of 1,2,4-triazole sorption onto PG and PGB1 copolymers. The two models mentioned above are described by the following equations:

- Lagergren model [58]:

$$q_t = q_e\left(1 - e^{-k_1 t}\right) \quad (8)$$

- Ho model [59]:

$$q_t = \frac{k_2 \cdot q_t^2 \cdot t}{1 + k_2 \cdot q_e \cdot t} \quad (9)$$

where q_e and q_t are the amounts of 1,2,4-triazole sorbed at equilibrium and at time t (mg/g), k_1 is the rate constant of the pseudo-first order sorption process (min^{-1}), and k_2 is the rate constant of pseudo-second order sorption process (g/mg·min).

Figure 13 presents the plots of the Lagergren and Ho models in the case of 1,2,4-triazole sorption ($C_{1,2,4\text{-triazole}} = 15 \times 10^{-3}$ g/mL) onto PG and PGB1 copolymers at T = 308 K, and Table 5 shows the parameter values corresponding to the Lagergren and Ho models.

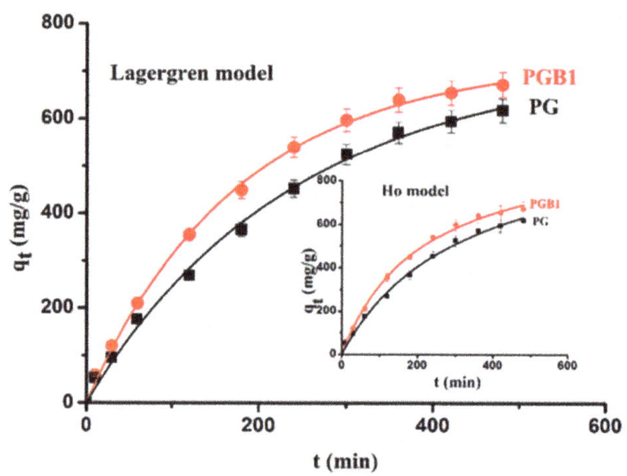

Figure 13. Graphical representation of Lagergren and Ho models in case of 1,2,4-triazole sorption onto PG and PGB1 copolymers at T = 308 K and $C_{1,2,4\text{-triazole}} = 15 \times 10^{-3}$ g/mL.

Table 5. The parameters corresponding to the Lagergren and Ho models used in the case of 1,2,4-triazole sorption onto PG and PGB1 copolymers.

$C_{1,2,4\text{-triazole}}$ (g/mL)		PG			PGB1		
		298	303	308	298	303	308
	$q_{e,exp}$ (mg/g)	325	394	473	412	496	587
				Lagergren model			
	$q_{e,calc}$ (mg/g)	346.29	415.91	491.95	428.82	511.76	600.94
	k_1 ($\times 10^3$ min^{-1})	2.19	3.01	3.31	3.83	3.92	4.14
	χ^2	1.738	2.263	1.119	1.119	1.682	1.103
3.6×10^{-4}	R^2	0.998	0.998	0.998	0.999	0.998	0.999
				Ho model			
	$q_{e,calc}$ (mg/g)	449.72	482.38	564.84	579.40	616.92	696.22
	k_2 [$\times 10^6$ (g/mg· min)]	1.26	1.62	2.43	2.89	3.51	3.92
	χ^2	4.014	4.686	5.976	2.411	5.499	5.040
	R^2	0.997	0.997	0.997	0.998	0.997	0.998

Table 5. *Cont.*

$C_{1,2,4\text{-triazole}}$ (g/mL)		PG			PGB1		
		298	303	308	298	303	308
	$q_{e,exp}$ (mg/g)	325	394	473	412	496	587
3.6 × 10^{-3}	$q_{e,exp}$ (mg/g)	432	491	593	507	583	659
		Lagergren model					
	$q_{e,calc}$ (mg/g)	473.63	523.96	622.21	531.87	606.53	679.53
	k_1 (×10^3 min^{-1})	3.10	3.52	4.01	4.13	4.27	4.87
	χ^2	1.456	1.446	1.569	1.312	1.852	1.786
	R^2	0.998	0.999	0.998	0.997	0.998	0.997
		Ho model					
	$q_{e,calc}$ (mg/g)	546.70	621.17	709.71	649.28	694.28	766.03
	k_2 [×10^6 (g/mg· min)]	2.19	2.46	3.11	3.07	3.72	4.01
	χ^2	2.071	2.539	2.390	2.772	2.083	2.310
	R^2	0.997	0.996	0.997	0.995	0.997	0.996
7 × 10^{-3}	$q_{e,exp}$ (mg/g)	449	504	606	519	594	667
		Lagergren model					
	$q_{e,calc}$ (mg/g)	463.71	527.91	622.32	533.86	621.43	681.07
	k_1 (×10^3 min^{-1})	3.42	3.96	4.43	4.60	4.90	5.10
	χ^2	1.722	1.949	1.654	1.034	1.326	1.764
	R^2	0.997	0.997	0.998	0.998	0.997	0.999
		Ho model					
	$q_{e,calc}$ (mg/g)	567.15	613.04	713.83	621.28	734.63	796.60
	k_2 [×10^6 (g/mg min)]	3.09	3.87	4.11	3.35	4.19	5.17
	χ^2	2.609	2.913	1.895	2.776	2.864	3.775
	R^2	0.996	0.997	0.996	0.997	0.997	0.998
15 × 10^{-3}	$q_{e,exp}$ (mg/g)	457	512	618	529	601	672
		Lagergren model					
	$q_{e,calc}$ (mg/g)	483.63	532.96	644.21	544.25	627.05	694.28
	k_1 (×10^3 min^{-1})	3.67	4.67	4.81	4.92	5.14	5.48
	χ^2	1.466	1.343	1.691	1.144	1.671	1.024
	R^2	0.997	0.999	0.998	0.998	0.999	0.999
		Ho model					
	$q_{e,calc}$ (mg/g)	568.17	624.11	754.83	635.77	729.27	783.34
	k_2 [×10^6 (g/mg min)]	3.85	4.38	5.83	4.23	4.97	5.73
	χ^2	3.050	3.208	3.903	3.887	4.321	3.678
	R^2	0.996	0.997	0.996	0.998	0.997	0.997

From data presented in Table 5, it is observed that the value of the sorption rate increases with the increase of the 1,2,4-triazole concentration. The highest amount of 1,2,4-triazole sorbed was obtained in the case of the PGB1 copolymer. The q_e values calculated by applying the Lagergren model are close to the experimental values compared to the q_e values calculated by applying the Ho model. Although the values of the correlation coefficients R^2 are higher when applying the two kinetic models, the χ^2 values are higher when applying the Ho model, which indicates that the Lagergren model describes better the experimental data. These results indicate that 1,2,4-triazole sorption onto PG and PGB1 copolymer is a physical process.

2.4. In Vitro Release Studies

In vitro release studies were performed on the PG-T and PGB1-T systems with the highest amount of immobilized 1,2,4-triazole. In vitro 1,2,4-triazole release studies were realized at pH = 1.2, and the release profiles are shown in Figure 14.

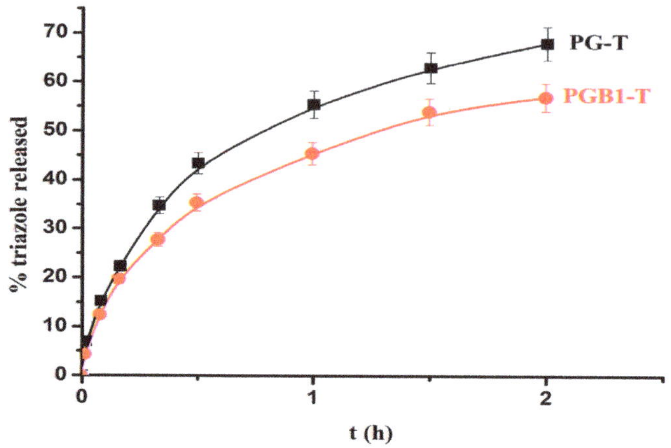

Figure 14. Release profiles of 1,2,4-triazole from PG-T and PGB1-T systems at pH = 1.2.

The kinetic studies and the release mechanism from PG-T and PGB1-T copolymers loaded with 1,2,4-triazole were examined using the following mathematical models:

- Higuchi model [60]:

$$Q_t = k_H \cdot t^{1/2} \quad (10)$$

where Q_t is the amount of drug released at time t; k_H is the Higuchi dissolution constant; and t is the time.

- Korsmeyer–Peppas model [61]:

$$\frac{M_t}{M_\infty} = k_r \cdot t^n \quad (11)$$

where M_t/M_∞ is the fraction of drug released at time t; k_r is the release rate constant that is characteristic to bioactive compound-polymer interactions; and n is the diffusion coefficient that is characteristic to the release mechanism. The values of the 1,2,4-triazole release parameters from PG-T and PGB1-T systems are presented in Table 6.

Table 6. Kinetic parameters of 1,2,4-triazole release from PG-T and PGB1-T systems.

Sample Code	Higuchi Model		Korsmeyer–Peppas Model		
	k_H ($h^{-1/2}$)	R^2	k_r (min^{-n})	n	R^2
PG-T	0.324	0.993	0.014	0.534	0.995
PGB1-T	0.294	0.994	0.017	0.634	0.996

The release exponent n from the Korsmeyer–Peppas equation is situated between 0.534 and 0.634. These results suggest that the release mechanism of 1,2,4-triazole from PG-T and PGB1-T systems was controlled by more than one process, namely diffusion and swelling.

3. Materials and Methods

3.1. General Information

All reactants and solvents were purchased from Merck Company KGaA (Darmstadt, Germany) and were used without purification. Elemental analyses were performed by Exeter Analytical CF-440 Elemental Analyzer (Coventry, United Kingdom). The bioactive compounds (1,4-disubstituted-thiosemicarbazides, 1,3,4-thiadiazoles, and 1,2,4-triazoles) were characterized by FT-IR spectroscopy (Bruker Vertex FT-IR spectrometer) as potassium bromide pellets in the range of 4000–400 cm^{-1} and at a resolution of 2 cm^{-1}. The ^{1}H NMR spectra were recorded in DMSO-d_{60} on Bruker ARX 400 Spectrometer (400 MHz). X-ray diffraction analysis was performed in a Rigaku Miniflex 600 diffractometer using CuKα-emission in the angular range 3–90° (2θ) with a scanning step of 0.01° and a recording rate of 5°/min. The surface morphologies of PG-T system, PGB1-T system, and 1,2,4-triazole in powder form were analyzed with an environmental scanning electron microscope type Quanta 200 at 25kV with secondary electrons in low vacuum.

3.2. Synthesis of Benzoxazole-2′-yl-mercapto-acetic Acid Hydrazide (Compound I)

It was obtained by treating benzoxazol-2′-yl-mercapto-acetic acid ethyl ester with 98% hydrazine hydrate in anhydrous ethanol at room temperature.

3.3. General Procedure for Synthesis of 1-(Benzoxazole-2′-yl-mercapto-acetyl)-4-aryl-thiosemicarbazide (Compounds II–VII)

A total of 0.005 moles of benzoxazole-2′-yl-mercapto-acetic acid hydrazide (I) dissolved in 30 mL absolute methyl alcohol was treated with 0.005 moles of aromatic isothiocyanate in 5 mL anhydrous methanol. The mixture was refluxed for 2 h. During heating, a crystalline precipitate appeared and became more and more abundant once the reaction was complete. After cooling, the precipitate was filtered and washed with anhydrous ethyl ether. Purification was carried out by recrystallization from anhydrous methanol.

1-(Benzoxazole-2′-yl-mercapto-acetyl)-4-phenyl-thiosemicarbazide (Compound II), Physical state: White, crystalline powder; M_p = 182–183 °C. FT-IR (KBr,ν, cm^{-1}): 2917, 3150 (NH); 1670 (C=O); 1400 (CH_2); 1145 (C=S); 755 (monosubstituted benzene nucleus); 700 (C-S).

1-(Benzoxazole-2′-yl-mercapto-acetyl)-4-(p-tolyl)-tiosemicarbazide (Compound III), Physical state: White-ivory powder, acicular crystals; M_p = 163–164 °C. FT-IR (KBr,ν, cm^{-1}): 2880, 3096 (NH); 1709 (C=O); 1424 (CH_2); 720 (C-S); 750 (para-disubstituted benzene nucleus). ^{1}H-NMR (DMSO-d_6, 400 MHz, δ ppm): 2.11 (t, 3H, CH_3); 4.39 (s, 2H, CH_2); 6.95–6.98 (d, 2H, CHAr); 7.13–7.18 (d, 2H, CHAr); 7.57–7.63 (d, 2H, CHAr); 7.76–7.79 (d, 2H, CHAr); 9.86 (s, 1H, NHCO); 10.19 (s, 1H, NH); 10.76 (s, 1H, NH).

1-(Benzoxazole-2′-yl-mercapto-acetyl)-4-(p-methoxytolyl) thiosemicarbazide (Compound IV), Physical state: White, crystalline powder; M_p = 186–188 °C. FT-IR (KBr,ν, cm^{-1}): 2918, 3096 (NH); 1623 (C=O); 1492 (CH_2); 1239 (C=S); 789 (monosubstituted benzene nucleus); 749 (C-S). ^{1}H-NMR (DMSO-d_6, 400 MHz, δ ppm): 3.38 (t, 3H, OCH_3); 4.38 (s, 2H, CH_2); 6.09–6.11 (d, 2H, CHAr); 7.11–7.13 (d, 2H, CHAr); 7.41–7.43 (d, 2H, CHAr); 7.84–7.92 (d, 2H, CHAr); 8.86 (s, 1H, NHCO); 10.26 (s, 1H, NH); 10.86 (s, 1H, NH).

1-(Benzoxazole-2′-yl-mercapto-acetyl)-4-(p-chlorophenyl) thiosemicarbazide (Compound V), Physical state: White-ivory, crystalline powder; M_p = 147–149 °C. FT-IR (KBr,ν, cm^{-1}): 3096 (NH); 1620 (C=O); 1482 (CH_2); 1249 (C=S); 749 (para-disubstituted benzene nucleus); 710 (C-S); 629 (C-Br). ^{1}H-NMR (DMSO-d_6, 400 MHz, δ ppm): 4.27 (d, 2H, CH_2); 7.32–7.35 (d, 2H, CHAr); 7.38–7.40 (d, 2H, CHAr); 7.52–7.56 (d, 2H, CHAr); 7.65–7.67 (d, 2H, CHAr); 9.67 (s, 1H, NHCO); 9.90 (s, 1H, NH); 10.49 (s, 1H, NH).

1-(Benzoxazole-2′-yl-mercapto-acetyl)-4-(p-chlorophenyl)-thiosemicarbazide (Compound VI), Physical state: White-ivory, crystalline powder; M_p = 150–151 °C. FT-IR (KBr,ν, cm^{-1}): 3080, 3124 (NH); 1620 (C=O); 1498 (CH_2); 1250 (C=S); 725 (C-Cl); 708 (C-S); 680 (para-disubstituted benzene nucleus). ^{1}H-NMR (DMSO-d_6, 400 MHz, δ ppm): 4.23 (d, 2H, CH_2); 6.92–7.04 (d, 2H, CHAr); 7.29–7.32 (d, 2H, CHAr); 7.39–7.41 (d, 2H, CHAr); 8.77 (s, 1H, NHCO); 10.11 (s, 1H, NH); 10.85 (s, 1H, NH).

1-(Benzoxazole-2′-yl-mercapto-acetyl)-4-(p-iodophenyl)-thiosemicarbazide (Compound VII). Physical state: White-yellow, crystalline powder; M_p = 172–179 °C. FT-IR (KBr,ν, cm^{-1}): 3097, 3140 (NH); 1620 (C=O); 1490 (CH_2); 1120 (C=S); 785 (para-disubstituted benzene nucleus); 712 (C-S); 630 (C-I). ^{1}H-NMR (DMSO-d_6, 400 MHz, δ ppm): 4.25 (d, 2H, CH_2); 6.91–7.03 (d,

2H, CHAr); 7.30–7.31 (d, 2H, CHAr); 7.40–7.42 (d, 2H, CHAr); 7.66–7.69 (d, 2H, CHAr); 8.74 (s, 1H, NHCO); 10.10 (s, 1H, NH); 10.31 (s, 1H, NH).

3.4. General Procedure for Synthesis of 1-(Benzoxazole-2′-yl-mercapto-methyl)-5-(aryl-amino)-1,3,4,thiadiazoles (Compounds VIII–XIII)

A total of 6 mL of concentrated sulfuric acid were added under stirring to 0.005 moles of 1-(benzoxazole-2′-yl-mercapto-acetyl)-4-aryl-thiosemicarbazide (compounds II–VII). The reaction mixture was stirred at room temperature for 45–50 min to complete the cyclization and then was poured on crushed ice when an abundant precipitate appeared. The mixture was neutralized with ammonium hydroxide, and, after 2 h, the microcrystals were filtered in vacuum and washed with distilled water until the washing water reached pH = 7. Then, the obtained compounds were dried in vacuum oven at 55–60 °C and were purified by recrystallization from ethyl alcohol.

2-(Benzoxazole-2′-yl-mercapto-methyl)-5-(phenyl-amino)-1,3,4-thiadiazole (Compound VIII), Physical state: White-yellow, crystalline powder; M_p = 194–196 °C. FT-IR (KBr,ν, cm^{-1}): 2945, 3256 (NH); 1496 (CH_2); 1454 (N=C-S); 1401, 1535 (thiadiazole nucleus) 1090 (C-S-C); 785, 815 (para-disubstituted benzene nucleus); 750 (C-S). ^{1}H-NMR (DMSO-d_6, 400 MHz, δ ppm): 4.26 (s, 2H, CH_2); 6.91 (t, 1H, CHAr); 6.98–7.00 (d, 2H, CHAr); 7.46–7.49 (d, 2H, CHAr); 8.35 (d, 2H, CHAr); 8.71–8.72 (d, 2H, CHAr); 10.48 (s, 1H, NH).

2-(Benzoxazole-2′-yl-mercapto-methyl)-5-(tolyl-amino)-1,3,4-thiadiazole (Compound IX), Physical state: Brown-red, amorphous powder; M_p = 154–156 °C. FT-IR (KBr,ν, cm^{-1}): 2945 (NH); 1480 (CH_2); 1450 (N=C-S); 1448, 1517 (thiadiazole nucleus); 1070 (C-S-C); 828 (para-disubstituted benzene nucleus); 730 (C-S). ^{1}H-NMR (DMSO-d_6, 400 MHz, δ ppm): 2.29 (t, 3H, CH_3); 4.49 (s, 2H, CH_2); 7.36–7.38 (d, 2H, CHAr); 7.66–7.69 (d, 2H, CHAr); 8.21–8.23 (d, 2H, CHAr); 8.39–8.42 (d, 2H, CHAr); 10.21 (s, 1H, NH).

2-(Benzoxazole-2′-yl-mercapto-methyl)-5-(p-methoxyphenyl-amino)-1,3,4-thiadiazole (Compound X), Physical state: Brown, crystalline powder; M_p = 116–118 °C. FT-IR (KBr,ν, cm^{-1}): 2500, 3000 (NH); 1475 (CH_2); 1438 (N=C-S); 1243, 1615 (thiadiazole nucleus); 1068 (C-S-C); 830 (para-disubstituted benzene nucleus); 710 (C-S). ^{1}H-NMR (DMSO-d_6, 400 MHz), δ ppm: 3.38 (t, 3H, OCH_3); 4.53 (s, 2H, CH_2); 7.31–7.34 (d, 2H, CHAr); 7.53–7.56 (d, 2H, CHAr); 8.50 (d, 2H, CHAr); 8.86 (d, 2H, CHAr); 10.37 (s, 1H, NH).

2-(Benzoxazole-2′-yl-mercapto-methyl)-5-(p-bromophenyl-amino)-1,3,4-thiadiazole (Compound XI), Physical state: Brown, crystalline powder; M_p = 197–197°C. FT-IR (KBr,ν, cm^{-1}): 2880, 2905 (NH); 1480 (CH_2); 1449 (N=C-S); 1240, 1518 (thiadiazole nucleus); 1070 (C-S-C); 828 (para-disubstituted benzene nucleus); 750 (C-Br). ^{1}H-NMR (DMSO-d_6, 400 MHz), δ ppm: 4.49 (s, 2H, CH_2); 7.21-7.23 (d, 2H, CHAr); 7.47-7.51 (d, 2H, CHAr); 8.19-8.21 (d, 2H, CHAr); 8.33-8.35 (d, 2H, CHAr); 10.98 (s, 1H, NH).

2-(Benzoxazole-2′-yl-mercapto-methyl)-5-(p-chlorophenyl-amino)-1,3,4-thiadiazole (Compound XII), Physical state: Brown-white, crystalline powder; M_p = 201–203 °C. FT-IR (KBr,ν, cm^{-1}): 3318 (NH); 1475 (CH_2); 1450 (N=C-S); 1240, 1518 (thiadiazole nucleus); 1071 (C-S-C); 830 (para-disubstituted benzene nucleus); 755 (C-Cl). 1H-NMR (DMSO-d_6, 400 MHz), δ ppm: 4.32 (s, 2H, CH_2); 7.17–7.19 (d, 2H, CHAr); 7.39–7.42 (d, 2H, CHAr); 8.22–8.24 (d, 2H, CHAr); 8.54–8.57 (d, 2H, CHAr); 10.31 (s, 1H, NH).

2-(Benzoxazole-2′-yl-mercapto-methyl)-5-(p-iodophenyl-amino)-1,3,4-thiadiazole (Compound XIII), Physical state: Brown, crystalline powder; M_p = 187 °C. FT-IR (KBr,ν, cm^{-1}): 2880, 2918 (NH); 1478 (CH_2); 1462 (N=C-S); 1438, 1580 (thiadiazole nucleus); 1070 (C-S-C); 825 (para-disubstituted benzene nucleus); 760 (C-I). ^{1}H-NMR (DMSO-d_6, 400 MHz), δ ppm: 4.25 (s, 2H, CH_2); 7.28–7.32 (d, 2H, CHAr); 7.40–7.42 (d, 2H, CHAr); 8.16–8.18 (d, 2H, CHAr); 8.47–8.50 (d, 2H, CHAr); 10.27 (s, 1H, NH).

3.5. General Procedure for Synthesis of 3-(Benzoxazole-2′-yl-mercapto-methyl)-4-aryl-5mercapto-1,2,4-triazoles (Compounds XIV–XIX)

A total of 0.0027 moles of 1-(Benzoxazole-2′-yl-mercapto-acetyl)-4-aryl-thiosemicarbazide (II-VII) were treated with 20 mL of 2N sodium hydroxide solution at room temperature. Then, the reaction mixture was heated under reflux for 60 min. After that, the mixture was cooled, diluted with distilled water (*v/v*), and treated with a diluted solution of hydrochloric acid (1:1) to pH = 4.5. The precipitates [mercapto-1,2,4-triazole (compounds XIV–XIX) were filtered, washed on filter with 500 mL of distilled water, and finally dried. After recrystallization from boiling ethyl alcohol, the finished products were crystalline.

3-(Benzoxazole-2′-yl-mercapto-methyl)-4-phenyl-5-mercapto-1,2,4-triazoles (Compound XIV), Physical state: White, crystalline powder; M_p = 239–241 °C. FT-IR (KBr,ν, cm^{-1}): 3095 (NH); 1622 (C=N); 1490 (C=S); 820 (monosubstituted benzene nucleus); 785 (S-CH_2). ^{1}H-NMR (DMSO-d_6, 400 MHz), δ ppm: 4.78 (s, 2H, CH_2); 6.99–7.01 (d, 2H, CHAr); 7.13–7.15 (d, 2H, CHAr); 7.22 (t, 1H, CHAr); 14.06 (s, 1H, NH).

3-(Benzoxazole-2′-yl-mercapto-methyl)-4-(p-tolyl)-5-mercapto-1,2,4-triazole (Compound XV), Physical state: White-yellow, crystalline powder; M_p = 198–200 °C. FT-IR (KBr,ν, cm^{-1}): 3036 (NH); 1623 (C=N); 1492 (C=S); 898 (para-disubstituted benzene nucleus); 782 (S-CH_2). ^{1}H-NMR (DMSO-d_6, 400 MHz), δ ppm: 2.34 (t, 3H, CH_3); 4.51 (s, 2H, CH_2); 7.07–7.09 (d, 2H, CHAr); 7.28–7.30 (d, 2H, CHAr); 7.51–7.53 (d, 2H, CHAr); 14.04 (s, 1H, NH).

3-(Benzoxazole-2′-yl-mercapto-methyl)-4-(p-methoxyphenyl)-5-mercapto-1,2,4-triazole (Compound XVI), Physical state: White-ivory, crystalline powder; M_p = 213–215 °C. FT-IR (KBr,ν, cm^{-1}): 3100 (NH); 1628 (C=N); 1492 (C=S); 837 (para-disubstituted benzene nucleus); 783 (S-CH_2). ^{1}H-NMR (DMSO-d_6, 400 MHz), δ ppm: 3.46 (t, 3H, OCH_3); 4.29 (s, 2H, CH_2); 7.09–7.11 (d, 2H, CHAr); 7.25–7.27 (d, 2H, CHAr); 7.53–7.55 (d, 2H, CHAr); 14.09 (s, 1H, NH).

3-(Benzoxazole-2′-yl-mercapto-methyl)-4-(p-bromoyphenyl)-5-mercapto-1,2,4-triazole (Compound XVII), Physical state: Brown-red, crystalline powder; M_p = 202–204 °C. FT-IR (KBr,ν, cm^{-1}): 3328 (NH); 1617 (C=N); 1494 (C=S); 914 (para-disubstituted benzene nucleus); 780 (S-CH_2). ^{1}H-NMR (DMSO-d6, 400 MHz), δ ppm: 4.17 (s, 2H, CH_2); 7.20–7.23 (d, 2H, CHAr); 7.47–7.50 (d, 2H, CHAr); 7.75–7.77 (d, 2H, CHAr); 14.21 (s, 1H, NH).

3-(Benzoxazole-2′-yl-mercapto-methyl)-4-(p-chloroyphenyl)-5-mercapto-1,2,4-triazole (Compound XVIII), Physical state: White-ivory, crystalline powder; M_p = 189–191 °C. FT-IR (KBr,ν, cm^{-1}): 2894 (NH); 1615 (C=N); 1493 (C=S); 899 (para-disubstituted benzene nucleus); 765 (S-CH_2); 746 (c-Cl). ^{1}H-NMR (DMSO-d_6, 400 MHz), δ ppm: 4.64 (s, 2H, CH_2); 7.31–7.33 (d, 2H, CHAr); 7.42–7.44 (d, 2H, CHAr); 7.73–7.75 (d, 2H, CHAr); 14.29 (s, 1H, NH).

3-(Benzoxazole-2′-yl-mercapto-methyl)-4-(p-iodoyphenyl)-5-mercapto-1,2,4-triazole (Compound XIX), Physical state: Brown-red, crystalline powder; M_p = 222–224 °C. FT-IR (KBr,ν, cm^{-1}): 3038 (NH); 1614 (C=N); 1495 (C=S); 915 (para-disubstituted benzene nucleus); 770 (S-CH_2); 748 (C-I). ^{1}H-NMR (DMSO-d_6, 400 MHz), δ ppm: 4.23 (s, 2H, CH_2); 7.35–7.37 (d, 2H, CHAr); 7.42–7.45 (d, 2H, CHAr); 7.59–7.61 (d, 2H, CHAr); 14.33 (s, 1H, NH).

3.6. Antibacterial Activity

The microbial strains (*Staphylococcus aureus* (ATCC-2593), *Bacillus subtilis* (ATCC-6638), *Bacillus cereus* (ATCC-10876), *Salmonella entiritidis* (P-1131) and *Escherichia coli* (ATCC-25922)) were grown on Mueller–Hinton agar, with incubation at 37 °C for 48 h. Previously, the tested substances were weighted and dissolved in dimethylsulphoxide (DMSO) to prepare the solutions with the following concentrations: 1 mg/mL, 0.5 mg/mL, and 0.25 mg/mL culture medium. The solutions thus prepared were included in 100 mL of culture medium, previously agarized, melted, and then homogenized. In parallel, the control sample was prepared consisting of glucose agar. The culture media were then distributed in the test tubes and sterilized for 20 min at 120 °C. The bacterial cultures were prepared according to the manufacture recommendations, using suspensions with a concentration of about $5.2 \cdot 10^7$ CFU/mL.

3.7. Immobilization of 1,2,4-Triazole

Immobilization of the 1,2,4-triazole was performed as follows: 0.2 g powder of PG and PGB1 copolymers were introduced into 50 mL conical flasks over which 20 mL of 1,2,4-triazole solution was added ($C_{1,2,4\text{-triazole}} = 3.6 \times 10^{-4} - 15 \times 10^{-3}$ g/mL). Then, the samples were placed in a thermostated water-bath shaker (Memmert MOO/M01, Schwabach, Germany) and shaken at 180 strokes/minute until equilibrium was reached. The 1,2,4-triazole immobilization studies on PG and PGB1 copolymers were performed at different temperatures: 25, 30, and 35 °C and for different contact times. The conical flasks were removed from the water-bath shaker, and the copolymers were centrifuged at 78 RCF for 10 min. Then, the samples were separated by filtration, and the amount of 1,2,4-triazole immobilized on PG and PGB1 copolymers was determined by UV-VIS spectrophotometry (SPEKOL 1300 spectrophotometer, Analytik Jena, Jena, Germany) at a wavelength of 226 nm, based on a calibration curve. The retained amount of 1,2,4-triazole was calculated by the difference between 1,2,4-triazole concentration in the supernatant before and after sorption process.

The amount of immobilized 1,2,4-triazole was calculated using the following equation:

$$q_e = \frac{(C_0 - C_e) \cdot V}{W} \tag{12}$$

where q_e is the amount of 1,2,4-triazole immobilized onto PG and PGB1 copolymers (mg/g), C_0 is the initial 1,2,4-triazole concentration (mg/mL), C_e is the equilibrium 1,2,4-triazole concentration (mg/mL), V is the volume of 1,2,4-triazole solution (mL), and W is the amount of copolymers. PG and PGB1 copolymers are in the form of powder consisting of irregularly shaped particles.

3.8. Release of 1,2,4-triazole

In vitro 1,2,4-triazole release studies were performed by immersing the PG-T and PGB1-T systems (0.1 g) in 10 mL of simulated gastric fluids of pH = 1.2 for 2 h at 37 °C. The samples were placed in a thermostated water-bath shaker under gentle stirring (50 strokes/minute). Samples (1 μL) of supernatant solution were collected at different time intervals using a microsyringe and then analyzed spectrophotometrically at a wavelength of 226 nm, using a UV-VIS spectrophotometer (Nanodrop ND 100, Wilmington, DE, USA). The amount of 1,2,4-triazole released was determined using the calibration curve.

4. Conclusions

The 2-mercapto-benzoxazole molecule was used to achieve a selective support for some heterocycles with 1,3,4-thiadiazole and 1,2,4-triazole structure, respectively. The addition reaction of benzoxazolyl-2-mercapto-acetic acid hydrazide to aromatic isothiocyanantes was used to obtain a new series of acyl-thiosemicarbazides (compounds II–VII). Applying the cyclization reaction, the 4-substituted acyl-thiosemicarbazides were converted in acid medium into thiadiazole derivatives with benzoxazole residue in the molecule (compounds VIII–XIII) and in basic medium in mercapto-triazole-3,4-disubstituted (compounds XIV–XIX). All new compounds were characterized using elemental and spectral analysis (FT-IR and ^{1}H-NMR). Among the active principles synthesized, 3-(benzoxazole-2′-yl-mercapto-methyl)-4-(p-methoxyphenyl)-5-mercapto-1,2,4 triazole was chosen to be immobilized onto a polymeric support due to its biological properties. The antimicrobial studies have shown that 1,2,4 triazole presents very good antimicrobial activities against several microbial strains, such as *Escherichia coli*, *Salmonella enteritidis*, and *Bacillus subtilis* compared to 1,3,4-thiadiazoles that show good antimicrobial activity against *Staphylococcus aureus*.

Two polymeric supports were used for the immobilization of 1,2,4-triazole derivative (compound XVI): grafted copolymer and grafted copolymer carrying betaine moieties based on gellan and N-vinyl imidazole, which were chosen due to the biocompatibility of the polysaccharide as well as of the polymer/grafts generated by the vinyl monomer and betaine function.

To describe the interactions that occur in sorption processes between 1,2,4-triazole and PG and PGB1 copolymers, Langmuir, Freundlich, and Dubinin–Radushkevich model isotherms were used. Kinetic sorption studies conclude that the Lagergren model best describes the experimental data and confirm that the sorption of 1,2,4-triazole on the grafting copolymers is physical in nature.

The thermodynamic study completed by the kinetic one led to the conclusion that the sorption process of 1,2,4-triazole on the betainized copolymer is more intense than in the case of the grafted copolymer only. The sorption process of 1,2,4-triazole is spontaneous and favored by the increase of temperature. The release of 1,2,4-triazole from polymeric supports proceeds through a complex mechanism controlled by both swelling and diffusion processes. Similar to the results obtained in case of immobilization of the cefotaxime sodium salt, these results confirm the potential of the grafted copolymer with betaine structure as a candidate for developing sustained/controlled drug delivery systems.

Author Contributions: Conceptualization, N.B. and S.R.; methodology, S.R., N.B., and S.V.; validation, V.S., M.P., and J.D.; formal analysis, N.B., A.M.M., and C.C.; investigation, C.L. and V.S.; data curation, S.R., N.B., and S.V.; writing—original draft preparation, N.B. and S.V.; writing—review and editing, M.P. and J.D., visualization and supervision, S.V., M.P., and J.D. All authors have read and agreed to the published version of the manuscript.

Funding: This research received no external funding.

Data Availability Statement: The data presented in this study are available on request from the corresponding author.

Acknowledgments: The authors thank Marius Zaharia and Ana-Lavinia Vasiliu from Petru Poni Institute of Macromolecular Chemistry, Iasi, for their great support and help in the XRD studies and SEM analysis.

Conflicts of Interest: The authors declare no conflict of interest.

Sample Availability: Samples of the compounds are not available from the authors.

References

1. Laraia, L.; Robke, L.; Waldmann, H. Bioactive compound collections: From design to target identification. *Chem* **2018**, *4*, 705–730. [CrossRef]
2. Liechty, W.B.; Kryscio, D.R.; Slaughter, B.V.; Peppas, N.A. Polymers for drug delivery systems. *Annu. Rev. Chem. Biomol. Eng.* **2010**, *1*, 149–173. [CrossRef] [PubMed]
3. Farshori, N.N.; Banday, M.R.; Ahmad, A.; Khan, A.U.; Rauf, A. Synthesis, characterization and in vitro antimicrobial activities of 5-alkenyl/hydroxyalkenyl-2-phenylamine-1,3,4-oxadiazoles and thiadiazoles. *Bioorg. Med. Chem. Lett.* **2010**, *20*, 1933–1938. [CrossRef] [PubMed]
4. Pintilie, O.; Profire, L.; Sunel, V.; Popa, M.; Pui, A. Synthesis and antimicrobial activity of some new 1,3,4-thiadiazole and 1,2,4-triazole compounds having a D,L-methionine moiety. *Molecules* **2007**, *12*, 103–113. [CrossRef] [PubMed]
5. Kadi, A.A.; El-Brollosy, N.R.; Al-Deeb, O.A.; Habib, E.E.; Ibrahim, T.M.; El-Emam, A.A. Synthesis, antimicrobial and antiinflammatory activities of novel 2-(1-adamantyl)-5-substituted-1,3,4-oxadiazoles and 2-(1-adamantylomini)-5-substituted-1,3,4-thiadiazoles. *Eur. J. Med. Chem.* **2007**, *42*, 235–242. [CrossRef] [PubMed]
6. Tan, C.X.; Shi, Y.X.; Weng, J.Q.; Liu, X.H.; Zhao, W.G.; Li, B.J. Synthesis and antifungal activity of novel 1,2,4-triazole derivatives containing 1,2,3-thiadiazole moiety. *J. Heterocycl. Chem.* **2014**, *51*, 690–694. [CrossRef]
7. Klip, N.T.; Capan, G.; Gursoy, A.; Uzun, M.; Satana, D. Synthesis, structure and antifungal evaluation of some novel 1,2,4-triazolylmercaptoacetyl thiosemicarbazide and 1,2,4-triazolylmercaptomethyl-1,3,4 thiadiazole analogs. *J. Enzyme Inhib. Med. Chem.* **2010**, *25*, 126–131. [CrossRef] [PubMed]
8. Xu, J.; Cao, Y.; Zhang, J.; Yu, S.; Zou, Y.; Chai, X.; Wu, Q.; Zhang, D.; Jiang, Y.; Sun, Q. Design, synthesis and antifugal activities of novel 1,2,4-triazole derivatives. *Eur. J. Med. Chem.* **2011**, *46*, 3142–3148. [CrossRef] [PubMed]
9. Gupta, D.; Jain, D.K. Synthesis, antifungal and antibacterial activity of novel 1,2,4-triazole derivatives. *J. Adv. Pharm. Technol. Res.* **2015**, *6*, 141–146. [CrossRef] [PubMed]

10. Solak, N.; Rollas, S. Synthesis and antituberculosis activity of 2-(aryl/alkylamino)-5-(4-aminophenyl)-1,3,4-thiadiazoles and their Schiff bases. *Arkivoc* **2006**, *12*, 173–181. [CrossRef]
11. El-Reedy, A.A.M.; Soliman, N.K. Synthesis, biological activity and molecular modeling study of novel 1,2,4 triazolo[4,3-b] [1,2,4,5] tetrazines and 1,2,4-triazolo [4,3-b] [1,2,4] triazines. *Sci. Rep.* **2020**, *10*, 6137. [CrossRef] [PubMed]
12. Amir, M.; Kumar, H.; Javed, S.A. Condensed bridgehead nitrogen heterocyclic system: Synthesis and pharmacological activities of 1,2,4-triazolo-[3,4-b]-1,3,4-thiadiazole derivatives of ibuprofen and biphenyl-4-yloxy acetic acid. *Eur. J. Med. Chem.* **2008**, *43*, 2056–2066. [CrossRef] [PubMed]
13. Kumar, H.; Javed, S.A.; Khan, S.A.; Amir, M. 1,3,4-Oxadiazole/thiadiazole and 1,2,4-triazole derivatives of bisphenyl-4-yloxy acetic acid: Synthesis and preliminary evaluation of biological properties. *Eur. J. Med. Chem.* **2008**, *43*, 2688–2698. [CrossRef] [PubMed]
14. Salgin-Goksen, U.; Gokhan-Kelekci, N.; Goktas, O.; Koysa, Y.; Kilic, E.; Isik, S.; Aktay, G.; Ozalp, M. 1-Acylthiosemicarbazides, 1,2,4-triazole-5(4*H*)-thiones, 1,3,4-thiadiazoles and hydrazones containing 5-methyl-2-benzoxazolinones: Synthesis, analgesic-antiinflammatory and antimicrobial activities. *Bioorg. Med. Chem.* **2007**, *15*, 5738–5751. [CrossRef] [PubMed]
15. Amir, M.; Khan, M.S.Y.; Zaman, M.S. Synthesis, characterization and biological activities of substituted oxadiazole, triazole, thiadiazole and 4-thiazolidinone derivatives. *Indian J. Chem.* **2004**, *43B*, 2189–2194. [CrossRef]
16. Clerici, F.; Pocar, D.; Guido, M.; Loche, A.; Perlini, V.; Brufani, M. Synthesis of 2-amino-5-sulfanyl-1,3,4-thiadiazole derivatives and evaluation of their antidepressant and anxiolytic activity. *J. Med. Chem.* **2001**, *44*, 931–936. [CrossRef] [PubMed]
17. Cressier, D.; Prouillac, C.; Hernandez, P.; Amourette, C.; Diserbo, M.; Lion, C.; Rima, G. Synthesis, antioxidant properties and radioprotective effects ofnew benzothiazoles and thiadiazoles. *Bioorg. Med. Chem.* **2009**, *17*, 5275–5284. [CrossRef] [PubMed]
18. Sharma, P.C.; Bansal, K.K.; Sharma, A.; Sharma, D.; Deep, A. Thiazole-containing compounds as therapeutic targets for cancer therapy. *Eur. J. Med. Chem.* **2020**, *188*, 112016. [CrossRef] [PubMed]
19. Gomha, S.M.; Abdelhamid, A.O.; Abdelrehem, N.A.; Kandeel, S.M. Efficient synthesis of new benzofuran-based thiazoles and investigation of their cytotoxic activity against human breast carcinoma cell lines. *J. Heterocycl. Chem* **2018**, *55*, 995–1001. [CrossRef]
20. Padmavathi, V.; Reddy, G.S.; Padmaja, A.; Kondaiah, P.; Shazia, A. Synthesis, antimicrobial and cytotoxic activities of 1,3,4-oxadiazoles, 1,3,4-thiadiazoles and 1,2,4-triazoles. *Eur. J. Med. Chem.* **2009**, *44*, 2106–2112. [CrossRef] [PubMed]
21. Gomha, S.M.; Kheder, N.A.; Abdelaziz, M.R.; Mabkhot, Y.N.; Alhajoj, A.M. A facile synthesis and anticancer activity of some novel thiazoles carrying 1,3,4-thiadiazole moiety. *Chem. Cent. J.* **2017**, *11*, 25. [CrossRef] [PubMed]
22. Choritos, G.; Trafalis, D.T.; Delezis, P.; Potamitis, C.; Sarli, V.; Zoumpoulakis, P.; Camoutsis, C. Synthesis and anticancer activity of novel 3,6-disubstituted 1,2,4-triazolo-[3,4-b]-1,3,4-thiadiazole derivatives. *Arabian J. Chem.* **2019**, *12*, 4784–4794. [CrossRef]
23. Jain, A.K.; Sharma, S.; Vaidya, A.; Ravichandran, V.; Agrawal, R.K. 1,3,4-Thiadiazole and itas derivatives: A review on recent progress in biological activities. *Chem. Biol. Drug Des.* **2013**, *81*, 557–576. [CrossRef] [PubMed]
24. Chen, J.; Sun, X.Y.; Chai, K.Y.; Lee, J.S.; Song, M.S.; Quan, Z.S. Synthesis and anticonvulsant evaluation of 4-(4-alkoxylphenyl)-3-ethyl-4H-1,2,4-triazoles as open-chain analogues of 7-alkoxyl-4,5-dihydro[1,2,4]triazolo[4,3-a]quinolines. *Bioorg. Med. Chem.* **2007**, *15*, 6775–6781. [CrossRef] [PubMed]
25. Datar, P.A.; Deokule, T.A. Development of thiadiazole as an antidiabetic agent-A review. *Mini Rev. Med. Chem.* **2014**, *14*, 136–153. [CrossRef] [PubMed]
26. Van der Nieuwendijk, A.M.C.H.; Pietra, D.; Heitman, L.; Goblyos, A.; IJzerman, A.P. Synthesis and biological evaluation of 2,3,5 substituted [1,2,4] thiadiazoles as allosteric modulators of adenosine receptors. *J. Med. Chem.* **2004**, *47*, 663–672. [CrossRef]
27. Leung-Toung, R.; Wodzinka, J.; Li, W.; Lowrie, J.; Kukreja, R.; Desilets, D.; Karimian, K.; Tam, T.F. 1,2,4 thiadiazole: A novel Cathepsin B inhibitor. *Bioorg. Med. Chem.* **2003**, *11*, 5529–5537. [CrossRef] [PubMed]
28. Sayeed, I.B.; Vishnuvardhan, M.V.P.S.; Nagarajan, A.; Kantevari, S.; Kamal, A. Imidazopyridine linked triazoles as tubulin inhibitors, effectively triggering apoptosis in lung cancer cell line. *Bioorg. Chem.* **2018**, *80*, 714–720. [CrossRef] [PubMed]
29. Desai, T.; Desai, V.; Shingade, S. In-vitro Anti-cancer assay and apoptotic cell pathway of newly synthesized benzaxazole-N-heterocyclic hybrids as potent tyrosine kinase inhibitors. *Bioorg. Chem.* **2020**, *94*, 103382. [CrossRef] [PubMed]
30. Yuan, X.; Yang, Q.; Liu, T.; Li, K.; Liu, Y.; Zhu, C.; Zhang, Z.; Li, L.; Zhang, C.; Xie, M.; et al. Design, synthesis and in vitro evaluation of 6-amide-2-aryl benzoxazole/benzimidazole derivatives against tumor cells by inhibiting VEGFR-2 kinase. *Eur. J. Med. Chem.* **2019**, *179*, 147–165. [CrossRef] [PubMed]
31. Belal, A.; Abdelgawad, M.A. New benzothiazole/benzoxazole-pyrazole hybrids with potential as COX inhibitors: Design, synthesis and anticancer activity evaluation. *Res. Chem. Intermed.* **2017**, *43*, 3859–3872. [CrossRef]
32. Padalkar, V.S.; Borse, B.N.; Gupta, V.D.; Phatangare, K.R.; Patil, V.S.; Umape, P.G.; Sekar, N. Synthesis and antimicrobial activity of novel 2-substituted benzimidazole, benzoxazole and benzothiazole derivatives. *Arab. J. Chem.* **2016**, *9*, S1125–S1130. [CrossRef]
33. Seth, K.; Garg, S.K.; Kumar, R.; Purohit, P.; Meena, V.S.; Goyal, R.; Banerjee, U.C.; Chakraborti, A.K. 2-(2-Arylphenyl)benzoxazole as a novel anti-inflammatory scaffold: Synthesis and biological evaluation. *ACS Med. Chem. Lett.* **2014**, *5*, 512–516. [CrossRef] [PubMed]
34. Sheng, C.; Xu, H.; Wang, W.; Cao, Y.; Dong, G.; Wang, S.; Che, X.; Ji, H.; Miao, Z.; Yao, J.; et al. Design, synthesis and antifungal activity of isosteric analogues of benzoheterocyclic N-myristoyltransferase inhibitors. *Eur. J. Med. Chem.* **2010**, *45*, 3531–3540. [CrossRef] [PubMed]

35. Ryu, S.K.; Lee, R.Y.; Kim, N.Y.; Kim, Y.H.; Song, A.L. Synthesis and antifungal activity of benzo[d]oxazole-4,7-diones. *Bioorg. Med. Chem. Lett.* **2009**, *19*, 5924–5926. [CrossRef]
36. Jauhari, P.K.; Bhavani, A.; Varalwar, S.; Singhal, K.; Raj, P. Synthesis of some novel 2-substituted benzoxazoles as anticancer, antifugal and antimicrobial agents. *Med. Chem. Res.* **2008**, *17*, 412–424. [CrossRef]
37. Haider, S.; Alam, M.S.; Hamid, H.; Shafi, S.; Nargotra, A.; Mahajan, P.; Nazreen, S.; Kalle, A.M.; Kharbanda, C.; Ali, Y.; et al. Synthesis of novel 1,2,3-triazole based benzoxazolinones: Their TNF-β based molecular docking with in-vivo anti-inflammatory, antinociceptive activities and ulcerogenic risk evaluation. *Eur. J. Med. Chem.* **2013**, *70*, 579–588. [CrossRef] [PubMed]
38. Kakkar, S.; Kumar, S.; Narasimhan, B.; Lim, S.M.; Ramasamy, K.; Mani, V.; Shah, S.A.A. Design, synthesis and biological potential of heterocyclic benzoxazole scaffolds as promising antimicrobial and anticancer agents. *Chem. Cent. J.* **2018**, *12*, 96. [CrossRef] [PubMed]
39. Duroux, R.; Renault, N.; Cuelho, J.E.; Agouridas, L.; Blum, D.; Lopes, L.V.; Melnyk, P.; Yous, S. Design, synthesis and evaluation of 2-aryl benzoxazoles as promising hit for the A_{2A} receptor. *J. Enzyme Inhib. Med. Chem.* **2017**, *32*, 850–864. [CrossRef] [PubMed]
40. Kale, M.; Chavan, V. Exploration of the biological potential of benzoxazoles: An overview. *Mini Rev. Org. Chem.* **2019**, *16*, 111–126. [CrossRef]
41. Padmavathi, V.; Kumara, C.P.; Venkatesh, B.C.; Padmaja, A. Synthesis and antimicrobial activity of amido linked pyrrolyl and pyrazolyl-oxazoles, thiazoles and imidazoles. *Eur. J. Med. Chem.* **2011**, *46*, 5317–5326. [CrossRef] [PubMed]
42. Racovita, S.; Baranov, N.; Macsim, A.M.; Lionte, C.; Cheptea, C.; Sunel, V.; Popa, M.; Vasiliu, S.; Desbrieres, J. New grafted copolymers carrying betaine units based on gellan and N-vinylimidazole as precursors for design of drug delivery systems. *Molecules* **2020**, *25*, 5451. [CrossRef] [PubMed]
43. Aparaschivei, R.; Holban, M.; Sunel, V.; Popa, M.; Desbrieres, J. Synthesis and characterization of new heterocyclic compounds with potential antituberculosis activity and their immobilization on polymer supports. *Cellul. Chem. Technol.* **2012**, *46*, 301–306.
44. Cheptea, C.; Sunel, V.; Desbrieres, J.; Popa, M. Synthesis and antimicrobial activity of new derivatives of 1,3,4-thiadiazoles and 1,2,4-triazoles with 5-nitroindazole as support. *J. Heterocycl. Chem.* **2013**, *50*, 366–372. [CrossRef]
45. Moise, M.; Sunel, V.; Profire, L.; Popa, M.; Desbrieres, J.; Peptu, C. 1,4-Disubstituted thiosemicarbazides with potential tuberculostatic action. *Bull. Inst. Polit. Iasi.* **2009**, *55*, 57–64.
46. Moise, M.; Sunel, V.; Profire, L.; Popa, M.; Desbrieres, J.; Peptu, C. Synthesis and biological activity of some new 1,3,4-thiadiazole and 1,2,4-triazole compounds containing a phenylalanine moiety. *Molecules* **2009**, *14*, 2621–2631. [CrossRef] [PubMed]
47. Hussain, S.; Sharma, J.; Amir, M. Synthesis and antimicrobial activities of 1,2,4-triazole and 1,3,4-thiadiazole derivatives of 5-amino-2-hydroxybenzoic acid. *J. Chem.* **2008**, *5*, 963–968. [CrossRef]
48. Cheptea, C.; Sunel, V.; Morosanu, C.; Dorohoi, D.O. Derivatives of 1,2,4-triazole-3,4-disubstituted based on aminoaxids with potential biological activity. *Rev. Chim.* **2020**, *71*, 211–221. [CrossRef]
49. Langmuir, I. The adsorption of gases on plane surface of glass, mica and platinum. *J. Am. Chem. Soc.* **1918**, *40*, 1361–1368. [CrossRef]
50. Weber, T.W.; Chakravot, R.K. Pore and solid diffusion models for fixed bed adsorbents. *AiChE J.* **1974**, *20*, 228–238. [CrossRef]
51. Foo, K.Y.; Hameed, B.H. Insights into the modeling of adsorption isotherm systems. *Chem. Eng. J.* **2010**, *156*, 2–10. [CrossRef]
52. Freundlich, H.M.F. Over the adsorption in solution. *J. Phys. Chem.* **1906**, *57*, 385–470.
53. Dubinin, M.M.; Zaverina, E.D.; Radushkevich, L.V. Sorption and structure of active carbons. I. Adsorption of organic vapors. *Zhurnal Fizicheskoi Khimii.* **1947**, *21*, 1351–1362.
54. Chabani, M.; Amrane, A.; Bensmaili, A. Kinetic modelling of the adsorption of nitrates by ion exchange resin. *Chem. Eng. J.* **2006**, *125*, 111–117. [CrossRef]
55. Yu, Y.; Zhuang, Y.Y.; Wang, Z.H.; Qiu, M.Q. Adsorption of water-soluble dyes onto modified resin. *Chemosphere* **2004**, *54*, 425–430. [CrossRef]
56. Rodriguez, A.; Garcia, J.; Ovejero, G.; Mestanza, M. Adsorption of anionic and cationic dyes on activated carbon from aqueous solutions: Equilibrium and kinetics. *J. Hazard. Mat.* **2009**, *172*, 1311–1320. [CrossRef]
57. Cigu, T.A.; Vasiliu, S.; Racovita, S.; Lionte, C.; Sunel, V.; Popa, M.; Cheptea, C. Adsorption and release studies of new cephalosporin from chitosan-g-poly(glycidyl methacrylate) microparticles. *Eur. Polym. J.* **2016**, *82*, 132–156. [CrossRef]
58. Lagergren, S.; Svenska, B.K. Zur theorie der sogenannten adsorption geloester stoffe. *Kungl. Sven. Vetenskapsakad. Handl.* **1898**, *24*, 1–3.
59. Ho, Y.S.; McKay, G. The kinetics of sorption of basic dyes from aqueous solutions by sphagnum moss peat. *Can. J. Chem. Eng.* **1998**, *76*, 822–827. [CrossRef]
60. Higuchi, W.I. Diffusional models useful in biopharmaceutics. *J. Pharm. Sci.* **1967**, *56*, 315–324. [CrossRef]
61. Korsmeyer, R.W.; Gurny, R.; Doelker, E.; Buri, P.; Peppas, N.A. Mechanisms of solute release from porous hydrophilic polymers. *Int. J. Pharm.* **1983**, *15*, 25–35. [CrossRef]

molecules

MDPI

Article

Effect of the Addition of Alginate and/or Tetracycline on Brushite Cement Properties

Claudia Morilla [1,2,3], Elianis Perdomo [4], Ana Karla Hernández [1], Ramcy Regalado [1], Amisel Almirall [1], Gastón Fuentes [1,2,*], Yaima Campos Mora [1,2], Timo Schomann [2,3], Alan Chan [3] and Luis J. Cruz [2]

1 Biomaterials Center, University of Havana, La Habana 10400, Cuba; C.Morilla_Espino@lumc.nl (C.M.); anakarla@estudiantes.fq.uh.cu (A.K.H.); ramcy@estudiantes.fq.uh.cu (R.R.); amisel@biomat.uh.cu (A.A.); mana@biomat.uh.cu (Y.C.M.)
2 Translational Nanobiomaterials and Imaging Group, Department of Radiology, Leiden University Medical Center, 2333 ZA Leiden, The Netherlands; T.Schomann@lumc.nl (T.S.); L.J.Cruz_Ricondo@lumc.nl (L.J.C.)
3 Percuros B.V., 2333 CL Leiden, The Netherlands; achan@percuros.com
4 Faculty of Automatic and Biomedical Engineering, Technological University of Havana, La Habana 11300, Cuba; elianisperdomo@gmail.com
* Correspondence: gastonfe@biomat.uh.cu or gastonfe@gmail.com

Abstract: Calcium phosphate cements have the advantage that they can be prepared as a paste that sets in a few minutes and can be easily adapted to the shape of the bone defect, which facilitates its clinical application. In this research, six formulations of brushite (dicalcium phosphate dihydrated) cement were obtained and the effect of the addition of sodium alginate was analyzed, such as its capacity as a tetracycline release system. The samples that contain sodium alginate set in 4 or 5 min and showed a high percentage of injectability (93%). The cements exhibit compression resistance values between 1.6 and 2.6 MPa. The drug was released in a range between 12.6 and 13.2% after 7 days. The antimicrobial activity of all the cements containing antibiotics was proven. All samples reached values of cell viability above 70 percent. We also observed that the addition of the sodium alginate and tetracycline improved the cell viability.

Keywords: brushite; calcium phosphate cements; physical-chemical properties; mechanical properties; microbiological properties; cell viability

Citation: Morilla, C.; Perdomo, E.; Hernández, A.K.; Regalado, R.; Almirall, A.; Fuentes, G.; Campos Mora, Y.; Schomann, T.; Chan, A.; Cruz, L.J. Effect of the Addition of Alginate and/or Tetracycline on Brushite Cement Properties. *Molecules* **2021**, *26*, 3272. https://doi.org/10.3390/molecules26113272

Academic Editors: Marcel Popa and Leonard Ionut Atanase

Received: 24 April 2021
Accepted: 14 May 2021
Published: 28 May 2021

Publisher's Note: MDPI stays neutral with regard to jurisdictional claims in published maps and institutional affiliations.

1. Introduction

Calcium phosphates are biomaterials well known to stimulate bone regeneration and have excellent biocompatibility and bioactivity, which has led to a substantial increase in their use in biomedical applications in the last three decades.

These materials are widely used in various applications of orthopedic and maxillofacial surgery, whether for alveolar ridge augmentation, filling of bone defects, middle ear implants, fusion of spinal vertebrae or in the coating of metal prostheses. They are applied in different ways: as granules, blocks or as cements [1]. Due to their excellent biocompatibility, bioactivity and osteoconductivity, they can be reabsorbed by new bone, by the action of bone cells (osteoclasts and osteoblasts) responsible for bone remodeling [2,3]. Since the first research conducted in the 1980s [4], calcium phosphate cements (CPC) have attracted significant interest as a bone substitute. Additionally, due to the malleability of CPC, they have the ability to adapt to bone defects and implant sites, and then harden in situ to provide stability and support [5]. Unlike other materials, these biomaterials can repair bone defects permanently [6], promoting the formation of new bone tissue during cement degradation [7] due to their osteoconductivity [8,9]. In addition, the characteristics of calcium phosphate cements make them an excellent alternative for the release of drugs and other active ingredients, including growth factors and cells [10,11].

Inorganic CPCs often have critical drawbacks that limit their possible clinical application, including a lack of injectability [12,13] that is generally characterized by phase separation during injection, low mechanical properties for the loading requirements of the implantation site [14,15] as well as a weak cohesion that results in the disintegration of the cement paste when in contact with physiological fluids [2,13].

Brushite cements (dicalcium phosphate dihydrated, DCPD, $CaHPO_4{\cdot}2H_2O$) are prepared by mixing water with a powder consisting of an acid calcium phosphate (monocalcium phosphate monohydrated, MCPM, $CaHPO_4{\cdot}H_2O$) and a basic calcium phosphate (β-tricalcium phosphate, β-TCP, β-$Ca_3(PO_4)_2$). The result of this mixture is a moldable paste that solidifies by an exothermic reaction, forming a hard material. These materials were first described in 1989 by Mirtchi and Lemaître [6,16,17].

The first studies performed showed that brushite cements, despite their biocompatibility, are difficult to handle, their setting time is too short (usually less than 30 s) and they have poor mechanical properties [2,12,13,18].

Different additives have been added to these cements to improve some of their properties like injectability [5], cohesion and mechanical properties [19], and the setting time [6]. Polymers have been proven to enhance the mechanical properties of cements due to the role they fulfill in the bone itself, which is a composite material made of an organic phase reinforced with hydroxyapatite crystals. Between the natural polymers, sodium alginate has been studied for many biomedical applications because it is biocompatible, biodegradable and able to form hydrogels. Sodium alginate hydrogels can be prepared under mild conditions by ionic crosslinking and shows a structural similarity to the extracellular matrices of living tissues, which leads to use in applications such as the administration of bioactive agents, the healing of wounds and in tissue engineering [13,20–26].

Other concerns in maxillofacial surgical operations include the risk of infections that demand the use of antibiotics. A large number of infections in bone implants are caused by bacteria, which are very common in the oral cavity and are related to periodontal diseases. However, antibiotics generally have a negative effect on the mechanical properties of cements due to two effects: the increase in the porosity and the inhibitory effect on the setting reaction [10]. One of the most widely used antibiotics in stomatology is tetracycline, which is known as a very effective antibiotic, with a broad spectrum against bacterial infections, generally related to periodontal diseases [27,28].

In our present study, several brushite or DCPD bone cements for maxillofacial applications made from MCPM and β-TCP with or without sodium alginate were prepared and evaluated as drug release systems for tetracycline. Although the use of CPC as a drug delivery system has been analyzed, the study of more complex formulations that include reinforcement materials, such as sodium alginate, could have a significant impact on the development of more efficient bone regenerative biomaterials with the capability to be used as a drug delivery system and an injectable restoration biomaterial.

2. Results

All the cements prepared for our study formed a malleable paste that set in about 2 to 3 min. According to thermochemical studies of calcium phosphate minerals, the dissolution of MCPM in calcium and phosphate ions must be exothermic. In the presence of water, the MCPM tends to hydrolyze in diphosphate and calcium ions following the exothermic reaction of Equation (1). Simultaneously, the exothermic dissolution of β-TCP occurs as a result of its exposure to the acidic medium (Equation (2)). Following the initial dissolution of the reagents, the cement undergoes an increase in pH as a result of the exothermic precipitation of brushite crystals (Equation (3)). Finally, the overall reaction of the cement (Equation (4) = Equation (1) + Equation (2) + 4 × Equation (3)) is exothermic and the brushite cements are usually slightly heated in the final set reaction [6].

$$Ca(H_2PO_4)_2{\cdot}H_2O = Ca^{2+} + 2H_2PO_4^{-} + H_2O \quad (1)$$

$$\beta\text{-}Ca_3(PO_4)_2 + 4H^+ = 3Ca^{2+} + 2H_2PO_4^{-} \quad (2)$$

$$Ca^{2+} + H_2PO_4^{-} + 2H_2O = CaHPO_4{\cdot}2H_2O + H^+ \quad (3)$$

$$\beta\text{-}Ca_3(PO_4)_2 + Ca(H_2PO_4)_2{\cdot}H_2O + 7H_2O = 4CaHPO_4{\cdot}2H_2O \quad (4)$$

It is known that DCPD is a precursor of hydroxyapatite in aqueous solutions, which is thermodynamically more stable, by a dissolution–reprecipitation mechanism. This compound has a relatively low solubility. Therefore, just the presence of water is not enough to trigger the reprecipitation mechanism. However, the aqueous medium in which the DCPD is immersed contains Ca^{2+} ions. In the presence of Ca^{2+} ions, the process occurs according to Equation (5) [29].

$$6\ CaHPO_4{\cdot}2H_2O + (4-x)\ Ca_3{}^{2+} = Ca_{10-X}(HPO_4) \times (PO_4)_{6-X}(OH)_{2-X} + (10+x)\ H_2O + (8-2x)\ H^+ \quad (5)$$

2.1. X-ray Diffraction (XRD)

Figure 1 shows the XRD pattern of the cement samples A0T0, A0T1, A2T0, A2T1, A5T0 and A5T1 after 72 h of immersion in Ringer's solution. The most significant peaks of the pattern were compared with the XRD pattern of DCPD, which is the expected product of the setting reaction of the cement, using the ICDD PDF 9-0077. A congruence in both position and intensity of the peaks for all samples were observed, which confirms the occurrence of the setting reaction and the precipitation of brushite crystals. It can be observed that the addition of sodium alginate and/or tetracycline do not affect the setting reaction of the cements.

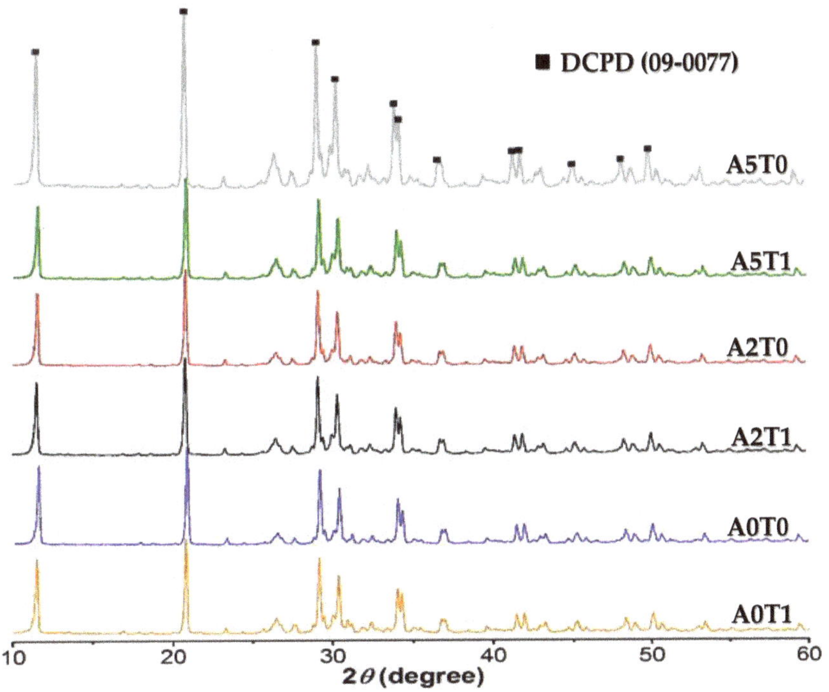

Figure 1. XRD pattern of samples. All peaks correspond to DCPD according to the ICDD PDF 9-0077 X-ray diffraction pattern. The six-maximum intensity DCPD peaks could be observed between 12° and 35° at 2 θ.

Table 1 shows the crystallite size of the cements. The addition of alginate causes a decrease in the size of the crystals while the incorporation of the drug shows the opposite behavior, but not enough to compensate the effect of the polymer. In none of the cases was there a significant difference between any of the values, whether analyzing the variation of sodium alginate or tetracycline.

Table 1. Crystallite size of the cements.

Samples	A0T0	A0T1	A2T0	A2T1	A5T0	A5T1
Crystallite Size (nm)	58 ± 7	57 ± 7	54 ± 4	56 ± 7	50 ± 4	58 ± 7

2.2. *Morphology by Electronic Microscopy*

2.2.1. Scanning Electronic Microscopy (SEM)

In Figure 2, the micrographs of the samples showed mostly the presence of small size pores, although the presence of macropores can also be observed. In the sample A2T0, we can detect that the presence of alginate causes a low porosity due to its uniform distribution among the calcium phosphate crystals. In A5T0, with the increase of the alginate concentration, the number of pores decreased, although there is a presence of a non-uniform and rough surface that was not observed in the A2T0.

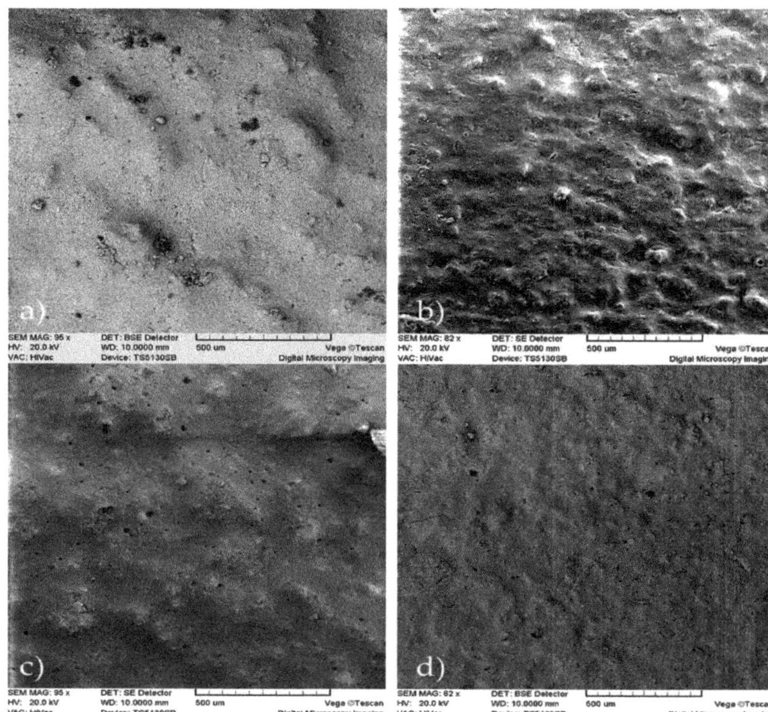

Figure 2. SEM micrographs of the samples: (**a**) A2T0, (**b**) A2T1, (**c**) A5T0 and (**d**) A5T1, (scale bar = 500 μm).

Comparing samples A2T0 and A2T1, it was determined that the presence of tetracycline causes a superficial uniformity, but at the same time there is an increased presence of macropores and deformities that are not observed in the sample lacking tetracycline. When comparing A5T0 with A5T1, a larger roughness of the surface with the presence of macropores is observed in A5T0 that was not formed in A5T1, which is characterized by a more uniform surface and is devoid of macropores.

In A2T1 and A5T1, both samples with tetracycline and different concentrations of alginate, a superficial roughness can be observed as well as a greater number of macropores in A2T1.

2.2.2. Transmission Electronic Microscopy (TEM)

In Figure 3, the transmission micrographs of the cements show agglomerates of spherical particles of around 50 nm in size. This particle size corresponds with the crystallite size calculated previously (Table 1), and is evidence that increases in the amount of alginate induces a decrease in the size of the crystals. The addition of tetracycline affects the particles size in the opposite way, but the influence is less marked, and cannot counteract the impact of the alginate.

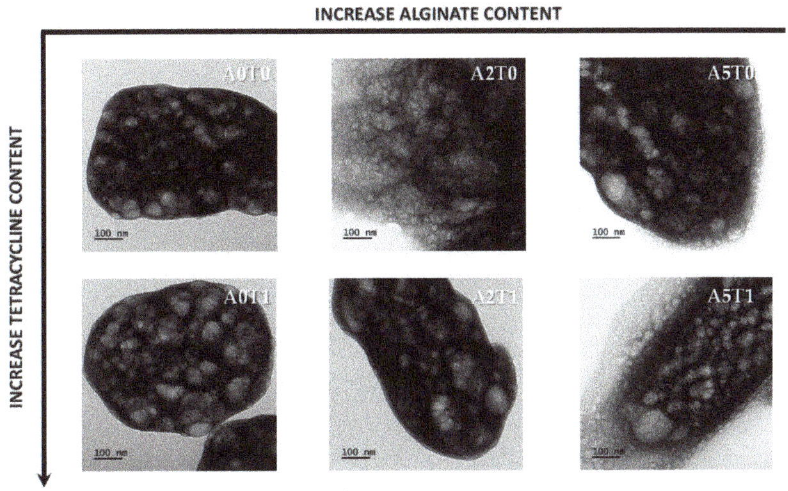

Figure 3. TEM micrographs of all samples at the same magnification.

2.3. Mechanical Properties

Table 2 shows the results of the mechanical property (compression strength) evaluation. Values between 1.6 and 2.6 MPa for the compressive strength of the samples were obtained, which corresponds to what is reported in the literature for this type of cement [13,16,21,30].

Table 2. Compression strength (after 72 h of incubation) and injectability of the samples.

Samples	Compressive Strength (MPa)	Injectability (%)
A0T0	1.6 ± 0.5	72.56
A0T1	1.9 ± 0.3	87.07
A2T0	2.6 ± 0.2	41.71
A2T1	1.9 ± 0.5	93.04
A5T0	2.3 ± 0.7	4.78
A5T1	1.9 ± 0.4	93.43

It is also observed that the samples with sodium alginate and without the presence of tetracycline in their formulations (A2T0 and A5T0) were the ones that reached the highest values. As expected, the addition of sodium alginate increases the mechanical properties of the material; this result corresponds with the effect reported in other studies [30].

2.4. Injectability

The results of the injectability process are shown in Table 2, where it can be seen that the percentage of injectability ranges between 4.78% and 93.43%. When samples do not contain tetracycline, the addition of sodium alginate does not improve the injectability but makes the cement paste more difficult to handle in the first few minutes and therefore decreases the injectability. In the case of sample A5T0, this effect is more marked by the high viscosity of the liquid phase that does not allow the preparation of the paste in the time required for the test. Thus, the measurement that is reported was made approximately 40 s later than the rest of the samples. In samples with tetracycline, the addition of sodium alginate improved the injectability.

2.5. Drug Release Study

In order to study the drug release mechanisms, the Peppas and Sahlin model [31] was used (Equation (6)). In this equation, the first term shows the fraction of drug release. K_1 is the kinetic constant related to the diffusion process and K_2 to the polymer chain relaxation process, and t is the time of release. The diffusional coefficient n is a measure of the diffusion type for a device of any geometric shape that exhibits a controlled release.

$$\frac{M_t}{M_\infty} = K_1 t^n + K_2 t^{2n} \tag{6}$$

The release profile for the first eight hours (around 75% of the release), as well as the calculated parameters for the Peppas and Sahlin Equation, are shown in Figure 3 and Table 3, respectively. In Figure 4, the inset shows the first eight hours of the release study that is governed by a diffusion mechanism.

Table 3. Values of the parameters of the Peppas and Sahlin Equation for each sample (inset Figure 3, first eight hours).

Plot	A0T1	A2T1	A5T1
K_1	0.0559 ± 0.0001	0.0483 ± 0.0002	0.0452 ± 0.0002
K_2	−0.00719 ± 0.00004	−0.00537 ± 0.00005	−0.00482 ± 0.00006
n	0.589 ± 0.003	0.610 ± 0.004	0.620 ± 0.006
R^2	99.99%	99.98%	99.97%

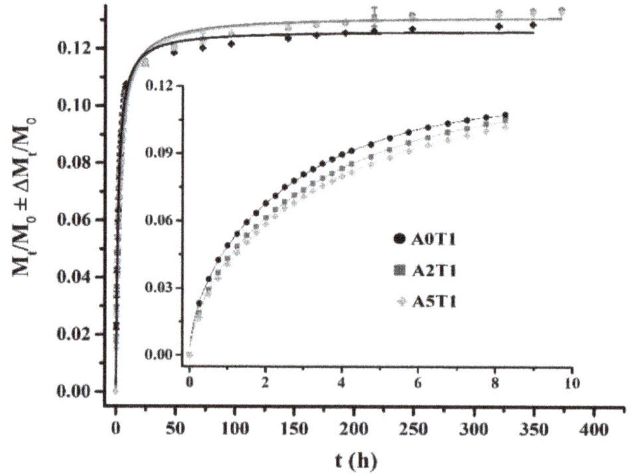

Figure 4. Tetracycline release profiles fitting to Equation (6) (the inset, first 8 h) and Equation (7), from the beginning to the end of the process.

It can be observed that as the amount of alginate increases, the system moves away from diffusion (n = 0.5) to enter a process where the diffusion governs in conjunction with the relaxation of the polymer chains (n > 0.5). The increase in the chain relaxation effect is also described by the increment of K_2, which is the part of the equation associated with the relaxation phenomenon, and the decrease of K_1 that represents the diffusion process (Table 3).

In the inset of Figure 4, it can be seen that the formulation A0T1 showed the highest release in that period of time due to the absence of sodium alginate in the sample, which causes a higher surface porosity and therefore facilitates the release of the drug. However, in the remaining days (full graphic, Figure 3) the other two formulations were the ones that were most released, since A0T1 had already expelled the highest amount of tetracycline in the first stages.

The generalized logistic function or curve, also known as the Richard's curve, is a mathematical function that appears in various models of population growth, and the spread of epidemic diseases and dissemination in social networks. This function constitutes an extension of the sigmoid function for the growth of one magnitude [32], and is considered one of the best options to adjust dissolution curves [33]:

$$y = A2 + \frac{A1 - A2}{1 + (\frac{x}{x_0})^p} \tag{7}$$

where y is the amount of tetracycline released, x is time in minutes, $A1$ is the lower asymptote, $A2$ is the upper asymptote, x_0 is the value of the central node and p is the growth rate.

The values of the parameters for each sample, based on the model of Equation (7), can be seen in Table 4. The asymptote $A1$ shows values very close to zero, while the upper asymptote $A2$ shows values that correspond to the maximum percentage of drug released by each sample. In addition, the p values for each sample are approximately 0.9. Coefficients of determination with a value greater than 99% were obtained for the fitting of all the release curves, which confirms that this model explains more than 99% of the variability in the amount of drug release.

Table 4. Values of the parameters of the logistic curve for each sample (full graphic, Figure 4, 15 days).

Plot	A0T1	A2T1	A5T1
$A1$	0.003 ± 0.002	0.00162 ± 0.00189	0.00142 ± 0.00144
$A2$	0.1266 ± 0.0008	0.1319 ± 0.0008	0.1319 ± 0.0006
x_0	1.70 ± 0.07	2.27 ± 0.09	2.54 ± 0.08
p	0.98 ± 0.03	0.93 ± 0.03	0.92 ± 0.02
R^2	99.58%	99.65%	99.80%

2.6. Microbiological Study

Figure 5 shows the inhibition zones where no bacterial growth occurred due to the release of the drug to the culture medium.

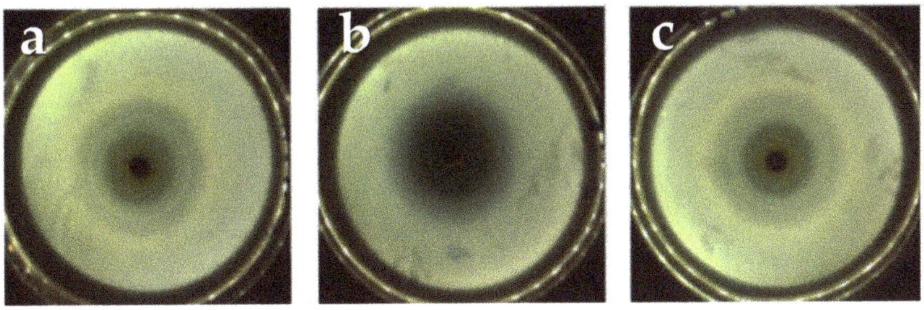

Figure 5. Diameters of the inhibition halos from the A2T1 sample at: (**a**) 24 h, (**b**) 48 h and (**c**) 72 h. The other sample (A5T1) shows a similar behavior.

The measurements of the halos at 24, 48 and 72 h respectively can be observed in Table 5. Here, it is important to notice that the sample A2T1 showed the greatest inhibition halo, which corresponds with the results obtained in the release study.

Table 5. Diameter of the growth inhibition halos (mm) on time.

Samples	24 h	48 h	72 h
A0T1	37 ± 2	37 ± 2	37 ± 2
A2T1	37.0 ± 0.6	38 ± 1	40.1 ± 0.5
A5T1	37.5 ± 0.5	37.6 ± 0.4	37.7 ± 0.3

2.7. pH Study

The pH values of the samples over time are shown in Table 6, where it can be observed that in the first minutes of the setting reaction, the pH remains basic but as the reaction progresses the medium acidifies in accordance with Equations (1)–(4). This effect is more significant with the increase in the sodium alginate contents of the formulations.

Table 6. Investigation of pH values of the samples.

t (min)	t (h)	A0T0	A0T1	A2T0	A2T1	A5T0	A5T1
5	0.08	7.24	7.24	7.26	7.22	7.19	7.23
60	1	7.06	6.87	7.13	7.16	7.11	7.13
120	2	6.92	6.37	7.01	6.69	6.84	6.54
185	3	6.84	6.28	6.79	6.63	6.35	6.42
240	4	6.78	5.93	6.74	6.44	5.98	6.33
300	5	6.74	5.81	6.50	6.22	5.70	6.17
360	6	6.72	5.71	6.41	6.18	5.58	6.12
420	7	6.67	5.68	6.36	6.10	5.49	6.08
1440	24	6.60	5.41	6.29	5.89	4.91	5.87
2304	96	6.47	5.01	6.05	5.60	4.74	5.45

2.8. Cell Viability Test

Figure 6 displays the cell viability percentages of the samples in the first 72 h of the initial setting. All samples show a percentage above 70, which means that the materials are not cytotoxic according to the standard ISO 10993-5. It also can be seen that the addition of sodium alginate and tetracycline improve the cell viability.

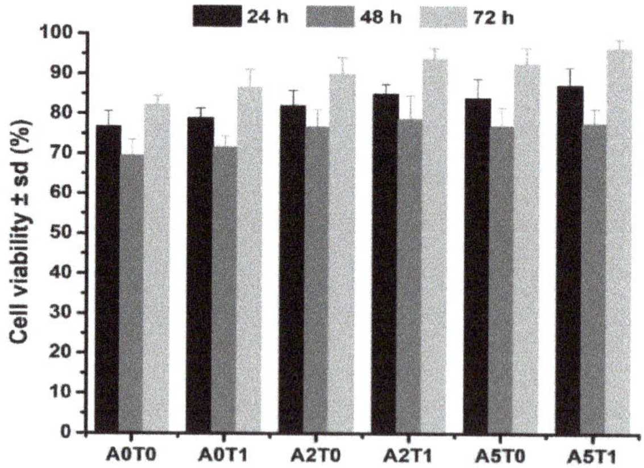

Figure 6. Cell viability assay of the samples at the 24, 48 and 72 h.

3. Discussion

The addition of sodium alginate interferes in the setting reaction, delaying the nucleation of the brushite crystals by making a substitution with the Ca^{2+} ions in the solution. This delay in the initial growth of the crystals causes a marked decrease in the final size of crystallites, as the amount of alginate increases. On the other hand, the presence of the drug also affects the initial growth of the crystals, as explained elsewhere. However, in the presence of alginate, the effect is opposite, but not enough to completely counteract the influence of the drug. The change to a higher value of the cation valence causes an increase in the stability of the alginate complexes that go from being inter- or intramolecular to only being intermolecular at higher valences. Compared with sodium (monovalent cation), the stability of the calcium chelate is remarkably higher than that of sodium, not only for the solubility or viscosity, but several thermodynamic evidences [34,35].

In this study, we showed that the addition of tetracycline in combination with the alginate causes a decrease in the compressive strength of CPC. Tetracycline tends to form chelates with Ca^{2+} ions, which causes a delay in the primary nucleation of the crystals [34,35]. This produces a greater porosity and an increase in setting time, which affects the mechanical properties of cement [10,16,21,23].

The decrease in compressive strength values is expected after 72 h of incubation. The solubility and high hydration capacity of MCPM and the biodegradability of β-TCP are conditioning factors for the low values of mechanical properties, even with the addition of alginate, widely reported and discussed, which should increase them, but competes with the drug's solubility that decreases it [10,13,21].

In the samples with tetracycline, the presence of alginate improved the injectability. Samples with a drug as a salt are more injectable, which is also evidence that the presence of antibiotics affects the setting reaction of the cement, causing it to be retarded. This allows the material to be more fluid for a longer period of time, promoting injectability, as has been reported in previous investigations [10,12,13].

The release of tetracycline in the matrices is guided by a diffusion mechanism in the first few hours, mainly due to the drug that is close to the edges of the matrix and according to the geometric shapes and the drug solubility. In the case of the samples that contain sodium alginate, the release is controlled by a diffusion mechanism in conjunction with the relaxation of the polymer chains. For a second stage, the release no longer depends only on the solubility of the drug, but also on other factors, such as the advance of the release front, the concentration of the drug and the diffusion medium added. In this case, the logistic function that fits the release profile manages to encompass not only the mechanisms of the second stage but also the diffusion mechanisms that govern the first stage, which is why it is able to describe the entire release process [10,30,32,33,36].

The effectivity of the cements as drug release systems was proven by the inhibition zones, within which there was no bacterial growth in the culture medium. The results of the microbiological study demonstrated that the antibiotic not only is released, but also maintains its pharmacological activity [30,37,38].

The pH study reveals that the increment of hydroxyl groups causes a drop in the pH around the cement due to the presence of alginate. The crosslinking among Ca^{2+} and alginate increases the viscosity of the medium in the first stages. In addition, the Ca^{2+} incorporated in the crosslinking process with alginate releases PO_4^{3-} into the medium, contributing to the possibility of a decrease of pH values [24,26,39]. This is an important factor to take into account when viability tests of cells are made, since it is a determinant for the survival of the cells in the in vitro tests.

The addition of sodium alginate, a biocompatible natural polymer, provides a substrate that promotes cell growth, resulting in an increase in the cell viability [12,25,26,40]. Moreover, tetracycline creates a safe nest, free from bacteria, which can positively affect the cell development in the implant.

4. Materials and Methods

4.1. Cement Preparation

All chemicals employed were of analytical grade, used as received. β-TCP was synthetized by a wet neutralization reaction using CaO and H_3PO_4 (both: Merck KGaA, Darmstadt, Germany), following the method described by Carrodeguas and de Aza [41].

For the preparation of the CPC, MCPM (Merck KGaA, Darmstadt, Germany) and β-TCP were mixed in the solid phase and, depending on the experiment, 2 or 5% of sodium alginate was added to the liquid phase as a reinforcement. In order to determine their possible uses as a drug release system, 1% tetracycline (Ningxia Qiyuan Pharmaceutical Co., Yinchuan, China) was added as an antibiotic. A solution of sodium citrate was used as a liquid phase of the cements and as a setting retarder. A comparative study to analyze the composition effect on compressive strength, injectability, drug release and antimicrobial activity was carried out. The effect of the independent variables, the quantity of β-TCP used and the addition or not of sodium alginate and/or tetracycline, was studied through the experiments described in Table 7.

Table 7. Experimental design, MCPM/β-TCP = 45/55%. Liquid phase = 0.5 mL/g.

Series	Sodium Alginate (% *w/w*)	Tetracycline (% *w/w*)
A0T0	0	0
A0T1	0	1
A2T0	2	0
A2T1	2	1
A5T0	5	0
A5T1	5	1

4.2. X-ray Diffraction (XRD)

Phase characterization was carried out by means of X-ray diffraction, in a Rigaku Rotaflex, RU200B diffractometer with Cu-K_α radiation (1.54056 nm). The scans were made in a 2 θ angular interval of 10–60° and a scanning speed of 0.02°/min. The results were interpreted using the X'Pert HighScore PANalytical program database, version 3.0 (PANalytical B. V. Almelo, The Netherlands). Crystal size was calculated using the Debye Scherrer tools of the software.

4.3. Electronic Microscopy

4.3.1. Scanning Electronic Microscopy (SEM)

The samples were coated with a 20 nm film of gold in a BAL-TEC MED 020 system and placed in a desiccator until analysis. A JEOL microscope, JSM-6360LV (Jeol Ltd., Tokyo, Japan) with Oxford EDX probe (Oxford Instruments, High Wycombe, UK), magnification of 5–300,000, resolution of 3 nm and acceleration voltage of 30 kV, was used for the microstructural analysis.

4.3.2. Transmission Electron Microscopy (TEM)

Samples were diluted in Milli-Q water. Subsequently, carbon-coated grids (Formvar/Carbon on 200 Mesh Copper; AGS162; Van Loenen Instruments; Zaandam, the Netherlands) were glow-discharged using the Emitech K950X Turbo Evaporator (Quorum Technologies; Ashford, UK) at 2×20^{-1} mbar and 20 mA for 1 min. Next, 3 μL of sample solution were applied on the freshly glow-discharged grid and allowed to adhere for 1 min. Afterwards, excess liquid was discarded by blotting onto a filter paper and the sample was air-dried for 10 min. Grids were mounted on a room temperature holder and examined using a FEI T12 Spirit BioTwin (FEI Company; Hillsboro, OR, USA) equipped with an OneView Camera Model 1095 (Gatan; Pleasanton, CA, USA) at a voltage of 120 kV. Digital images were acquired and stored using DigitalMicrograph 3.4 (Gatan, Pleasanton, CA, USA).

4.4. Mechanical Characterization

For the compressive strength of the material, 12 mm height and 6 mm diameter specimens were prepared. The samples were immersed in Ringer's solution at 37 °C and tested after 24 h of the cement preparation, immediately after being extracted in order to maintain hydration. The study was carried out in a universal testing machine (TestCom-5,

IBERTEST, Madrid, Spain) with a load cell of 200 N and at 1 mm min^{-1} load application speed. The compressive strength (σ_c) in MPa was determined by the following formula:

$$\sigma_c = \frac{F}{A_0} = \frac{4P}{\pi d^2} \cdot 10^{-6} \quad (8)$$

where P is the maximum breaking load (N) and d is the diameter of the specimen (m). Five specimens were tested for each formulation.

4.5. Injectability Study

The injectability of the samples was determined by extruding a certain quantity of the paste placed in a commercial plastic syringe of 5 mL capacity and with an exit diameter in the nozzle of 2 mm [42]. The extrusion was performed by placing the syringe in a universal testing machine (TestCom-5, IBERTEST, Madrid, Spain) using a compression speed of 15 mm/min until reaching a maximum load of 100 N [43]:

$$\%Injectability = \frac{mass\ of\ injected\ material}{total\ mass\ of\ material} \cdot 100\% \quad (9)$$

4.6. Drug Release Study

Test specimens of the cements of 6 mm in height and 12 mm in diameter loaded with tetracycline were used. The samples were immersed in 10 mL of Ringer's solution in glass bottles at (37.0 ± 0.5) °C throughout the study. The solution in contact with the specimens was completely extracted at the established times and replaced with 10 mL of fresh solution. The extractions were made every half hour until 5 h of the cement preparation and, after that, at 24 h up to seven days. Five specimens of each formulation were prepared and evaluated. The determination of the antibiotic released to the solution was carried out in a UV-Visible Spectrophotometer (Shimadzu, Kyoto, Japan) at a wavelength of 276 nm and the results were reported as a cumulative amount of the tetracycline released versus time [44].

4.7. Microbiological Study

For the microbiological study, specimens of 6 mm in height and 12 mm in diameter were prepared. In order to evaluate the antimicrobial susceptibility of the composites, strains of Staphylococcus aureus Agar Tripton were used, at a strain concentration adjusted with a turbidimetric method employing as a reference a 0.5 MacFarland standard (1×10^8 CFU mL^{-1}). Subsequently, 500 µL of a previously prepared culture medium of Mueller-Hinton agar (Merck KGaA, Darmstadt, Germany) was inoculated in Petri dishes. After a period of 20 min the test specimens were placed on top of the plates containing the culture medium and the bacterial suspension and were incubated at 37 ± 1 °C for a period of 72 h. Three specimens were tested for each formulation and the inhibition zone was measured with the software SCAN 500 Automatic Colony Counter Version 6.

4.8. pH Study

For the pH study, the samples of 300 mg approximately, were immersed in 10 mL of PBS at 37 °C of temperature and the pH was measure (HI-83300 pH-meter, Hanna Instruments, Woonsocket, RI, USA) over 7 h of the first day and then at the 24 and 96 h.

4.9. Cell Viability

MTS assay. To further corroborate the cell viability results, an MTS assay was performed. This is a colorimetric technique in which (3-(4,5-dimethylthiazol-2-yl)-5-(3-carboxymethoxyphenyl)-2-(4-sulfophenyl)-2H-tetrazolium), in the presence of phenazine methosulfate (PMS), produces a formazan product that has an absorbance maximum at 490 nm in PBS. Scaffold samples with dimensions like a viability assay were loaded with osteoblastic MC3T3-E1 cells (density: 10^4 per well; 500 µL of cell suspension), and then incubated for 24, 48 and 72 h; 100 μL of the supernatant solution was extracted to a 96-well plate for reading into a tunable, spectrophotometric microplate reader (VersaMax, Molecular Devices, San José, CA, USA with Program Softmax Pro) and the absorbance (λ = 490 nm) was measured.

4.10. Statistical Calculations

Graphs and statistics were performed with OriginPro 2021 (OriginLab Corp., Northampton, MA, USA). Data are reported as mean ± standard deviation (SD), unless stated otherwise. Error bars represent the SD calculated from tests of triplicate measurements for each scaffold.

5. Conclusions

Six DCPD formulations were obtained in which the influence of the addition of sodium alginate and/or tetracycline on magnitudes such as mechanical properties, release capacity, injectability, microbiological response and cell proliferation was studied. The addition of sodium alginate caused an increase in mechanical properties and cell proliferation, as well as release in the final stages. The injectability and the pH values decreased, as well the release in the first stage, due to its dependence on diffusion and where the viscosity provided by the sodium alginate interferes with the process.

The addition of tetracycline had less marked effects. In this case, all the magnitudes increased their value except the case of pH, results that agree with the state of the art of the subject. Certainly, when both substances coincided in the formulation, the values of the properties broke the trend, an unequivocal sign of materials science, where magnifying one property leads to sacrificing the values of another. However, the materials obtained proved to be a promising option in the restoration of bone tissue with the added functionality of a controlled drug release system.

Author Contributions: Conceptualization and methodology: C.M., A.A., G.F., L.J.C.; Software: C.M., G.F.; Investigation, data curation, formal analysis and visualization: C.M., E.P., A.K.H., R.R., Y.C.M., T.S., G.F.; Resources: A.A., T.S., L.J.C.; Writing—original draft preparation: C.M., A.A., G.F.; Writing—review and editing: C.M., G.F., T.S., A.C., L.J.C.; Supervision: G.F., L.J.C.; Project administration and Funding acquisition: A.C., L.J.C. All authors have read and agreed to the published version of the manuscript.

Funding: We would like to acknowledge the financial support from the European Union through Erasmus PLUS doctoral fellowship, project code 2015-1-NL01-KA 107-008639 (C. Morilla and Y. Campos). This work was also supported by project grants from the European Commission H2020-MSCA-RISE (644373—PRISAR), H2020-MSCA-RISE (777682—CANCER), H2020-WIDESPREAD-05-2017-Twinning (807281—ACORN), H2020-WIDESPREAD-2018-03 (852985—SIMICA), H2020-MSCA-RISE-2016 (734684—CHARMED), (952520-BIOSAFETY) (872391-CONCRETE) and MSCA-ITN-2015-ETN (860173, RISE-WELL) (675743—ISPIC), 861190 (PAVE), 857894 (CAST), 859908 (NOVA-MRI); 872860 (PRISAR2). This work was also financially supported by the VIDI personal grant (project number 723.012.110) (Luis J. Cruz).

Data Availability Statement: The data presented in this study are available on request from the corresponding author.

Acknowledgments: We want to express our gratitude to the specialists of the Center for the State Control of Medicines, Equipment and Medical Devices, Havana, Cuba, especially Raisi Morales, Dainik Alvarez and Armando Pérez Peña for their invaluable help in microbiological and mechanical tests.

Conflicts of Interest: The authors declare no conflict of interest.

References

1. Dorozhkin, S.V. Calcium Orthophosphate-Based Bioceramics. *Materials* **2013**, *6*, 3840–3942. [CrossRef]
2. Le Ferrec, M.; Mellier, C.; Lefèvre, F.-X.; Boukhechba, F.; Janvier, P.; Montavon, G.; Bouler, J.-M.; Gauthier, O.; Bujoli, B. In vivo resorption of injectable apatitic calcium phosphate cements: Critical role of the intergranular microstructure. *J. Biomed. Mater. Res. Part B Appl. Biomater.* **2020**, *108*, 367–376. [CrossRef] [PubMed]

3. Mills, D.K. The Role of Polymer Additives in Enhancing the Response of Calcium Phosphate Cement. In *Orthopedic Biomaterials: Progress in Biology, Manufacturing, and Industry Perspectives*; Li, B., Webster, T., Eds.; Springer International Publishing: Cham, Switzerland, 2018; pp. 345–379. [CrossRef]
4. Brown, W.E.; Chow, L.C. Dental Restorative Cement Pastes. U.S. Patent 4518430A, 21 May 1985.
5. O'Neill, R.; McCarthy, H.; Montufar, E.; Ginebra, M.-P.; Wilson, D.; Lennon, A.; Dunne, N. Critical Review: Injectability of Calcium Phosphate Pastes and Cements. *Acta Biomater.* **2016**, *50*, 1–19. [CrossRef] [PubMed]
6. Tamimi, F.; Sheikh, Z.; Barralet, J. Dicalcium phosphate cements: Brushite and monetite. *Acta Biomater.* **2012**, *8*, 474–487. [CrossRef] [PubMed]
7. Sheikh, Z.; Abdallah, M.-N.; Hanafi, A.A.; Misbahuddin, S.; Rashid, H.; Glogauer, M. Mechanisms of in Vivo Degradation and Resorption of Calcium Phosphate Based Biomaterials. *Materials* **2015**, *8*, 7913–7925. [CrossRef] [PubMed]
8. Ko, C.-L.; Chen, J.-C.; Hung, C.-C.; Wang, J.-C.; Tien, Y.-C.; Chen, W.-C. Biphasic products of dicalcium phosphate-rich cement with injectability and nondispersibility. *Mater. Sci. Eng. C* **2014**, *39*, 40–46. [CrossRef]
9. Alge, D.L.; Goebel, W.S.; Gabriel Chu, T.-M. In vitro degradation and cytocompatibility of dicalcium phosphate dihydrate cements prepared using the monocalcium phosphate monohydrate/hydroxyapatite system reveals rapid conversion to HA as a key mechanism. *J. Biomed. Mater. Res. Part B Appl. Biomater.* **2012**, *100*, 595–602. [CrossRef]
10. Ginebra, M.-P.; Canal, C.; Espanol, M.; Pastorino, D.; Montufar, E.B. Calcium phosphate cements as drug delivery materials. *Adv. Drug Deliv. Rev.* **2012**, *64*, 1090–1110. [CrossRef]
11. Vorndran, E.; Geffers, M.; Ewald, A.; Lemm, M.; Nies, B.; Gbureck, U. Ready-to-use injectable calcium phosphate bone cement paste as drug carrier. *Acta Biomater.* **2013**, *9*, 9558–9567. [CrossRef]
12. Hurle, K.; Oliveira, J.M.; Reis, R.L.; Pina, S.; Goetz-Neunhoeffer, F. Ion-doped Brushite Cements for Bone Regeneration. *Acta Biomater.* **2021**, *123*, 51–71. [CrossRef]
13. Luo, J.; Engqvist, H.; Persson, C. A ready-to-use acidic, brushite-forming calcium phosphate cement. *Acta Biomater.* **2018**, *81*, 304–314. [CrossRef] [PubMed]
14. Cui, X.; Huang, C.; Chen, Z.; Zhang, M.; Liu, C.; Su, K.; Wang, J.; Li, L.; Wang, R.; Li, B.; et al. Hyaluronic acid facilitates bone repair effects of calcium phosphate cement by accelerating osteogenic expression. *Bioact. Mater.* **2021**, *6*, 3801–3811. [CrossRef] [PubMed]
15. Fathi, M.; Kholtei, A.; EL Youbi, S.; El Idrissi, B.C. Setting Properties of Calcium Phosphate Bone Cement. *Mater. Today Proc.* **2019**, *13*, 876–881. [CrossRef]
16. Dorozhkin, S.V. Self-Setting Calcium Orthophosphate Formulations. *J. Funct. Biomater.* **2013**, *4*, 209–311. [CrossRef]
17. Mirtchi, A.A.; Lemaitre, J.; Terao, N. Calcium phosphate cements: Study of the β-tricalcium phosphate—Monocalcium phosphate system. *Biomaterials* **1989**, *10*, 475–480. [CrossRef]
18. Guo, H.; Pujari-Palmer, M.; Yu, Y.; Stevensson, B.; Engqvist, H.; Edén, M. Quantitative phase analyses of biomedical pyrophosphate-bearing monetite and brushite cements by solid-state NMR and powder XRD. *Ceram. Int.* **2020**, *46*, 11000–11012. [CrossRef]
19. Engstrandn, J.; Persson, C.; Engqvist, H. The effect of composition on mechanical properties of brushite cements. *J. Mech. Behav. Biomed. Mater.* **2014**, *29*, 81–90. [CrossRef]
20. Ucar, S.; Bjørnøy, S.H.; Bassett, D.C.; Strand, B.L.; Sikorski, P.; Andreassen, J.-P. Nucleation and Growth of Brushite in the Presence of Alginate. *Cryst. Growth Des.* **2015**, *15*, 5397–5405. [CrossRef]
21. Dabiri, S.M.H.; Lagazzo, A.; Aliakbarian, B.; Mehrjoo, M.; Finocchio, E.; Pastorino, L. Fabrication of alginate modified brushite cement impregnated with antibiotic: Mechanical, thermal, and biological characterizations. *J. Biomed. Mater. Res. Part A* **2019**, *107*, 2063–2075. [CrossRef]
22. Dabiri, S.M.H.; Lagazzo, A.; Barberis, F.; Farokhi, M.; Finochio, E.; Pastorino, L. Characterization of alginate-brushite in-situ hydrogel composites. *Mater. Sci. Eng. C* **2016**, *67*, 502–510. [CrossRef]
23. Lee, H.-J.; Kim, B.; Padalhin, A.R.; Lee, B.-T. Incorporation of chitosan-alginate complex into injectable calcium phosphate cement system as a bone graft material. *Mater. Sci. Eng. C* **2019**, *94*, 385–392. [CrossRef] [PubMed]
24. Rödel, M.; Teßmar, J.; Groll, J.; Gbureck, U. Highly flexible and degradable dual setting systems based on PEG-hydrogels and brushite cement. *Acta Biomater.* **2018**, *79*, 182–201. [CrossRef]
25. Roy, M.; DeVoe, K.; Bandyopadhyay, A.; Bose, S. Mechanical property and in vitro biocompatibility of brushite cement modified by polyethylene glycol. *Mater. Sci. Eng. C* **2012**, *32*, 2145–2152. [CrossRef]
26. Schröter, L.; Kaiser, F.; Stein, S.; Gbureck, U.; Ignatius, A. Biological and mechanical performance and degradation characteristics of calcium phosphate cements in large animals and humans. *Acta Biomater.* **2020**, *117*, 1–20. [CrossRef] [PubMed]
27. Bottino, M.C.; Münchow, E.A.; Albuquerque, M.T.P.; Kamocki, K.; Shahi, R.; Gregory, R.L.; Chu, T.-M.G.; Pankajakshan, D. Tetracycline-incorporated polymer nanofibers as a potential dental implant surface modifier. *J. Biomed. Mater. Res. Part B Appl. Biomater.* **2017**, *105*, 2085–2092. [CrossRef]
28. Shahi, R.G.; Albuquerque, M.T.P.; Münchow, E.A.; Blanchard, S.B.; Gregory, R.L.; Bottino, M.C. Novel bioactive tetracycline-containing electrospun polymer fibers as a potential antibacterial dental implant coating. *Odontology* **2017**, *105*, 354–363. [CrossRef]
29. Tas, A.C.; Bhaduri, S.B. Preparation of brushite powders and their in vitro conversion to nanoapatites. In *Bioceramics: Materials and Applications V: Proceedings of the 106th Annual Meeting of the American Ceramic Society, Indianapolis, Indiana, USA*; Sundar, V., Rusin, R.P., Rusiter, C., Eds.; American Chemical Society: Westerville, OH, USA, 2004; pp. 119–128.
30. Morilla, C.; Lima, Y.M.; Fuentes, G.; Almirall, A. Synthesis and evaluation of a collagen–brushite cement as a drug delivery system. *Int. J. Mater. Res.* **2019**, *110*, 367–374. [CrossRef]

31. Peppas, N.A.; Sahlin, J.J. A simple equation for the description of solute release. III. Coupling of diffusion and relaxation. *Int. J. Pharm.* **1989**, *57*, 169–172. [CrossRef]
32. Richards, F.J. A Flexible Growth Function for Empirical Use. *J. Exp. Bot.* **1959**, *10*, 290–301. [CrossRef]
33. Adams, E. Non-linear mixed effects models for the evaluation of dissolution profiles. *Int. J. Pharm.* **2002**, *240*, 37–53. [CrossRef]
34. El-Shatoury, S.; Hassan, R.; Said, A. Alginate Polyelectrolyte Ionotropic Gels-XVI. Kinetics and Chemical Equilibria Studies for Heterogeneous Ion Exchange of Polyvalent Metal Ions in Alginate Gel Complexes. *High Perform. Polym.* **1992**, *4*, 173–179. [CrossRef]
35. Hassan, R.; Makhlouf, M.T.; Summan, A.; Awad, A. Influence of frequency on specific conductance of polyelectrolyte gels with special correlation between strength of chelation and stability of divalent metal alginate ionotropic gels. *Eur. Polym. J.* **1989**, *25*, 993–996. [CrossRef]
36. Reed, L.J.; Berkson, J. The Application of the Logistic Function to Experimental Data. *J. Phys. Chem.* **1929**, *33*, 760–779. [CrossRef]
37. van Staden, A.D.; Brand, A.M.; Dicks, L.M.T. Nisin F-loaded brushite bone cement prevented the growth of Staphylococcus aureus in vivo. *J. Appl. Microbiol.* **2012**, *112*, 831–840. [CrossRef]
38. Li, X.; Li, G.; Zhang, K.; Pei, Z.; Zhao, S.; Li, J. Cu-loaded Brushite bone cements with good antibacterial activity and operability. *J. Biomed. Mater. Res. Part B Appl. Biomater.* **2021**, *109*, 877–889. [CrossRef]
39. Şahin, E.; Çiftçioğlu, M. Compositional, microstructural and mechanical effects of NaCl porogens in brushite cement scaffolds. *J. Mech. Behav. Biomed. Mater.* **2021**, *116*, 104363. [CrossRef]
40. Wang, L.; Shelton, R.M.; Cooper, P.R.; Lawson, M.; Triffitt, J.T.; Barralet, J.E. Evaluation of sodium alginate for bone marrow cell tissue engineering. *Biomaterials* **2003**, *24*, 3475–3481. [CrossRef]
41. Carrodeguas, R.G.; De Aza, S. α-Tricalcium phosphate: Synthesis, properties and biomedical applications. *Acta Biomater.* **2011**, *7*, 3536–3546. [CrossRef]
42. Delgado, J.; Harr, I.; Almirall, A.; del Valle, S.; Planell, J.A.; Ginebra, M. Injectability of a macroporous calcium phosphate cement. *Key Eng. Mater.* **2005**, *284–286*, 157–160. [CrossRef]
43. La Serna, A.A. Desarrollo y Caracterización de Cementos Macroporosos de Fosfatos de Calcio. Ph.D. Thesis, Universidad de La Habana, La Habana, Cuba, 2006.
44. Peppas, N.A.; Narasimhan, B. Mathematical models in drug delivery: How modeling has shaped the way we design new drug delivery systems. *J. Control. Release* **2014**, *190*, 75–81. [CrossRef]

molecules

MDPI

Review

Recent Progress in Transdermal Nanocarriers and Their Surface Modifications

Zhixi Yu [1,†], Xinxian Meng [1,†], Shunuo Zhang [1], Yunsheng Chen [1,*], Zheng Zhang [1,*] and Yixin Zhang [1,2,*]

1 Department of Plastic and Reconstructive Surgery, Shanghai Ninth People's Hospital, School of Medicine, Shanghai Jiao Tong University, 639 Zhizaoju Rd, Shanghai 200011, China; caroleyu@126.com (Z.Y.); m_sharon@sina.com (X.M.); feltfans2013@163.com (S.Z.)
2 Shanghai National Engineering Research Center for Nanotechnology, 245 Jiachuan Road, Shanghai 200237, China
* Correspondence: Yunshengchen@126.com (Y.C.); Zhangzheng958@163.com (Z.Z.); Zhangyixin6688@163.com (Y.Z.)
† These authors contributed equally to this work.

Abstract: Transdermal drug delivery system (TDDS) is an attractive method for drug delivery with convenient application, less first-pass effect, and fewer systemic side effects. Among all generations of TDDS, transdermal nanocarriers show the greatest clinical potential because of their non-invasive properties and high drug delivery efficiency. However, it is still difficult to design optimal transdermal nanocarriers to overcome the skin barrier, control drug release, and achieve targeting. Hence, surface modification becomes a promising strategy to optimize and functionalize the transdermal nanocarriers with enhanced penetration efficiency, controlled drug release profile, and targeting drug delivery. Therefore, this review summarizes the developed transdermal nanocarriers with their transdermal mechanism, and focuses on the surface modification strategies via their different functions.

Keywords: transdermal drug delivery; transdermal nanocarrier; surface modification; enhanced penetration efficiency; controlled release; targeting delivery

Citation: Yu, Z.; Meng, X.; Zhang, S.; Chen, Y.; Zhang, Z.; Zhang, Y. Recent Progress in Transdermal Nanocarriers and Their Surface Modifications. *Molecules* **2021**, *26*, 3093. https://doi.org/10.3390/molecules26113093

Academic Editors: Marcel Popa, Leonard Ionut Atanase and Roberta Cassano

Received: 27 April 2021
Accepted: 18 May 2021
Published: 21 May 2021

1. Introduction

Transdermal drug delivery systems (TDDS) have been developed as an attractive drug administration method in clinical dermatology [1]. Compared to the conventional oral route, intravenous, or subcutaneous injection, the TDDS provide an easier application without the first-pass effect and systemic side effects [2]. Nowadays, TDDS are widely applied in treatment for many dermatological diseases, such as psoriasis, contact dermatitis, and skin cancer. Moreover, TDDS is also suitable for long-term administration, especially for insulin delivery and some analgesic agents [3]. However, TDDS need to conquer the skin barriers, including stratum corneum (SC), surrounding lipid bilayer, and dermal tissue [4].

Therefore, great efforts have been made to achieve a better transdermal delivery efficiency during the past decades. Many transdermal delivery enhancement technologies have been developed from the first-generation to the fourth-generation, such as iontophoresis, electroporation, ultrasound, microneedles, and most recently, the nanocarriers [5]. Among them, nanocarriers show the greatest potentials in clinical applications, because of their independence on extra equipment, low skin irritation, and no damage on the intact skin barrier. In addition, nanocarriers can encapsulate macromolecules and hydrophilic agents such as non-steroidal drugs, photosensitizers, and some chemotherapy agents, as well as increase the drug stability and retention [6–9].

Various nanocarriers have been developed and investigated as useful TDDS. By optimizing the components and formulation, the nanocarriers can achieve ideal morphological and physico-chemical characteristics (such as size and surface charge). Although they display the potentials in transdermal delivery, their therapeutic applications suffer from their limited penetration ability, drug encapsulation and release, and so on [10]. Recently, based on the well-developed nanocarriers, many studies concentrated on surface modification to further improve their therapeutic applications. The flexible structures and exposed functional groups provide the opportunity for this procedure.

Herein, this review aims to present recent progress on different transdermal nanocarriers as well as their functionalization with different surface modifications. Furthermore, the functional transdermal nanocarriers can overcome the barriers in transdermal delivery, and their surface modification strategies provide feasibility ideas to promote their clinical applications.

2. Overview of Transdermal Drug Nanocarriers

In the past decades, lots of nanocarriers have been developed and applied for transdermal drug delivery. Their structure and transdermal delivery mechanism vary with different components. The most investigated transdermal nanocarriers, including lipid-based nanovesicles, lipid nanoparticles, polymeric nanoparticles, some inorganic nanoparticles, and others, are summarized below (Table 1 and some transmission electron microscopy (TEM) images shown in Figure 1).

Table 1. Summary of typical components, structure, and transdermal mechanisms of transdermal nanocarriers.

Classification		Typical Components	Structure	Transdermal Delivery Mechanism	Ref.
Lipid-based nanovesicles	Liposomes	Phospholipid, cholesterol	Spherical vesicles with one or more lipid bilayers and aqueous inner core	The phospholipid component interacts with the lipids of the SC.	[16]
	Ethosomes	Phospholipid, ethanol (high concentration up to 20–50% *w/w*)		Ethanol increases the fluidity of phospholipid bilayers and disrupts the membrane barrier of SC.	[17]
	Transfersomes	Phospholipid, edge activators		The edge activators increase the flexibility and deformability for passing through the narrow pores.	[18]
	Niosomes	Nonionic surfactants, cholesterol		Similar to liposome The nonionic surfactants enhance the drug encapsulation efficiency.	[19]
	Glycerosomes	Phospholipid, glycerol, cholesterol		Similar to transfersome The glycerol improves the elasticity and deformability.	[20]
	Invasomes	Phospholipid, ethanol (low concentration as 3% *w/w*), terpenes		Ethanol and terpenes disrupt the SC lipid structure and increase the membrane elasticity.	[21]
Lipid nanoparticles	Solid lipid nanoparticles	Solid lipids	Solid particles with non-aqueous core	The lipid nanoparticles form a mono-layer lipid film to enlarge the inter-keratinocyte gap. The lipid components and incorporated surfactant could disrupt skin structure and increase the intercellular space.	[11,22]
	Nanostructured lipid carriers	Solid and liquid lipids			
Polymeric nanoparticles		Natural polymers and synthetic polymers	Solid colloidal carriers	They create a drug concentration gradient to enhance the drug permeation. Some small or positively charged polymeric nanoparticles can penetrate through epidermal barrier through paracellular route and hair-follicle route.	[13,22,23]

Table 1. *Cont.*

Classification		Typical Components	Structure	Transdermal Delivery Mechanism	Ref.
Inorganic nanoparticles	Gold nanoparticles	Au	Solid and rigid particles	Some small nanoparticles can penetrate the skin through the lipidic matrix of the stratum corneum and through hair follicle orifices.	[24,25]
	Fe_3O_4 nanoparticles	Fe_3O_4		They penetrate into deeper dermis via transfollicular route.	[26]
	CuS nanoparticles	CuS		The near-infrared absorption induces the localized thermal ablation of SC and facilitate the penetration.	[27]
Dendrimers		Poly (amidoamine) and other polymers	High-branched polymeric nanocarriers	Serving as penetration enhancer through interaction with skin lipid bilayers	[28]
Micelles		Amphiphilic polymers, surfactants	Spherical or irregular monolayer structure	Improving the water solubility of drugs	[29–31]
Nanoemulsions		Water, oil, surfactants	Dispersions of water and oil	Disrupting the skin lipid bilayers; Increasing solubility for both liposoluble and water-soluble drugs	[22]
Nanogels		Polymers	Cross-linked network structure	The nanocarrier dispersions prolong the topical contact duration and increase localized drug concentration. Cationically charged nanogels interact with epidermis.	[32–34]

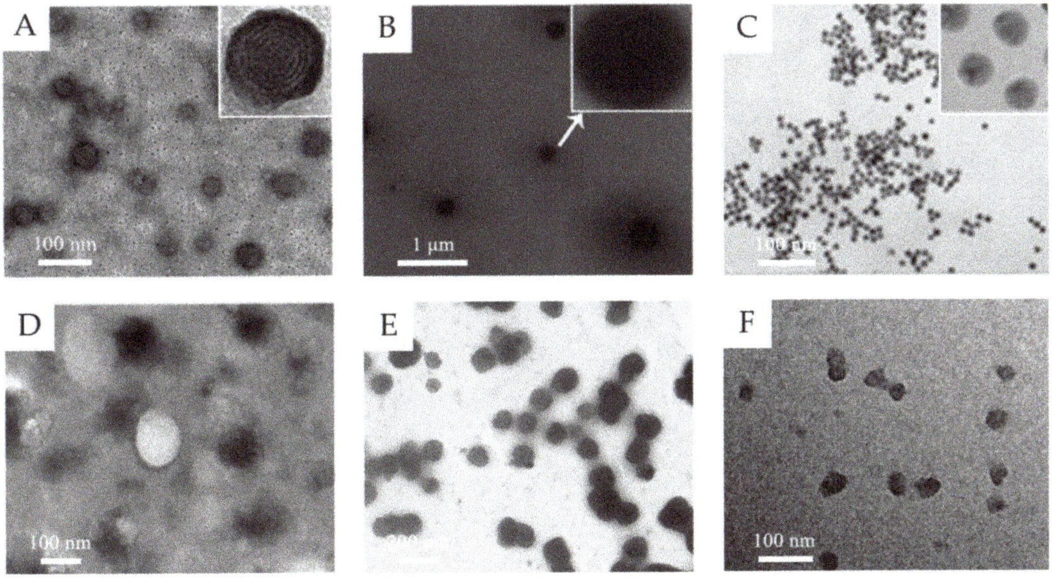

Figure 1. (**A**) TEM images of 5-aminolevulinic acid-loaded ethosomal vesicles. Adapted from Reference [8] with permission of The Royal Society of Chemistry. (**B**) TEM images of triptolide-loaded nanostructured lipid carriers (NLC). Adapted from Reference [11] with permission. (**C**) TEM images of the citrate-capped gold nanoparticles. Adapted from Reference [12] with permission. Copyright (2017) American Chemical Society. (**D**) TEM images of silk fibroin nanoparticles. Reproduced from Reference [13] Copyright (2018), with permission from Elsevier. (**E**) TEM images of water-in-oil (w/o) nanoemulsions of caffeine. Reproduced from Reference [14]. Copyright (2010), with permission from Elsevier. (**F**) TEM images of hyaluronic acid nanogels. Reproduced from Reference [15]. Copyright (2015), with permission from Elsevier.

2.1. Lipid-Based Nanovesicles

Lipid-based nanovesicles are defined as spherical vesicles with one or more lipid bilayers and aqueous inner core, in which hydrophilic drugs can be encapsulated in internal core and hydrophobic drugs can be inserted in external lipid bilayers. Liposomes, the classic and most mature lipid-based nanovesicles, are generally formed with phospholipid molecules and cholesterol as a stabilizer [35]. The phospholipid component could interact with the lipids of the SC, namely fusion mechanism, and thus achieve the transdermal effect. Some deformable liposomes are also reported to penetrate through SC with their intact structure [16]. However, its penetration in the deeper skin layer is limited and the accumulation was mainly observed in the epidermis because of its relatively low fluidity [36]. Therefore, great efforts have been made to improve the skin penetration.

Ethosomes and transfersomes are the most investigated innovative lipid-based nanovesicles, with increased softness, deformability, and elasticity [37]. Ethosomes, composed of phospholipid and ethanol, are generally formed as multilamellar nanovesicles. Ethanol can increase the fluidity of phospholipid bilayers, disrupts the membrane barrier of SC, and thus enhances the penetration ability [17]. Transfersomes are composed of phospholipid and edge activators (such as Tween 80, Span 80, and sodium cholate). The edge activators, as the membrane-softening agent, endow transfersomes with remarkable flexibility and deformability for passing through the narrow pores [18].

Some other lipid-based nanovesicles are also developed with different compositions. Niosomes are synthesized with cholesterol and nonionic surfactants, which replace the phospholipid. The nonionic surfactants enhance the drug encapsulation efficiency and cholesterol increases the rigidity and stability [19]. Glycerosomes, as the term implies, are composed of glycerol, which serves as the same role of edge activators in transfersomes. The glycerol improves the elasticity and deformability of the vesicle with a dose-dependent effect [20]. Terpenes are introduced to form the invasomes because they can disrupt the SC lipid structure and increase the membrane elasticity at a low concentration as 1% *w/v* [21].

2.2. Lipid Nanoparticles

Different from lipid-based nanovesicles, lipid nanoparticles are solid particles, mainly encapsulating drug molecules in a non-aqueous core. They include solid lipid nanoparticles (SLN), composed of solid lipids, and nanostructured lipid carriers (NLC), as the second generation, composed of solid and liquid lipids. Compared to SLNs, liquid lipids in NLCs avoid recrystallization of solid lipids and thus increase the stability. Some surfactants are introduced to reduce the interfacial tension between the aqueous phase and hydrophobic lipid structure and to improve the formulation stability [38].

The transdermal delivery mechanism of lipid nanoparticles remains unclear. The possible mechanisms are as follows: the lipid nanoparticles demonstrate skin adhesive property and form a mono-layer lipid film, which results in the "occlusion effect", to avoid the water evaporation, enlarge the inter-keratinocyte gap, and thus enhance the drug penetration [22]. It is also reported that the lipid components and incorporated surfactant could disrupt skin structure and increase the intercellular space by their interaction with skin lipid layer, especially in SC. Gu et al. observed the skin structure after being treated with triptolide-loaded lipid nanoparticles [11]. The SEM images and histopathological analysis demonstrated the inflated SC, loose texture, and dilated epidermis, indicating the interaction between lipid nanoparticles and skin.

2.3. Polymeric Nanoparticles

Polymeric nanoparticles can be prepared by both natural polymers and synthetic polymers. The natural polymers, such as chitosan, gelatin, and albumin, are easily biodegradable and biocompatible. Chitosan, one of the most frequently used natural materials, is positively charged, which is suitable for encapsulating negative-charged drugs via electrostatic interaction and facilitates the cellular internalization [39]. Meanwhile, the functional groups of chitosan provide great potential for further surface modification [40].

Compared to natural polymers, the synthetic polymers show better purity and consistency. Synthetic polymers are biodegradable, such as polylactides and poly (lactic-*co*-glycolic acid) (PLGA) copolymers, or non-biodegradable, like poly (methyl methacrylate) and polyacrylates. The nanoparticles formed from synthetic polymers are frequently used for delivering the hydrophobic drugs [39].

The size of polymeric nanoparticles is generally too large to pass through the SC, where they can form as a drug reservoir and create a drug concentration gradient to enhance the drug permeation [22]. Tomoda et al. introduced the iontophoresis to enhance the skin penetration of negatively charged PLGA nanoparticles [41]. Takeuchi et al. developed positively charged chitosan-coated PLGA nanoparticles for iontophoretic transdermal delivery of positively charged drugs and observed that the nanoparticles penetrated through the follicle pathway [23]. Recently, Takeuchi et al. synthesized 40-nm silk fibroin nanoparticles which can reach the deeper dermis through both paracellular route and hair-follicle route [13]. It is indicated that polymeric nanoparticles smaller than a certain size can penetrate through the epidermal barrier.

2.4. *Inorganic Nanoparticles*

Inorganic nanostructures are widely applied in drug delivery for cancer therapy with their probable bio-imaging and phototherapy potential. Some nanoparticles with positive charge, high surface lipophilicity, and small size have transdermal capability by passively penetrating SC [2]. Because of their great stability and high potential for surface functionalization, some studies have also evaluated the transdermal drug delivery efficiency of some inorganic nanoparticles.

Gold nanoparticles are well explored as TDDS, because of their low cytotoxicity and controllable particle size [42]. As reported, the skin penetration amount and rate increase with the decreasing particle size; moreover, the accumulation of 15-nm size particles are observed in deep dermis [24]. Furthermore, the surface ligands bind with the gold nanoparticles with Au-S bonding, which means various ligands could be conveniently conjugated on the surface of particles by pre-thiolation [12]. This specific characteristic of gold nanoparticles enables the application in gene delivery. Zheng et al. conjugated gold nanoparticles with thiolated epidermal growth factor receptor (EGFR) siRNA duplexes (SNA-NCs) via Au-S bonding [25]. The SNA-NCs penetrated through whole skin layer and successfully abolished EGFR expression in mouse model.

Some other metallic particles are also observed for transdermal application. Fe_3O_4 nanoparticles with a pH-sensitive amide bond successfully penetrate into deeper dermis via transfollicular route [26]. Photothermal CuS nanoparticles with near-infrared irritation can induce localized thermal ablation of SC and facilitate the penetration of particles [27].

2.5. *Other Nanocarriers*

Dendrimers are high-branched polymeric nanocarriers used as TDDS for transport hydrophobic agents and macromolecules, such as photosensitizers and chemotherapeutic agents. Conventional dendrimers with composition of poly (amidoamine) could not effectively cross the skin barrier but show penetration enhancer potential because of the interaction with skin lipid bilayers [28]. Recently, some studies concentrate on the size and surface charge of dendrimers and then developed the second-generation dendrimers with smaller size and skin penetration capability [43].

Micelles, also one type of polymeric nanocarriers, are composed of amphiphilic polymers or surfactants agglomerating in an aqueous medium [29]. Their flexible formulation endows micelles with drug co-delivery ability. Generally, the lipophilic drugs are loaded in core, while hydrophilic drugs are located mainly in the shell [30]. As TDDS, the micelles can increase the skin penetration by improving the water solubility of drugs. The polymer components are blocked by intact SC, and thus have less improvement for low-permeability drugs [31].

Nanoemulsions (NE) are dispersions of water and oil stabilized by surfactants, forming nanometric droplets of oil in water (O/W), water in oil (W/O), and bi-continuous emulsions. These different systems are suitable for delivering both hydrophobic and hydrophilic drugs. The oil phase and surfactants can disrupt the skin lipid bilayers and SC and enhance the drug permeation. Furthermore, the unique bi-phase structure provides excellent solubility for both liposoluble and water-soluble drugs, which leads to an increased drug concentration gradient for better penetration [22]. Shakeel et al. prepared W/O nanoemulsions of hydrophilic caffeine and 5-fluorouracil for transdermal anti-cancer therapy [14,44]. The W/O nanoemulsions significantly increased the drug permeation through para-cellular transportation. The limitation of nanoemulsions is long-term stability with the risk of Ostwald ripening, and overdosing surfactants may result in additional cytotoxicity [45]. Therefore, an ideal formulation with a suitable surfactant ratio still needs to be developed.

Nanogels are chemically or physically cross-linked polymers in nanosize swelling in medium. The cross-linked network structure demonstrates unique stimuli-responsive nature for drug loading and modulated releasing [15]. Nanogels could enhance the drug penetration mainly by following two aspects. Firstly, the stable nanocarrier dispersions in medium prolong the topical contact duration and form a localized high concentration of loaded drugs. Secondly, cationic polymers could be introduced in the network structure to interact with negatively charged keratinocytes [32]. Jayakumar et al. developed curcumin and 5-fluorouracil-loaded chitin nanogels for melanoma treatment [33,34]. The drug retention in the dermis of cationically charged chitin nanogels increased 4- to 5-fold as compared to the control solution. As observed in histological studies, the thinner keratin layer and fragmentation of SC revealed the interaction of chitin and epidermis.

3. Surface Modifications and Their Functions

Although various types of nanocarriers have been developed as TDDS, they still face some challenges, such as limited drug penetration efficiency, burst release, and lack of targeting. Recently, some studies focused on surface modifications to improve the transdermal nanocarriers to overcome their limitations (Table 2).

Table 2. Summary of introduced surface modifications via their functions.

Functions	Surface Modifier		Nanocarrier	Achieved Improvement	Ref.
Enhanced penetration efficiency	Oleic acid		Niosomes; polymeric nanoparticles	Increasing penetration depth	[46,47]
	Ceramides		Vesicular phospholipid system	Increasing drug deposition in the skin	[48]
	Polymers	PEG	PDA NPs	Preventing aggregation; Increasing penetration depth	[49]
		PIPP	Cellulose nanocrystal (CNC)-stabilized emulsions	Increasing surface hydrophobicity and stability; Increasing penetration depth	[50]
	Cell-penetrating peptide		Lipid-based vesicles	Increasing cell uptake and internalization; Improving skin penetration and permeation	[51–53]

Table 2. *Cont.*

Functions		Surface Modifier	Nanocarrier	Achieved Improvement	Ref.
Controlled release	Sustained drug release	Oleic acid	Polymeric nanoparticles	Controlled release lasting 72 h	[54]
		Dabco surfactants	Liposomal system	Sustained release of 50% loaded drug in 12 h	[55]
		Oxidized chitosan	Nanofibers	Sustained release of 50% loaded drug in 9 h	[56]
	Activated modulated release	Thermosensitive poly-N-isopropylacrylamide (PNIPAAm)	Liposomes	Drug release and skin permeation begin from 37 °C	[57]
		Thermosensitive polyglycerol (tPG)	Nanogels	Drug release and skin permeation begin from 40 °C	[58]
		Cationic hydroxyethyl cellulose (JR400)	MoS_2 nanoparticles	Irritated by near-infrared 808 nm laser	[59]
Targeting drug delivery		Hyaluronic acid and derivatives	Niosomes; Cdot-Ce6	Cancer cell targeting	[60,61]
		Epidermal growth factor	Fusion peptide carrier	Melanoma cell targeting	[62]
		BQ-788 (endothelin ETB receptor ligand)	ZnO quantum dots	Melanocyte targeting	[63]
		Integrin $\alpha v\beta 3$ ligand	*Escherichia coli* derived outer membrane vesicles	Melanoma cell targeting	[64]

3.1. Enhanced Penetration Efficiency

The penetration efficiency, including drug penetration depth and drug permeation amount, is the most critical parameter of the transdermal nanocarriers [65]. The drug transdermal diffusion pathways are mainly divided into two pathways: transappendageal pathway and transepidermal pathway [4]. The transappendageal pathway is known as a shunt route, because the presence of skin appendages, such as hair follicles and sweat glands, provide natural openings for drug penetration [66]. The transepidermal pathway, including intercellular pathway, is a tortous route through intercellular bilayered matrix, and the transcellular pathway is where drugs diffuse directly through corneocytes. Although the transappendageal pathway has natural opening with high penetration efficiency, it constitutes only 0.1% of skin surface area [4,22], and even does not exist in some skin lesions and scars. Therefore, the improvement generally concentrates on intercellular pathway and transcellular pathway. The current main strategies are divided into two aspects, disturbing cell membrane and cell junction with lipid components and some surface surfactants, or increasing cell uptake with cationic ligands and cell penetrating agents.

Oleic acid is generally used as penetration enhancer for drug passive transdermal penetration by increasing the fluidity of skin lipids [67]. Pando et al. investigated the effect of oleic and linoleic acids on the niosomes entrapping resveratrol, a low water-soluble and photosensitive drug [46]. The results indicated the oleic and linoleic acids can enhance the penetration of poorly water-soluble drugs. Punit et al. developed polymeric bilayered nanoparticles with PLGA inner core and chitosan as an outer coat for simultaneous delivery of spantide II (SP) and ketoprofen (KP) for inflammatory skin disorders [47]. The nanoparticles are modified with succinimidyl glutarate ester of PEGylated oleic acid by the formed covalent amide bond between chitosan and PEG derivative. The in vitro penetration depth of the modified nanoparticles has been enhanced up to 240 µm and the cumulative amount after 24 h of SP and KP were 4.1- and 3.1-fold higher, respectively.

Ceramides, the major component of SC, has been investigated for improving the skin penetration. Abdelgawad et al. doped the ceramide into a vesicular phospholipid system, called "cerosomes", which entrapped the tazarotene for the psoriasis treatment [48]. The in vitro experiments revealed that the "cerosomes" increased the drug deposition in the skin compared to the marketed product Acnitaz®. They also evaluated the clinical efficiency of the tazarotene loaded "cerosomes" and Acnitaz®. After 8 weeks of treatment, the lesions tropically treated with "cerosomes" were significantly reduced and had marked improvement. It should be noted that, as reported, the presence of ceramide resulted in a shape transformation from spherical to the tubular morphology, which still needs more research.

Some polymers have also been used to enhance the penetration and permeation of transdermal nanocarriers. Poly (ethylene glycol) (PEG), one of the most commonly used polymers, shows great improvement in the transdermal penetration efficiency. As Mahmoud et al. reported, the PEG coating could prevent the aggregation by providing steric repulsion [68]. Sunoqrot et al. have further investigated PEG with different termini for the melanin-mimetic polydopamine nanoparticles (PDA NPs) [49]. The PDA NP itself accumulated in the Strat-M membrane, while the PEGylated NPs could penetrate to the dermis. Among different various termini on the NPs surface, the anionic PEGylated NPs obtained up to 78% drug accumulation, with great colloidal stability. Hiranphinyophat et al. introduced a polymer, poly (2-isopropoxy-2-oxo-1,3,2-dioxaphospholane) (PIPP), as a surface modifier for the cellulose nanocrystal (CNC)-stabilized emulsions [50]. The PIPP bound with CNC via ring opening polymerization and the PIPP increased the surface hydrophobicity, emulsifying efficiency, and the particle stabilization. The in vitro study demonstrated that the PIPP-modified particle-stabilized emulsions could penetrate through all skin layers with increasing drug accumulation.

Cell-penetrating peptide (CPP), a kind of short, water soluble, cationic peptide, with cell penetration ability, was recently investigated as the surface modifier. Several studies demonstrated its improvement in skin penetration and permeation because it could destabilize SC, interact with cell membrane via electrostatic interaction, and increase cell uptake [69]. Experimental results reveal that the content of cationic amino acids is crucial for CPP internalization and the arginine residues have the greatest impact on internalization. The 7–15 residues of polyarginine provide most optimal improvement [70]. Kwon et al. conjugated the CPPs to DOPC liposomes via a thiol-maleimide reaction for delivering *Polygonum aviculare* extract [51]. Compared to the typical liposomes, this system improved skin penetration ability and increased the cell uptake. Wang et al. reported a CPP conjugated lipid vesicle with SPDP cross-linker, demonstrating a higher storage stability, increased drug delivery amount and enhanced penetration ability [52]. Jiang et al. have synthesized CPP-modified transfersomes for delivering the paclitaxel, a chemotherapy drug of melanoma [53]. With the CPP modification, they successfully sent the paclitaxel to the xenografted tumor tissues via transdermal route in a mouse model.

3.2. Controlled Release

Traditional topical medicines, such as topical corticosteroids, generally result in local side effects with long-term application and could lead to the systemic side effects with an extreme overdose and over-absorption [71]. With development of nanocarrier encapsulation structure, some studies concentrated on the transdermal controlled release to overcome the mentioned problems. The reported controlled release system of transdermal nanocarriers could be divided into two release patterns, the sustained release and the activated modulated release.

In sustained drug release, the drug release is prolonged over a period of time, which provides a decreased application frequency, a prolonged treatment effect, and a relatively stable drug concentration. Silva et al. conjugated the oleic acid on the polymeric nanoparticles to deliver betamethasone [54]. In vitro drug release study demonstrated a burst release of 50% loaded corticoid in the initial 10 h with a following controlled release lasting 72 h. In addition, with encapsulation in the nanoparticles, the drug degradation was reduced and the skin penetration was enhanced by oleic acid as reported before.

Pashirova et al. synthesized a type of surfactant containing natural moiety quinuclidine and 1,4-diazabicyclo [2.2.2]octane (Dabco) presenting antimicrobial activity [55].

They modified liposomal system with this surfactant and reported a more sustained release of loaded drug with 50% of release in 12 h. The embedded surfactants increased the rigidity and stability of the liposomal structure, which provided the controlled release ability, with significant influence on the skin penetration ability.

Chitosan, a biodegradable and biocompatible polymer with mucoadhesive property, is also an ideal surface modifier for increasing the surface rigidity for nanocarriers [40]. Shekh et al. conjugated the oxidized chitosan (OC) on electrospun polyacrylonitrile nanofibers (PAN NFs) with ethylene diamine (EDA) as cross-linker for the transdermal delivery of acyclovir (ACY) [56]. The surface anchored OC provided abundant unreacted CHO for facilitating the binding of drug molecules via Schiff's reaction. In vitro drug release in PBS presented that the bared PAN NFs released over 90% of ACY in 5–7 h, while the PAN-EDA-OC NFs released 50% of ACY with a controlled rate in 9 h.

In activated modulated release, the drug release is triggered by certain physical or chemical process, meaning that, different from the sustained drug release pattern, activated modulated release requires a response element. The temperature difference is one of the most notable features of skin barrier, therefore, Fujimoto et al. developed a temperature-responsive liposome (TR liposome) by incorporating a thermosensitive polymer, named poly-*N*-isopropylacrylamide (PNIPAAm) [57]. The drug release and skin permeation began at 37 °C and presented a temperature-dependent effect. Although the PNIPAAm incorporation led to a larger particle size and slightly decreased the skin absorption, with the assistance of fractional laser irradiation, both drug release and transdermal permeation were enhanced.

Rancan et al. introduced a linear thermosensitive polyglycerol (tPG) to dendritic polyglycerol (dPG)-based nanogels for inflammatory skin diseases [58]. The skin penetration in disrupted skin samples, which mimicked the inflammatory situation, was enhanced but still mainly limited in the SC. After irritation with infrared lamp to achieve 40 °C of topical temperature, the drug release was triggered and the tissue penetration could reach the deeper dermis.

Photothermal nanocarriers also show the potential for triggered drug release. Zhang et al. synthesized a molybdenum disulfide (MoS_2) nanoparticle modified with cationic hydroxyethyl cellulose (JR400) [59]. The JR400 provided the stability of drug encapsulation without irritation and reduced the cytotoxicity of MoS_2 NPs. JR400-MoS_2 NPs irritated by near-infrared 808 nm laser demonstrated a significant photothermal effect and an increased drug release. However, the synthesized nanoparticles themselves do not have transdermal properties, and thus the application of this drug delivery system is limited.

3.3. *Targeting Drug Delivery*

As long-term application of many chemotherapeutic agents and steroid hormones has severe local and even systemic side effects, development of tissue or cell targeting drug delivery is critical. The nanocarrier systems show great potential in targeting delivery because of their flexible surface modifications [72]. Different from the intravenous administration, the transdermal route bypasses the vascular system, which leads to absence of enhanced permeability and retention (EPR, a passive targeting delivery) effect. Therefore, the improvement of the transdermal nanocarriers concentrates on the active targeting strategies, such as some ligands for targeting specific receptors [73].

Hyaluronic acid (HA), a biocompatible polymer with transdermal effect, also could specifically bind to CD44 receptors, which overexpress in various cancer cells [74]. Kong et al. modified the niosomes with HA derivatives to develop the transdermal tumor targeting nanocarrier [60]. They reported an enhanced endocytosed amount of nanocarrier by 4T1 cells with the presence of HA derivatives. Beack et al. conjugated HA to the chlorin e6 (Ce6)-carbon dot (Cdot) by EDC/NHS cross-linking for the melanoma photodynamic therapy [61]. The Cdot-Ce6-HA conjugate demonstrated improved tissue penetration and specific cancerous skin accumulation. With laser irritation, the tumor growth of subcutaneous grafted B16F10 cells was significantly suppressed, while the tumor volume of free Ce6 group and Cdot-Ce6 group increased.

Epidermal growth factor (EGF), a specific ligand for epidermal growth factor receptor (EGFR), which is also overexpressed in cancer cells, showed its potential for target delivery. Ruan et al. designed a skin-penetrating and cell-entering peptide (SPACE) nanocarrier modified with EGF for siRNA delivery to melanoma cells [62]. The in vitro experiments indicated that the EGF gave the melanoma cell targeting ability without influence on the tissue penetration function of SPACE.

In recent years, some other ligands are screened for selective cell targeting strategies. Huang et al. chose the BQ-788, an antagonist selectively binding to the endothelin ET_B receptors for melanocytes, to functionalize the ZnO quantum dots (ZnO QDs), for transdermal delivery of a TYR inhibitor [63]. They reported a specific cell uptake by health melanocytes and no observed bind on the human keratinocyte line. Meanwhile, the BQ-788/ZnO QDs in vitro delivered the TYR inhibitor. Gu et al. synthesized a high-affinity ligand for integrin $\alpha v\beta 3$ (named PEP), an overexpressed cell adhesion protein in cancer cells [64]. They constructed an *Escherichia coli* (*E. coli*)-derived outer membrane vesicle (TEVs) modified with PEP for transdermal and tumor-targeting delivery. Both in vitro and in vivo experiments showed a specific accumulation of PEP-TEVs in the melanoma cells. As reported, the PEP-TEVs delivered the plasmid-DNA and ICG efficiently, indicating its potential as a multi-functional vector.

4. Conclusions and Outlook

In this review, various transdermal nanocarriers are introduced with their physicochemical characteristics, applications as well as delivery mechanism. Based on the present research, the lipophilic components can interact with skin lipid layer, disrupt skin structure, increase the intercellular space, and thus facilitate the skin penetration. Both nanovesicles and solid nanoparticles can serve as TDDS. The bilayer membrane structure endows nanovesicles with remarkable deformability and elasticity, while non-aqueous core provides solid nanoparticles with better stability. In general, nanovesicles encapsulate hydrophilic drugs in aqueous interior cores and load hydrophobic drugs in external lipid bilayers. On the contrary, solid nanoparticles shield hydrophobic drugs in lipid core and absorb hydrophilic drugs in outer aqueous phase. Moreover, smaller size is recommended for higher specific surface area, which probably increases non-specific cell uptake and facilitates transcellular penetration. However, the correlation of size and penetration efficiency is not absolute, and that encapsulation efficiency of drugs depends on the size of nanoparticles. In addition, the influence of surface charge still remains controversial. Although the positive surface charge is generally considered to promote the skin penetration via enhanced electrostatic interaction with the cell membrane, some researches have investigated a higher transdermal efficiency the negatively charged nanoparticles. It indicates that excessive cell internalization and skin retention will result in reduced skin penetration. Therefore, the ideal range of particle size and surface charge needs to be explored.

Furthermore, their surface modification strategies are also reviewed to overcome the challenges of limited penetration, uncontrolled release, and lack of targeting. As mentioned above, we summarized the investigated surface modifiers according to their functions. Conjugation of penetration enhancer increases the transdermal penetrability and thus allows therapeutic agents to penetrate thick skin tissue in dermatitis and dense fibrous tissue in scars. The realization of both sustained and activated modulated drug release makes it possible to design the transdermal nanocarrier according to optimal tissue drug concentration. Moreover, within selected specific ligands for target receptors, the transdermal nanocarriers achieve the active drug targeting delivery capability, which is crucial especially for skin cancer treatment. These surface modifications improve and enlarge the clinical application of transdermal nanocarriers.

Recently, systemic delivery via transdermal route shows great potential, especially for insulin, vaccine, and analgesic agents, which can be only achieved by microneedle systems at present. However, the existing transdermal nanocarriers are still limited in systemic administration. To achieve a systemic delivery, deeper tissue penetration, transvascular transport ability, and stable release rate are necessary. In addition, there are inadequate in vivo studies about clear transdermal delivery mechanism and safety of long-term administration, which are critical for further clinical translation. Moreover, ideal formulation, high synthetic reproducibility, and good stability for storage are also necessary for pharmaceutical aspect.

Author Contributions: Writing—original draft preparation, Z.Y.; resources, X.M. and S.Z.; conceptualization, Y.C.; writing—review and editing, Z.Z. and Y.Z.; supervision, Y.C. and Y.Z. All authors have read and agreed to the published version of the manuscript.

Funding: This review was funded by the National Natural Science Foundation of China (grant numbers: 81772098 and 81801917), the Outstanding Professional and Technical Leader Program of the Shanghai Municipal Science and Technology Commission (18XD1423700), the Shanghai Clinical Research Project of Health Industry (20204Y0443), and the Cross Research Project of Ninth People's Hospital, Shanghai Jiao Tong University School of Medicine (JYJC202009), Shanghai Municipal Education Commission-Gaofeng Clinical Medicine Grant Support (Grant Number: 20152227).

Acknowledgments: This review was supported by the National Natural Science Foundation of China (grant numbers: 81772098 and 81801917), the Outstanding Professional and Technical Leader Program of the Shanghai Municipal Science and Technology Commission (18XD1423700), the Shanghai Clinical Research Project of Health Industry (20204Y0443), and the Cross Research Project of Ninth People's Hospital, Shanghai Jiao Tong University School of Medicine (JYJC202009), Shanghai Municipal Education Commission-Gaofeng Clinical Medicine Grant Support (Grant Number: 20152227).

Conflicts of Interest: The authors declare no conflict of interest.

References

1. Li, J. Microneedle Patches as Drug and Vaccine Delivery Platform. *Curr. Med. Chem.* **2017**, *24*, 2413–2422. [CrossRef] [PubMed]
2. Wang, M.; Marepally, S.; Vemula, P.; Xu, C. Inorganic Nanoparticles for Transdermal Drug Delivery and Topical Application. In *Nanoscience in Dermatology*; Elsevier: Amsterdam, The Netherlands, 2016; pp. 57–72.
3. Xu, B.; Jiang, G.; Yu, W.; Liu, D.; Zhang, Y.; Zhou, J.; Sun, S.; Liu, Y. H2O2-Responsive mesoporous silica nanoparticles integrated with microneedle patches for the glucose-monitored transdermal delivery of insulin. *J. Mater. Chem. B* **2017**, *5*, 8200–8208. [CrossRef] [PubMed]
4. Pegoraro, C.; MacNeil, S.; Battaglia, G. Transdermal drug delivery: From micro to nano. *Nanoscale* **2012**, *4*, 1881–1894. [CrossRef] [PubMed]
5. Seah, B.C.; Teo, B.M. Recent advances in ultrasound-based transdermal drug delivery. *Int. J. Nanomed.* **2018**, *13*, 7749–7763. [CrossRef] [PubMed]
6. Kumar, L.; Verma, S.; Singh, M.; Chalotra, T.; Utreja, P. Advanced Drug Delivery Systems for Transdermal Delivery of Non-Steroidal Anti-Inflammatory Drugs: A Review. *Curr. Drug Deliv.* **2018**, *15*, 1087–1099. [CrossRef] [PubMed]
7. Zhang, Z.; Liu, Y.; Chen, Y.; Li, L.; Lan, P.; He, D.; Song, J.; Zhang, Y. Transdermal Delivery of 5-Aminolevulinic Acid by Nanoethosome Gels for Photodynamic Therapy of Hypertrophic Scars. *ACS Appl. Mater. Interfaces* **2019**, *11*, 3704–3714. [CrossRef] [PubMed]
8. Zhang, Z.; Chen, Y.; Xu, H.; Wo, Y.; Zhang, Z.; Liu, Y.; Su, W.; Cui, D.; Zhang, Y. 5-Aminolevulinic acid loaded ethosomal vesicles with high entrapment efficiency for in vitro topical transdermal delivery and photodynamic therapy of hypertrophic scars. *Nanoscale* **2016**, *8*, 19270–19279. [CrossRef]
9. Chen, Y.; Zhang, Z.; Xin, Y.; Yu, Z.; Meng, X.; Zhang, Y.; He, D.; Zhang, Y. Functional Transdermal Nanoethosomes Enhance Photodynamic Therapy of Hypertrophic Scars via Self-Generating Oxygen. *ACS Appl. Mater. Interfaces* **2021**, *13*, 7955–7965. [CrossRef] [PubMed]
10. Uchida, N.; Yanagi, M.; Hamada, H. Physical Enhancement? Nanocarrier? Current Progress in Transdermal Drug Delivery. *Nanomaterials* **2021**, *11*, 335. [CrossRef] [PubMed]
11. Gu, Y.; Yang, M.; Tang, X.; Wang, T.; Yang, D.; Zhai, G.; Liu, J. Lipid nanoparticles loading triptolide for transdermal delivery: Mechanisms of penetration enhancement and transport properties. *J. Nanobiotechnol.* **2018**, *16*, 68. [CrossRef]
12. Chen, Y.; Wu, Y.; Gao, J.; Zhang, Z.; Wang, L.; Chen, X.; Mi, J.; Yao, Y.; Guan, D.; Chen, B.; et al. Transdermal Vascular Endothelial Growth Factor Delivery with Surface Engineered Gold Nanoparticles. *ACS Appl. Mater. Interfaces* **2017**, *9*, 5173–5180. [CrossRef]
13. Takeuchi, I.; Shimamura, Y.; Kakami, Y.; Kameda, T.; Hattori, K.; Miura, S.; Shirai, H.; Okumura, M.; Inagi, T.; Terada, H.; et al. Transdermal delivery of 40-nm silk fibroin nanoparticles. *Colloids Surf. B Biointerfaces* **2019**, *175*, 564–568. [CrossRef]

14. Shakeel, F.; Ramadan, W. Transdermal delivery of anticancer drug caffeine from water-in-oil nanoemulsions. *Colloids Surf. B Biointerfaces* **2010**, *75*, 356–362. [CrossRef]
15. Yang, C.; Wang, X.; Yao, X.; Zhang, Y.; Wu, W.; Jiang, X. Hyaluronic acid nanogels with enzyme-sensitive cross-linking group for drug delivery. *J. Control. Release* **2015**, *205*, 206–217. [CrossRef]
16. Siler-Marinkovic, S. Liposomes as Drug Delivery Systems in Dermal and Transdermal Drug Delivery. In *Percutaneous Penetration Enhancers Chemical Methods in Penetration Enhancement*; J.B. Metzler: Stuttgart, Germany, 2016; pp. 15–38.
17. Touitou, E.; Dayan, N.; Bergelson, L.; Godin, B.; Eliaz, M. Ethosomes - novel vesicular carriers for enhanced delivery: Characterization and skin penetration properties. *J. Control. Release* **2000**, *65*, 403–418. [CrossRef]
18. Opatha, S.A.T.; Titapiwatanakun, V.; Chutoprapat, R. Transfersomes: A Promising Nanoencapsulation Technique for Transdermal Drug Delivery. *Pharmaceutics* **2020**, *12*, 855. [CrossRef]
19. Ge, X.; Wei, M.; He, S.; Yuan, W.E. Advances of Non-Ionic Surfactant Vesicles (Niosomes) and Their Application in Drug Delivery. *Pharmaceutics* **2019**, *11*, 55. [CrossRef]
20. Manca, M.L.; Zaru, M.; Manconi, M.; Lai, F.; Valenti, D.; Sinico, C.; Fadda, A.M. Glycerosomes: A new tool for effective dermal and transdermal drug delivery. *Int. J. Pharm.* **2013**, *455*, 66–74. [CrossRef]
21. Babaie, S.; Bakhshayesh, A.R.D.; Ha, J.W.; Hamishehkar, H.; Kim, K.H. Invasome: A Novel Nanocarrier for Transdermal Drug Delivery. *Nanomaterials* **2020**, *10*, 341. [CrossRef]
22. Zhou, X.; Hao, Y.; Yuan, L.; Pradhan, S.; Shrestha, K.; Pradhan, O.; Liu, H.; Li, W. Nano-formulations for transdermal drug delivery: A review. *Chin. Chem. Lett.* **2018**, *29*, 1713–1724. [CrossRef]
23. Takeuchi, I.; Takeshita, T.; Suzuki, T.; Makino, K. Iontophoretic transdermal delivery using chitosan-coated PLGA nanoparticles for positively charged drugs. *Colloids Surf. B Biointerfaces* **2017**, *160*, 520–526. [CrossRef] [PubMed]
24. Sonavane, G.; Tomoda, K.; Sano, A.; Ohshima, H.; Terada, H.; Makino, K. In vitro permeation of gold nanoparticles through rat skin and rat intestine: Effect of particle size. *Colloids Surf. B Biointerfaces* **2008**, *65*, 1–10. [CrossRef] [PubMed]
25. Zheng, D.; Giljohann, D.A.; Chen, D.L.; Massich, M.D.; Wang, X.Q.; Iordanov, H.; Mirkin, C.A.; Paller, A.S. Topical delivery of siRNA-based spherical nucleic acid nanoparticle conjugates for gene regulation. *Proc. Natl. Acad. Sci. USA* **2012**, *109*, 11975–11980. [CrossRef]
26. Rao, Y.F.; Chen, W.; Liang, X.G.; Huang, Y.Z.; Miao, J.; Liu, L.; Lou, Y.; Zhang, X.G.; Wang, B.; Tang, R.K.; et al. Epirubicin-loaded superparamagnetic iron-oxide nanoparticles for transdermal delivery: Cancer therapy by circumventing the skin barrier. *Small* **2015**, *11*, 239–247. [CrossRef]
27. Ramadan, S.; Guo, L.; Li, Y.; Yan, B.; Lu, W. Hollow copper sulfide nanoparticle-mediated transdermal drug delivery. *Small* **2012**, *8*, 3143–3150. [CrossRef] [PubMed]
28. Venuganti, V.V.; Perumal, O.P. Poly(amidoamine) dendrimers as skin penetration enhancers: Influence of charge, generation, and concentration. *J. Pharm. Sci.* **2009**, *98*, 2345–2356. [CrossRef]
29. Atanasova, D.; Staneva, D.; Grabchev, I. Textile Materials Modified with Stimuli-Responsive Drug Carrier for Skin Topical and Transdermal Delivery. *Materials* **2021**, *14*, 930. [CrossRef] [PubMed]
30. Lapteva, M.; Möller, M.; Kalia, Y.N. Polymeric Micelles in Dermal and Transdermal Delivery. In *Percutaneous Penetration Enhancers Chemical Methods in Penetration Enhancement*; J.B. Metzler: Stuttgart, Germany, 2016; pp. 223–240.
31. Yotsumoto, K.; Ishii, K.; Kokubo, M.; Yasuoka, S. Improvement of the skin penetration of hydrophobic drugs by polymeric micelles. *Int. J. Pharm.* **2018**, *553*, 132–140. [CrossRef] [PubMed]
32. Sivaram, A.J.; Rajitha, P.; Maya, S.; Jayakumar, R.; Sabitha, M. Nanogels for delivery, imaging and therapy. *Wiley Interdiscip. Rev. Nanomed. Nanobiotechnol.* **2015**, *7*, 509–533. [CrossRef]
33. Sabitha, M.; Sanoj Rejinold, N.; Nair, A.; Lakshmanan, V.K.; Nair, S.V.; Jayakumar, R. Development and evaluation of 5-fluorouracil loaded chitin nanogels for treatment of skin cancer. *Carbohydr. Polym.* **2013**, *91*, 48–57. [CrossRef]
34. Mangalathillam, S.; Rejinold, N.S.; Nair, A.; Lakshmanan, V.K.; Nair, S.V.; Jayakumar, R. Curcumin loaded chitin nanogels for skin cancer treatment via the transdermal route. *Nanoscale* **2012**, *4*, 239–250. [CrossRef]
35. Chacko, I.A.; Ghate, V.M.; Dsouza, L.; Lewis, S.A. Lipid vesicles: A versatile drug delivery platform for dermal and transdermal applications. *Colloids Surf. B Biointerfaces* **2020**, *195*, 111262. [CrossRef]
36. Elsayed, M.M.; Abdallah, O.Y.; Naggar, V.F.; Khalafallah, N.M. Lipid vesicles for skin delivery of drugs: Reviewing three decades of research. *Int. J. Pharm.* **2007**, *332*, 1–16. [CrossRef]
37. Natsheh, H.; Touitou, E. Phospholipid Vesicles for Dermal/Transdermal and Nasal Administration of Active Molecules: The Effect of Surfactants and Alcohols on the Fluidity of Their Lipid Bilayers and Penetration Enhancement Properties. *Molecules* **2020**, *25*, 2959. [CrossRef]
38. Souto, E.B.; Baldim, I.; Oliveira, W.P.; Rao, R.; Yadav, N.; Gama, F.M.; Mahant, S. SLN and NLC for topical, dermal, and transdermal drug delivery. *Expert Opin. Drug Deliv.* **2020**, *17*, 357–377. [CrossRef]
39. Zhang, Z.; Tsai, P.C.; Ramezanli, T.; Michniak-Kohn, B.B. Polymeric nanoparticles-based topical delivery systems for the treatment of dermatological diseases. *Wiley Interdiscip. Rev. Nanomed. Nanobiotechnol.* **2013**, *5*, 205–218. [CrossRef]
40. Fathi, M.; Sahandi Zangabad, P.; Majidi, S.; Barar, J.; Erfan-Niya, H.; Omidi, Y. Stimuli-responsive chitosan-based nanocarriers for cancer therapy. *Bioimpacts* **2017**, *7*, 269–277. [CrossRef]
41. Tomoda, K.; Terashima, H.; Suzuki, K.; Inagi, T.; Terada, H.; Makino, K. Enhanced transdermal delivery of indomethacin-loaded PLGA nanoparticles by iontophoresis. *Colloids Surf. B Biointerfaces* **2011**, *88*, 706–710. [CrossRef]

42. Ghosh, P.; Han, G.; De, M.; Kim, C.K.; Rotello, V.M. Gold nanoparticles in delivery applications. *Adv. Drug Delivery. Rev.* **2008**, *60*, 1307–1315. [CrossRef]
43. Yang, Y.; Sunoqrot, S.; Stowell, C.; Ji, J.; Lee, C.W.; Kim, J.W.; Khan, S.A.; Hong, S. Effect of size, surface charge, and hydrophobicity of poly(amidoamine) dendrimers on their skin penetration. *Biomacromolecules* **2012**, *13*, 2154–2162. [CrossRef]
44. Shakeel, F.; Haq, N.; Al-Dhfyan, A.; Alanazi, F.K.; Alsarra, I.A. Chemoprevention of skin cancer using low HLB surfactant nanoemulsion of 5-fluorouracil: A preliminary study. *Drug Deliv.* **2015**, *22*, 573–580. [CrossRef]
45. Singh, Y.; Meher, J.G.; Raval, K.; Khan, F.A.; Chaurasia, M.; Jain, N.K.; Chourasia, M.K. Nanoemulsion: Concepts, development and applications in drug delivery. *J. Control. Release* **2017**, *252*, 28–49. [CrossRef]
46. Pando, D.; Matos, M.; Gutierrez, G.; Pazos, C. Formulation of resveratrol entrapped niosomes for topical use. *Colloids Surf. B Biointerfaces* **2015**, *128*, 398–404. [CrossRef]
47. Shah, P.P.; Desai, P.R.; Singh, M. Effect of oleic acid modified polymeric bilayered nanoparticles on percutaneous delivery of spantide II and ketoprofen. *J. Control. Release* **2012**, *158*, 336–345. [CrossRef]
48. Abdelgawad, R.; Nasr, M.; Moftah, N.H.; Hamza, M.Y. Phospholipid membrane tubulation using ceramide doping "Cerosomes": Characterization and clinical application in psoriasis treatment. *Eur. J. Pharm. Sci.* **2017**, *101*, 258–268. [CrossRef]
49. Sunoqrot, S.; Mahmoud, N.N.; Ibrahim, L.H.; Al-Dabash, S.; Raschke, H.; Hergenroder, R. Tuning the Surface Chemistry of Melanin-Mimetic Polydopamine Nanoparticles Drastically Enhances Their Accumulation into Excised Human Skin. *ACS Biomater. Sci. Eng.* **2020**, *6*, 4424–4432. [CrossRef]
50. Hiranphinyophat, S.; Otaka, A.; Asaumi, Y.; Fujii, S.; Iwasaki, Y. Particle-stabilized oil-in-water emulsions as a platform for topical lipophilic drug delivery. *Colloids Surf. B Biointerfaces* **2021**, *197*, 111423. [CrossRef]
51. Kwon, S.S.; Kim, S.Y.; Kong, B.J.; Kim, K.J.; Noh, G.Y.; Im, N.R.; Lim, J.W.; Ha, J.H.; Kim, J.; Park, S.N. Cell penetrating peptide conjugated liposomes as transdermal delivery system of *Polygonum aviculare* L. extract. *Int. J. Pharm.* **2015**, *483*, 26–37. [CrossRef]
52. Wang, S.; Zeng, D.; Niu, J.; Wang, H.; Wang, L.; Li, Q.; Li, C.; Song, H.; Chang, J.; Zhang, L. Development of an efficient transdermal drug delivery system with TAT-conjugated cationic polymeric lipid vesicles. *J. Mater. Chem. B* **2014**, *2*, 877–884. [CrossRef]
53. Jiang, T.; Wang, T.; Li, T.; Ma, Y.; Shen, S.; He, B.; Mo, R. Enhanced Transdermal Drug Delivery by Transfersome-Embedded Oligopeptide Hydrogel for Topical Chemotherapy of Melanoma. *ACS Nano* **2018**, *12*, 9693–9701. [CrossRef]
54. Silva, C.O.; Rijo, P.; Molpeceres, J.; Figueiredo, I.V.; Ascensao, L.; Fernandes, A.S.; Roberto, A.; Reis, C.P. Polymeric nanoparticles modified with fatty acids encapsulating betamethasone for anti-inflammatory treatment. *Int. J. Pharm.* **2015**, *493*, 271–284. [CrossRef] [PubMed]
55. Pashirova, T.N.; Sapunova, A.S.; Lukashenko, S.S.; Burilova, E.A.; Lubina, A.P.; Shaihutdinova, Z.M.; Gerasimova, T.P.; Kovalenko, V.I.; Voloshina, A.D.; Souto, E.B.; et al. Synthesis, structure-activity relationship and biological evaluation of tetracationic gemini Dabco-surfactants for transdermal liposomal formulations. *Int. J. Pharm.* **2020**, *575*, 118953. [CrossRef] [PubMed]
56. Shekh, M.I.; Amirian, J.; Stadler, F.J.; Du, B.; Zhu, Y. Oxidized chitosan modified electrospun scaffolds for controllable release of acyclovir. *Int. J. Biol. Macromol.* **2020**, *151*, 787–796. [CrossRef]
57. Fujimoto, T.; Ito, M.; Ito, S.; Kanazawa, H. Fractional laser-assisted percutaneous drug delivery via temperature-responsive liposomes. *J. Biomater. Sci. Polym. Ed.* **2017**, *28*, 679–689. [CrossRef]
58. Rancan, F.; Giulbudagian, M.; Jurisch, J.; Blume-Peytavi, U.; Calderon, M.; Vogt, A. Drug delivery across intact and disrupted skin barrier: Identification of cell populations interacting with penetrated thermoresponsive nanogels. *Eur. J. Pharm. Biopharm.* **2017**, *116*, 4–11. [CrossRef]
59. Zhang, K.; Zhuang, Y.; Zhang, W.; Guo, Y.; Liu, X. Functionalized MoS2-nanoparticles for transdermal drug delivery of atenolol. *Drug Deliv.* **2020**, *27*, 909–916. [CrossRef]
60. Kong, M.; Park, H.; Feng, C.; Hou, L.; Cheng, X.; Chen, X. Construction of hyaluronic acid noisome as functional transdermal nanocarrier for tumor therapy. *Carbohydr. Polym.* **2013**, *94*, 634–641. [CrossRef]
61. Beack, S.; Kong, W.H.; Jung, H.S.; Do, I.H.; Han, S.; Kim, H.; Kim, K.S.; Yun, S.H.; Hahn, S.K. Photodynamic therapy of melanoma skin cancer using carbon dot–chlorin e6–hyaluronate conjugate. *Acta Biomater.* **2015**, *26*, 295–305. [CrossRef]
62. Ruan, R.; Chen, M.; Sun, S.; Wei, P.; Zou, L.; Liu, J.; Gao, D.; Wen, L.; Ding, W. Topical and Targeted Delivery of siRNAs to Melanoma Cells Using a Fusion Peptide Carrier. *Sci. Rep.* **2016**, *6*, 29159. [CrossRef]
63. Huang, X.; Chen, C.; Zhu, X.; Zheng, X.; Li, S.; Gong, X.; Xiao, Z.; Jiang, N.; Yu, C.; Yi, C. Transdermal BQ-788/EA@ZnO quantum dots as targeting and smart tyrosinase inhibitors in melanocytes. *Mater. Sci. Eng. C Mater. Biol. Appl.* **2019**, *102*, 45–52. [CrossRef]
64. Gu, T.W.; Wang, M.Z.; Niu, J.; Chu, Y.; Guo, K.R.; Peng, L.H. Outer membrane vesicles derived from *E. coli* as novel vehicles for transdermal and tumor targeting delivery. *Nanoscale* **2020**, *12*, 18965–18977. [CrossRef]
65. Ramadon, D.; McCrudden, M.T.C.; Courtenay, A.J.; Donnelly, R.F. Enhancement strategies for transdermal drug delivery systems: Current trends and applications. *Drug Deliv. Transl. Res.* **2021**. [CrossRef]
66. Patzelt, A.; Lademann, J. Recent advances in follicular drug delivery of nanoparticles. *Expert Opin. Drug Deliv.* **2020**, *17*, 49–60. [CrossRef]
67. Valjakka-Koskela, R.; Hirvonen, J.; Monkkonen, J.; Kiesvaara, J.; Antila, S.; Lehtonen, L.; Urtti, A. Transdermal delivery of levosimendan. *Eur. J. Pharm. Sci.* **2000**, *11*, 343–350. [CrossRef]
68. Mahmoud, N.N.; Al-Qaoud, K.M.; Al-Bakri, A.G.; Alkilany, A.M.; Khalil, E.A. Colloidal stability of gold nanorod solution upon exposure to excised human skin: Effect of surface chemistry and protein adsorption. *Int. J. Biochem. Cell Biol.* **2016**, *75*, 223–231. [CrossRef]
69. Nasrollahi, S.A.; Taghibiglou, C.; Azizi, E.; Farboud, E.S. Cell-penetrating peptides as a novel transdermal drug delivery system. *Chem. Biol. Drug Des.* **2012**, *80*, 639–646. [CrossRef]

70. Gao, S.; Tian, B.; Han, J.; Zhang, J.; Shi, Y.; Lv, Q.; Li, K. Enhanced transdermal delivery of lornoxicam by nanostructured lipid carrier gels modified with polyarginine peptide for treatment of carrageenan-induced rat paw edema. *Int. J. Nanomed.* **2019**, *14*, 6135–6150. [CrossRef]
71. Ference, J.D.; Last, A.R. Choosing topical corticosteroids. *Am. Fam. Physician* **2009**, *79*, 135–140.
72. Kebebe, D.; Liu, Y.; Wu, Y.; Vilakhamxay, M.; Liu, Z.; Li, J. Tumor-targeting delivery of herb-based drugs with cell-penetrating/tumor-targeting peptide-modified nanocarriers. *Int. J. Nanomed.* **2018**, *13*, 1425–1442. [CrossRef]
73. Lestini, B.J.; Sagnella, S.M.; Xu, Z.; Shive, M.S.; Richter, N.J.; Jayaseharan, J.; Case, A.J.; Kottke-Marchant, K.; Anderson, J.M.; Marchant, R.E. Surface modification of liposomes for selective cell targeting in cardiovascular drug delivery. *J. Control. Release* **2002**, *78*, 235–247. [CrossRef]
74. Choi, K.Y.; Chung, H.; Min, K.H.; Yoon, H.Y.; Kim, K.; Park, J.H.; Kwon, I.C.; Jeong, S.Y. Self-assembled hyaluronic acid nanoparticles for active tumor targeting. *Biomaterials* **2010**, *31*, 106–114. [CrossRef] [PubMed]

molecules

MDPI

Review

Polysaccharide-Based Drug Delivery Systems for the Treatment of Periodontitis

Nicolae Baranov [1], Marcel Popa [1,2,*], Leonard Ionut Atanase [3,*] and Daniela Luminita Ichim [3]

1 Faculty of Chemical Engineering and Protection of the Environment, "Gheorghe Asachi" Technical University, 700050 Iasi, Romania; baranov_nicolae@yahoo.com
2 Academy of Romanian Scientists, 50085 Bucharest, Romania
3 Faculty of Medical Dentistry, "Apollonia" University of Iasi, 700511 Iasi, Romania; danielaluminitaichim@yahoo.com
* Correspondence: marpopa@ch.tuiasi.ro (M.P.); leonard.atanase@univapollonia.ro (L.I.A.)

Abstract: Periodontal diseases are worldwide health problems that negatively affect the lifestyle of many people. The long-term effect of the classical treatments, including the mechanical removal of bacterial plaque, is not effective enough, causing the scientific world to find other alternatives. Polymer–drug systems, which have different forms of presentation, chosen depending on the nature of the disease, the mode of administration, the type of polymer used, etc., have become very promising. Hydrogels, for example (in the form of films, micro-/nanoparticles, implants, inserts, etc.), contain the drug included, encapsulated, or adsorbed on the surface. Biologically active compounds can also be associated directly with the polymer chains by covalent or ionic binding (polymer–drug conjugates). Not just any polymer can be used as a support for drug combination due to the constraints imposed by the fact that the system works inside the body. Biopolymers, especially polysaccharides and their derivatives and to a lesser extent proteins, are preferred for this purpose. This paper aims to review in detail the biopolymer–drug systems that have emerged in the last decade as alternatives to the classical treatment of periodontal disease.

Keywords: periodontitis; antibacterial properties; polysaccharides; drug delivery systems; hydrogels; microparticles; nanoparticles; films; electrospun fibers; gels

Citation: Baranov, N.; Popa, M.; Atanase, L.I.; Ichim, D.L. Polysaccharide-Based Drug Delivery Systems for the Treatment of Periodontitis. *Molecules* **2021**, *26*, 2735. https://doi.org/10.3390/molecules26092735

Academic Editor: Roberta Cassano

Received: 21 March 2021
Accepted: 5 May 2021
Published: 6 May 2021

1. Introduction

In the last 30 years, the field of controlled drug delivery systems, particularly the study of polymers as drug carriers, has received great interest. The polymer acts either as a support by protecting the bioactive agent during its transfer through the body until release or by its properties controlling the release kinetics. Such a controlled release system contains the bioactive principle loaded into the polymeric matrix or chemically bound (ionic, covalent) to the base chain, being administered orally, parenterally, transdermally, or surgically implanted in the body [1]. The drug will then be released at the site of the disease by diffusion, by hydrolysis of the chemical bonds between it and the support, or by erosion of the surface of the polymeric matrix [2].

The polymeric matrix must have an optimal combination of binding properties for compounds that work in or in contact with the human body, which makes biopolymers (proteins, polysaccharides) ideal candidates. Although the mechanisms of action of the release systems thus obtained are not yet fully elucidated and their technologies are not yet optimized, the results of some clinical experiments attest to the special value of some natural macromolecular compounds, and to a lesser extent of synthetic polymers, as adjuvants for the conditioning and immobilization of biologically active substances.

The main constraints that a polymer for bio-applications must meet are the (i) biocompatible and biodegradable. This assumes that the polymer must degrade in vivo into compounds that are easily eliminated from the body. Otherwise, the polymeric system must be surgically removed after completely or almost completely releasing the drug. (ii) The degradation products must be non-toxic and not create an inflammatory response in the body. (iii) Degradation of the polymer must occur within a reasonable period of time. In addition to these constraints, the choice of polymeric matrix also takes into account several criteria: (i) Molecular weight and polydispersity of molar masses. (ii) Basic chain geometry: linearity or cyclicity, branching, reticulation, etc. (iii) The chemical nature of the polymer, such as its chemical composition, aqueous solubility, and ionic charge. (iv) The polymer–drug relationship, such as the type of chemical bond that the partners can form (covalent, ionic, or coordinative); the physical interactions between partners, if no chemical bonds are involved; location of the pharmacologically active agent in the polymer (per chain, as a substituent, or included in the matrix). (v) The stereochemical phenomena.

The mentioned restrictions are fulfilled by few synthetic polymers, but mostly by biopolymers. Of this large category, the most used are polysaccharides and proteins; this review focuses only on polysaccharides and their derivatives. Given the specific diseases of the oral cavity, among which periodontal disease ranks first, the classes of drugs associated with different types of polysaccharide-based supports are those with anti-inflammatory and antimicrobial action on the microbial flora that causes periodontitis.

Numerous forms of the presentation of polymer–drug systems are currently known, these being chosen depending on the nature of the disease, the administration mode, the type of polymer used, etc. One of them is hydrogels (in the form of films, micro-/nanoparticles, implants, etc.), in which the drug can be included, encapsulated, or adsorbed on the surface. Another possibility is the direct association of the biologically active compound with the polymer chains by covalent or ionic binding (polymer–drug conjugates).

The literature abounds with a huge amount of articles dealing with the problem of classifying polymer–drug systems according to the way partners associate, how they are formulated, how they respond to external stimuli, the nature of the support polymer, the mechanism and kinetics of the release process, etc. This review will not refer to these aspects but will only investigate the polymer–drug systems with applications in the treatment of periodontitis. Moreover, the commercial applications/clinical trials are out of the scope of this review.

Apart from trauma produced by injuries, the most common diseases of the oral cavity are tooth decay, periodontal disease, oral infections, and cancer. It is estimated, for example, that approximately 48% of the US adult population is affected by periodontitis, but similar results have been reported in other countries. Periodontitis is therefore a chronic inflammatory disease of the periodontal tissue caused by pathogenic microorganisms and characterized by the destruction of the supporting structures of the teeth [3]. The inflammation is first located in the gum but then penetrates deep and in the absence of treatment creates bags colonized by anaerobic bacteria that erode the supporting ligaments of the tooth until it is lost; the progress of the inflamed bags can take several years. Several inflammatory and degenerative stages of the tissue surrounding the tooth are highlighted, the gum, periodontal ligaments, enamel, and alveolar bone being affected in turn; these stages in the evolution of the disease are suggestively presented in Figure 1 [4].

The understanding of the etiology and pathogenesis of periodontal disease has made significant progress in recent decades. In all forms of the disease, the strength of the microbial attack depends on the virulence of the microorganisms, the amount and composition of the bacterial plaque, their ability to invade the tissues, and their metabolic products [5]. The periodontal microbiota is a complex community of microorganisms, many of which are still difficult to isolate in the laboratory [6–9].

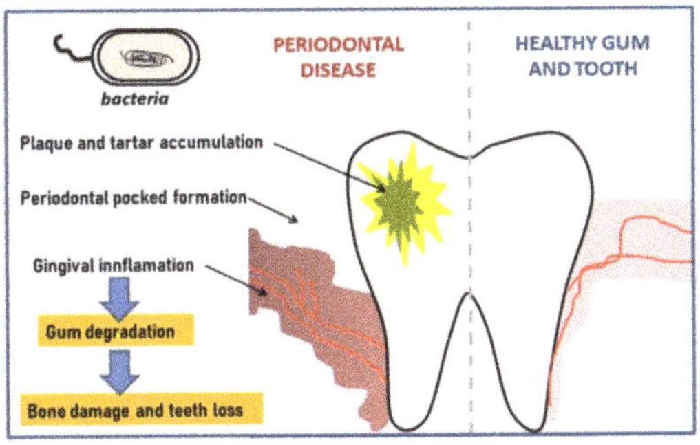

Figure 1. Schematic illustration of the stages of evolution of periodontal disease. Reprinted from ref. [4].

Many species function as pathogens, and those that function as pathogens in a site may exist in small amounts and in healthy sites. Bacterial accumulations cause vascular changes typical of inflammatory reactions, the result of a loss of vascular fluid and the migration of polymorphonuclear leukocytes from blood vessels into tissues and gingival sulcus. There are also early losses of apical collagen from the junction epithelium that begin during the time the bacterial plaque is deposited. The stage of chronic gingivitis is installed but limited to the tissues located very close to the sulcular and junctional epithelium. The supragingival plaque, if left to grow normally, accumulates and will cause gingival inflammation with the appearance of tissue products that serve as nutrients for Gram-negative anaerobes. At the same time, inflammation together with bacterial enzymes open a gateway to the tissue of high molecular weight products from bacteria that penetrate tissues, ultimately causing periodontal disease.

Among the bacteria that activate the destructive processes of the host tissues, the most important are: Porphyromonas gingivalis, some types of Prevotella intermedia, Tannerella forsythia, Aggregatibacter actinomycetemcomitans, Fusobacterium nucleatum, etc. [5].

Periodontal disease refers to all pathological manifestations with a multifactorial etiology, which appear as a consequence of interactions between microbial factors (determinants) and those of the host but also possible environmental influences (favorable factors) that affect the supporting periodontium and lead to tissue destruction, the experience of chronic evolution (with periods of activity and inactivity), and the requirement of a complex and long-term treatment [10,11]. It is characterized by gingival inflammation, the formation of periodontal pockets, and the destruction of the periodontal ligament and alveolar bone, as well as by the mobilization of teeth [12].

In fact, the prevention of periodontal disease is very important, and the best way is to remove its causes, which is the formation of bacterial plaque. Treatment often involves the systemic administration of the free drug, with the effectiveness of the treatment being reduced as the concentration of the active principle decreases rapidly over time to the sub-therapeutic level, when a new dose is required. The trend of recent years in the treatment of this disease has shifted to the local release of antibiotics, antimicrobials, and anti-inflammatory drugs, and this is through the local administration of drug delivery systems. An ideal system of this type consists of a biodegradable, bioadhesive support, based on natural and rarely synthetic polymers, capable of releasing the active principle relatively controlled and for as long as possible. Several types of such systems were investigated to date, such as electrospun fibers, tapes, inserts/implants, films, gels, hydrogels, polymer–drug conjugates, dendrimers, micelles, nanocapsules, micro- and nano-spheres, and particularly micro-/nano-capsules and liposomes.

Local administration has a number of apparent advantages over the systemic one, directly targeting the affected area, but even in this case repeated doses of the active ingredient are required to maintain a relatively constant concentration of the active ingredient, minimizing systemic side effects [13,14]. The most common formulations for local use in buccal cavity are semi-solid or liquid, especially because they are both easy to administer and borne by the patient. However, the disadvantage is the poor retention in the oral cavity, and, therefore, the achievement of a therapeutic effect is far below the optimal one. As a result, ways have been sought to solve these disadvantages, one of which is the association with polymers [15]. The use of a local system of release of the drug when placed in the periodontal pocket allows—and, therefore, maintains—a high and relatively constant concentration in time of the drug, which is more difficult to remove by salivary secretion and, therefore, cannot be swallowed by the patient, as well as preventing unwanted side effects that the free drug can generate [16].

2. Polymer–Drug Systems

An ideal mechanism for the release of drugs from polymer matrices should be characterized by zero-order kinetics (constant velocity) or at most of the first order.

Depending on the formulation and the application method, the duration of the drug release can vary from a few hours to a few days, months, or even years (in the case of implants). While the advantages of using such systems are significant, the potential disadvantages cannot be ignored: possible toxicity of the materials used; unwanted degradation of the products formed; surgical action required to implant and/or to remove the implant; the possibility that the system causes patient discomfort; and the cost of preparing and processing controlled release systems, which can be considerably higher compared to traditional pharmaceutical forms.

In the following, a detailed review of the different types of drug delivery systems based on polysaccharides and their derivatives and anti-inflammatory, antibacterial, etc. drugs, with potential application in the treatment of periodontitis that have been reported to date in the literature are presented.

2.1. Gels

Gels are nonfluid colloidal networks or polymer networks that are expanded throughout their whole volume by a fluid [17]. The network may be covalent in nature, formed by cross-linking the linear chains of a polymer, by nonlinear polymerization, or formed through the physical aggregation of polymer chains, caused by ionic bonds, coordinating bonds, hydrogen bonds, production of coils, hydrogen bonds or hydrophobic interactions, crystallization, or helix formation, that results in regions of local order acting as the network junction points. If the regions of local order are thermally reversible, the resulting swollen network may be termed a "thermoreversible gel". A special category of gels is the hydrogels, which will be discussed in the Section 2.2.

Due to the nature of the constituent polymers, the gels that can be administered in the oral cavity are mucoadhesive systems [18], which are easy to prepare. They are applied sublingually or in the periodontal pocket using a cannula or syringe [19] but have the disadvantage of a quick release of the active ingredient. Gelling occurs instantly through various mechanisms, which involve chemical or physical factors. For example, when using ionic polymers, the presence of polyions of the opposite charge to the ionic polymer used can cause gelling. The process occurs in situ when the polymer solution is instilled (polyanionite, for example) in the periodontal pocket, under the action of electrolytes, such as Ca^{2+}, Mg^{2+}, and Na^{+}, which are inevitably found in the fluids of the oral cavity and cause a change of the ionic strength of the polymer solution. Polymers that can be used to form gels in situ through this mechanism include alginates, HA, and GG. The gelling speed is determined by the osmotic gradient around the gel surface and the change in ionic strength when instilling the solution produces gelling.

2.1.1. Chitosan-Based Gels

Chitosan (CS) fulfills all of the aforementioned constraints for polymers usable as drug carriers, to which its strong mucoadhesive character and its intrinsic antimicrobial activity is added. CS-based gels, starting from solutions of different concentrations, are able to include and release tetracycline hydrochloride and metronidazole benzoate. The optimal concentration of the CS solution has been shown to be 3%, which allows the modulation of the dose of the drug substance in an optimal way for the local therapy of periodontitis [20]. Atorvastatin is an anti-inflammatory drug commonly used to treat inflammation in the oral cavity, but its limited solubility in water makes it less bioavailable. CS-based gels loaded with various anti-inflammatory drugs, including atorvastatin, have recently been reported by Özdoğan et al. In presence of atorvastatin, the bioadhesive property of the formulations was found to increase, which would retain the delivery system at the application site and maintain drug levels at the desired amount for a longer period of time [21]. The results showed that combining the drug with CS improves and enhanced the anti-inflammatory effect of atorvastatin as well as bone and tissue healing in vivo in periodontitis-induced rats [22]. In order to increase the bioavailability of the drug, the authors have prepared solid dispersions of the drug in some polymers (PEG, Pluronic F-68, CS) by using different drug/polymer ratios. CS gels incorporated with particles of atorvastatin exhibited suitable viscosity and a bioadhesive property for a high retention time of the drug at the application site. These atorvastatin-loaded formulations exert anti-inflammatory activity enhanced by the presence of CS [23].

Anti-inflammatory and antimicrobial CS-based nanogels using triclosan and flurbiprofen were obtained. Flurbiprofen was directly loaded into the nanogel, and Triclosan was prepared as nanoparticles using poly(ε-caprolactone) (PCL) [24]. The two systems showed a sustained and rapid in vitro release of drugs. The in vivo tests on rats proved an improved anti-inflammatory action of gels and reduction of bacterial plaque.

CS gels loaded with *Schinus molle L.* essential oil were evaluated for activity against bacteria associated with periodontal disease in dogs [25]. These gels showed their potential as a delivery system for *Schinus molle L.* and presented ideal physicochemical and rheological properties required for application on buccal tissue, such as good stability and antimicrobial activity against Gram-positive and Gram-negative bacteria (*Staphylococcus* spp., *Streptococcus, Corynebacterium* spp., *Pseudomonas* spp.) associated with canine periodontitis.

Thermoresponsive gels are attractive; they are liquid at low temperature and thus can be easily injected into deep pockets, and then they can form gels under body temperature. In the formulations reported so far, CS is associated with Poloxamer, a triblock copolymer with a hydrophobic central segment. Thermosensitive and mucoadhesive polymer-based sustained release moxifloxacin in situ gels for the treatment of periodontal diseases were reported recently by Shashala et al., [26]. The gel formulation consists of a mixture of Poloxamer and CS, which is formed during the injection in the periodontal bag at a temperature of 37 °C. The mucoadhesive nature of the polymers aids the gel to adhere to mucosa in the periodontal pocket for a prolonged time and releases the drug in a sustained manner. In vitro drug release studies demonstrated a sustained release for 8 h, and the antimicrobial studies proved a promising antimicrobial efficiency against *Aggregatibacter actinomycetemcomitans* and *Streptococcus mutans.* A combination of CS and Poloxamer 407 generates a heat-sensitive mucoadhesive gel that releases levofloxacin and metronidazole for 48 h [27,28].

2.1.2. Gels Based on Miscellaneous Polysaccharides

Microbial polysaccharides (gellan) or cellulosic derivatives have high solubility in water, the ability to form gels at higher concentrations, bioadhesive properties, and are being used especially in mixtures with other polysaccharides or synthetic polymers.

Formulations containing hydroxypropyl methylcellulose (HPMC), CMCNa, Carbopol 940, and propolis extract were reported. The antibacterial activity of the most promising formulation against *Porphyromonas gingivalis* was investigated by the disk diffusion method and it appeared that it was efficient in the treatment of periodontitis, which recommends it for clinical evaluation [29]. Doxycycline and metronidazole were incorporated recently into a hydroxyethylcellulose (HEC) poly(vinylpyrrolidone) (PVP)/calcium polycarbophil gel [30]. The antimicrobial activity was tested against *A. actinomycetemcomitans*, *S. sanguinis*, *P. micra*, and *E. corrodens*, and it was noticed to have an inhibitory effect during the first 24 h, which was kept constant until the 13th day. The in vivo test suggests that the formulated gel is effective against bacteria that are already present or will colonize the periodontal pocket.

This category of polysaccharides, especially in mixtures with Poloxamer, can generate heat-sensitive gels capable of sustained and constant release of the included drug. A mixture of Poloxamer 407 and methylcellulose generates a thermosensitive gel at room temperature, releasing simvastatin in a controlled manner for 10 days [31].

Recently, a gel based on Poloxamer 407, Carbopol 934 P, and gellan gum (thermo- and ion-sensitive), which is formed in situ, was prepared for the local release of moxifloxacin hydrochloride in periodontal pockets, showing a high in vitro antibacterial activity against *Staphylococcus aureus* and *Escherichia coli*, with prospects for clinical trials [32].

2.2. *Hydrogels*

Hydrogels are a derived form of gels, representing a three-dimensional solid resulting from the hydrophilic polymer chains being held together by cross-linking. The cross-linking methods that bond the polymers of a hydrogel fall under two general categories: physical and chemical. Hydrogels are networks capable of absorbing large amounts of water (over 90%, sometimes up to over 10,000% in the case of superabsorbent hydrogels) with the network remaining intact, unlike gels that can be disintegrated and fluidized by dilution. Being able to include high amounts of water as well as substances dissolved in it, hydrogels often possess physicochemical properties close to those of the native extracellular matrix [33,34].

Given the presence of functional groups, such as amines, carboxylic, and hydroxyl, in their structure, hydrogels can respond to various physical or chemical stimuli and are considered in many cases as "smart materials". Owing to this feature, they can encapsulate and especially release various chemical compounds, including drugs. A recent review discusses the "smart" hydrogels, which are able to respond to various external stimuli, such as temperature, pH, light, glucose, enzymes, pressure, magnetic field, electric field, or ultrasound, with the potential to be used in the treatment of periodontal disease [35]. Table 1 summarizes the latest information in the literature on obtaining hydrogels based on polysaccharides for inclusion in drugs, with applications in the treatment of periodontal disease, the table is organized by the cross-linking method and the nature of the polysaccharide. The main characteristics of the hydrogel and the effects on periodontal disease are presented.

Table 1. Hydrogels based on polysaccharides with uses in the treatment of periodontal disease.

Cross-Linking Method	Polysaccharide	Cross-Linking Agent	Drug	Features/Administration Route	Biological Activity/Application	Ref.
Physical (ionic)	- CS	β-glycerophosphate disodium salt and gelatin	aspirin and erythropetin	- fast gelation induced by gelatin - drug release in vitro/in vivo—up to 21 days	- effective in anti-inflammation and periodontium regeneration	[36]
		β-glycerophosphate disodium salt	atorvastatin and lovastatin (nano-emulsions)	- thermosensitive hydrogel	-treatment of *Porphyromonas gingivalis* infected oral epithelial cells and gingival fibroblasts - decreased significantly pro-inflammatory markers expression (TNF-a and IL1ß) and pro-osteoclastic RANKL	[37]
		β-glycerophosphate disodium salt	dental pulp stem cell-derived exosomes		- suppressing periodontal inflammation - modulating the immune response	[38]
	- CS - quaternized CS	β-glycerophosphate disodium salt	chlorhexidine	- thermosensitive hydrogel - fast gelation at physiological temperature (6 min after insertion into the periodontal pocket) - release during 18 h	*Porphyromonas gingivalis, Prevotella intermedia, Actinobacillus actinomycetemcomitan*	[39]
	- carboxymethyl-hexanoyl CS	β-glycerophosphate disodium salt	naringin	- thermosensitive hydrogel - subgingivally administration	- therapeutic effect evidenced with micro-CT imaging, histology, the expression of inflammation-associated genes, myeloid differentiation primary response gene-88, and tumor necrosis factor-alpha	[40]
			N-phenacylthia-zolium bromide	- thermosensitive hydrogel - drug release in vivo—up to 30 days	- delay the initiation and facilitates the in vivo recovery from periodontitis	[41]
	- gellan gum - chlorhexidine - nanohydroxya-patite	ionic gelation with $CaCl_2$	bone marrow mesenchymal stem	- injectable hydrogel	- inhibited *Enterococcus faecalis* - treating infectious bone defects caused by refractory periradicular periodontitis.	[42]
	- curdlan - polydopamine	ionic gelation	chlorhexidine acetate	- injectable hydrogel	- periodontal antibacterial treatment by combining photothermal effect and antimicrobial simultaneously - bacteriostatic rate until 99.9%.	[43]
Chemical (covalent)	- CS	GA	doxycycline hyclate	- fast sweeling in aqueous media with drug loading - release of 40 μg/mL drug after 24 h.	- periodontosis treatment and periodontal regeneration.	[44]
	- CS/hydroxyapatite	GA	recombinant human amelogenin	- mesoporous structure 7 nm in diameter of pores) - high surface area (33.95 m^2/g) - high drug loading efficiency	- antibacterial effects against *Fusobacterium nucleatum* and *Porphyromonas gingivalis* - formation of bone and cementum-like tissue.	[45]

Table 1. *Cont.*

Cross-Linking Method	Polysaccharide	Cross-Linking Agent	Drug	Features/Administration Route	Biological Activity/Application	Ref.
	- CS - PEO	GA	metronidazole glucose oxidase	- sensitivity to glucose (release of antimicrobial drug in response to the environmental glucose stimulus)	- drug release at higher glucose concentration - great capacity to inhibit *Porphyromonas gingivalis*	[46]
	- cellulose - cellulose nanofibers - κ-carrageenan oligosaccharide	- epichlorohydrin	surfactin and herbmedotcin	- 75% of the drug was released in vitro after 24 h.	- strong antibacterial activity against *Streptococcus mutans*, *Porphyromonas gingivalis*, *Fusobacterium nucleatum* and *Pseudomonas aeruginosa* - reduced the reactive oxygen species (ROS) generation, transcription factor, and cytokines production in human gingival fibroblast cells (HGF) under inflammatory conditions.	[47]
	- methyl cellulose	- xyloglucan from tamarind seeds	metronidazole	- high mucoadhesive property - superior injectability properties at 25 °C	- local drug carriers for periodontitis therapy	[48]
	- CS methacrylate	- photo-polymerization	metronidazole	- sensitivity to glucose (release of drug in response to the glucose stimulus)	- porous structure allows the control of the loaded metronidazole release with the glucose concentration rising. - local drug carriers for diabetics' periodontitis therapy	[49]
double cross-linking (ionic and covalent)	- oxidized HA - collagen - oligomeric proanthocyanidins	- tricalcium phosphate - carbonylated hyaluronic acid	tetracycline	- macroporous morphology with interconnected pores whose diameter varies between 50 and 250 µm continuous release with an effectiveness of 93% after 5 days.	- local release of the drug into the periodontal pocket for the treatment of advanced chronic periodontitis	[50]
	- CS decorated metronidazole microcapsules included in a PVA matrix	4-carboxyphenyl-boronic acid	metronidazole	- bioadhesive - injectable directly in the periodontal pocket	- antibacterial effect for 14 days (in vitro tests) and for a week (in vivo) tests on a rat model of periodontitis	[51]

2.2.1. Physical Hydrogels

It is well known that these networks are made by the interaction of the polysaccharide polyanion with at least divalent ions of opposite charge.

CS is by far the most widely used polysaccharide for obtaining hydrogels. CS derivatives or mixtures thereof with other biocompatible polymers, particularly with mineral powders (hydroxyapatite), are also reported in the recent literature as being able to generate hydrogels usable not only for the loading/releasing of anti-inflammatory or antimicrobial drugs with applications in the treatment of periodontitis but also for alveolar tissue regeneration. The presence of amino groups that are quaternized in an acidic environment makes the ionic interaction with polyvalent anions possible (sulfate anion, polyphosphate, etc.).

Recently, CS and CS derivative-based thermosensitive hydrogels have gained great attention. They are formed in situ, at a physiological temperature, when the liquid mixture of its components is introduced into the periodontal pocket.

Representative of these are hydrogels based on CS and cross-linked derivatives with β-glycerophosphat discodic salt and rapid formation, which are capable of releasing the bioactive agent (anti-inflammatory, antibacterial, antibiotic) in fluids that simulate saliva depending on its nature (after 18 h or even a month [41]); the advantages over the classical, periodic administration being obvious. Hydrogel administration is done by direct injection into the periodontal pocket [40] and, in addition to the anti-inflammatory and antibacterial action, such systems can exhibit a positive effect on the endogenous repair of alveolar bone [52] and can be used for periodontal treatment in an experimental periodontitis model, such as ligature-induced periodontitis [37,38].

Microbial polysaccharides such as gellan or curdlan can generate hydrogels by ionic gelling, in the presence of bivalent metal ions (Ca^{2+}, for example). Gellan and nano-hydroxyapatite-based biocomposites loaded with chlorhexidine (50 µg/m) and bone marrow mesenchymal stem inhibit *Enterococcus faecalis* in a concentration-dependent manner and are recommended for treating infectious bone defects caused by refractory periradicular periodontitis [42]. A hydrogel based on curdlan and polydopamine loaded with clorhexcidine acetate is a potential candidate for periodontal antibacterial treatment by combining photothermal and antimicrobial effects simultaneously [43].

2.2.2. Chemical Hydrogels

Chemical cross-linking exploits the presence of functional groups of the polysaccharide -OH, $-NH_2$, and -COOH in reaction with bifunctional compounds with complementary functions (dialdehydes, epichlorohydrin) to create the three-dimensional structure. Another way consists of the functionalization of the polysaccharide with polymerizable groups, so that by chemically, thermally, or photochemically initiated polymerization, hydrogel networks are obtained.

Although considered toxic, glutaraldehyde (GA) can be used, in small amounts, as a chemical cross-linker. For example, a hydrogel based on CS cross-linked with AG, loaded with doxycycline showing a release of 40 µg/mL after 24 h can be used for periodontal regeneration [44]. Interestingly, a local drug carrier for diabetics' periodontitis therapy is a glucose-sensitive CS-poly(ethylene oxide) (PEO) hydrogel of a semi-interpenetrated type, which can release metronidazole at a higher glucose concentration and has a great capacity to inhibit *Porphyromonas gingivalis* [46]. Using the same cross-linking agent, mesoporous hydroxyapatite/CS biocomposite hydrogels loaded with recombinant human amelogenin were obtained; they can inhibit the growth of periodontal pathogens and promote the formation of bone and cementum-like tissue [47].

Cellulose and its derivatives can be cross-linked with epichlorohydrin, biocomposites containing cellulose nanofibers, and κ-carrageenan oligosaccharide. The biocomposites loaded with surfactin and herbmedotcin exhibit antibacterial activity against *Streptococcus mutans*, *Porphyromonas gingivalis*, *Fusobacterium nucleatum*, and *Pseudomonas aeruginosa* [48]. Xyloglucan from tamarind seeds was used as a cross-linker to obtain an injectable and mucoadhesive methyl cellulose-based hydrogel, obtained directly in the periodontal pocket for the in situ release of metronidazole [48].

Interestingly, hydrogels obtained by in situ cross-linking through different modes of activation have been reported relatively recently by different groups of researchers. Light curing is one of the methods used for this purpose. The polysaccharide is pre-functionalized by introducing unsaturated groups as a substituent on the base chain. A hydrogel based on CS-methacrylate loaded with metronidazole containing glucose oxidase immobilized on its surface is able to rapidly and correspondingly adjust its inner pore structure to control the loaded metronidazole release film with the rising glucose concentration, so it is an important candidate for diabetics' periodontitis therapy [49]. Another variant is the use of glycidyl methacrylate as a cross-linker, which in the presence of CS and carboxymethyl CS generates a hydrogel by photopolymerization, whose gelling time decreases with the photoinitiator concentration [53].

2.2.3. Double Cross-Linked Hydrogels

Double cross-linking can be a method of obtaining hydrogels, which is applied in order to partially replace the covalent cross-linker with an ionic cross-linker, and thus, to reduce the toxicity of the obtained product. However, a minimum amount of covalent cross-linker must be maintained to ensure the required structural and mechanical stability of the hydrogel.

Wei et al. prepared a biocomposite based on collagen cross-linked with oxidized HA and with oligomeric proanthocyanidins, integrated with tricalcium phosphate [50], with the intention to include tetracycline for the local release of the drug into the periodontal pocket. The obtained hydrogels show a macroporous morphology with interconnected pores, whose diameter varies between 50 and 250 μm (Figure 2). The biocomposite exhibits a good mechanical strength, as well as a good biocompatibility in promoting the proliferation of MG-63 cells. The release of the antimicrobial drug is produced with a "burst effect" (50% released drug) after 24 h and continues with an effectiveness of 93% after 5 days.

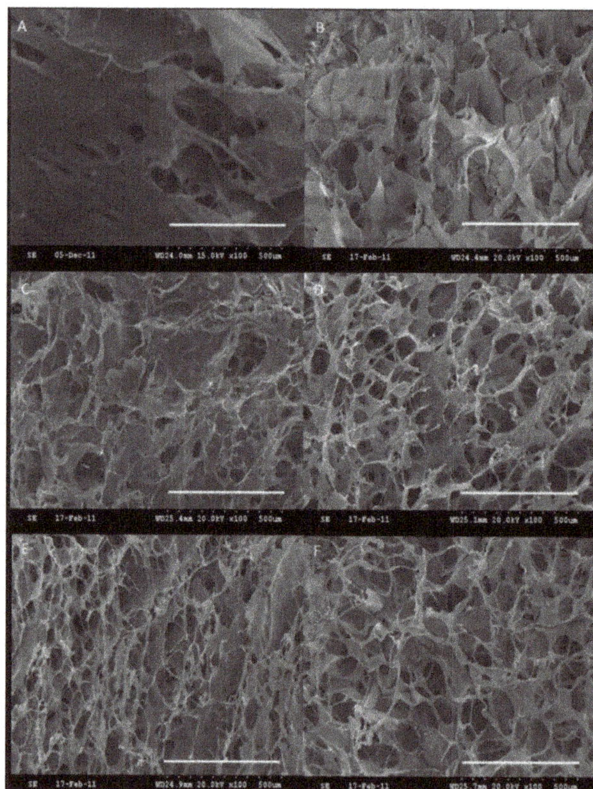

Figure 2. SEM morphology (scale bar = 500 μm) of (**A**) lyophilized collagen and oxi-HA/collagen hydrogel with the connective pores clearly shown in (**B**) CH–10%; (**C**) CH–20%; (**D**) CH–30%; (**E**) CH–35%; and (**F**) CH–40% (percentages indicate the concentration of the polymer mixture in the solution from which the hydrogel was obtained after cross-linking and lyophilization). Reprinted from ref. [50].

An interesting injectable hydrogel consisting of a PVA matrix that included CS-decorated metronidazole microcapsules was obtained by dual ionic and covalent cross-linking, with the network being created by 4-carboxyphenylboronic acid bridges [51]. The hydrogel is bioadhesive, injectable, and administered directly in the periodontal pocket where it exerts an antibacterial effect for 14 days according to in vitro tests and for a week according to in vivo tests on a rat model of periodontitis.

The summary examples presented are important findings in favor of the use of injectable hydrogels, especially of those sensitive to external factors, which can be formed by cross-linking in situ, e.g., directly in the periodontal pocket.

2.3. Films

The films are matrix systems similar to nanofibers and bands, with the drug being dispersed throughout their mass and the release being achieved predominantly by diffusion but also by erosion or even dissolution of the matrix. Their mucoadhesive behavior in most cases, as well as the flexibility and ease of preparation, make them preferable compared to other formulations. In addition, buccal films can protect the wound surface, thus reducing aches and treating the oral diseases more effectively. The size and shape of the film can be easily shaped so that it fits perfectly into the periodontal pocket into which it is inserted. Larger films can be applied even to the mucous membrane of the cheek but can also be cut into smaller pieces that can be placed directly in the periodontal pocket [54].

The mucoadhesive character of CS, to which its intrinsic antibacterial activity is added, determines its preferential use in making oral films. Ganjoo et al. obtained medium molecular weight CS-based films and loaded them with lincomycin hydrochloride and inserted them directly into the periodontal pouch, using solvent casting technology [55]. Although there are no strong chemical or physical interactions between the polymer matrix and the drug, its release occurs without a burst effect, practically linear up to 100 h when total release does not yet occur.

CS-based films charged with cyclohexidine have been shown to be effective against *Porphyromonas gingivalis*, the efficacy being higher even compared to free cyclohexidine [56]. The same polymer generates biodegradable films in which metronidazole and levifloxacin can be loaded and ensures a slow and steady release over time [57]. Combining the antimicrobial character of CS with the ability to load drugs with an antibacterial action, films were obtained based on this polysaccharide or in combination with HPMC, MC, HEC, or PVA loaded with a cetylpyridinium active principle, which has a bactericidal activity against some Gram-positive bacteria such as *Streptococcus mutans* and even against some Gram-negative bacteria in higher concentrations [58]. The films with PVA showed an antimicrobial activity comparable to Cetylpyridiunium chloride, but it is preferable due to the longer residence time in the periodontal pocket that ensures a longer action on *Streptococcus mutans*.

Labib and co-workers obtained films based on CS, HPMC, and Carbopol 934 loaded with pentoxifylline and metronidazole, which proved to be advantageous in clinical trials [59].

CS- and PVP-based films were optimized in terms of composition to obtain a high swelling capacity in aqueous media and mechanical properties suitable for use as release systems in the oral cavity [60]. The main goal was to optimize the oral mucoadhesive properties but also the loading capacity of tenoxicam. The in vivo salivary pharmacokinetic study of the optimized film revealed rapid and sustained salivary tenoxicam delivery in the buccal cavity, so we can conclude that the polymer–drug system can be an alternative to oral therapy for the treatment of chronic periodontitis and can provide a means to overcome the off-targeted side effects of an orally delivered drug.

Layered films based on CS and PCL, both polymers being biodegradable, were loaded with metronidazole, proving good mechanical properties and a slow release of the drug [61].

Monoadhesive films loaded with biologically active compounds with action against bacterial plaque have been reported using other types of polysaccharides. For example, ciprofloxacin-loaded buccal films were prepared using mucoadhesive polymers as sodium cellulose and sodium alginate, using the solvent evaporation method [62]. The films showed good physicochemical properties, mucoadhesion, high drug loading capacity, and high ex vivo drug release after 12 h. The significant reduction in the bacterial count in a periodontitis model proved a potent anti-periodontitis activity. Another study reports the preparation and characterization of films based on PCL and alginic acid, loaded with metronidazole and doxycycline [63]. The mechanical properties (tensile strength) are affected by the presence of the drug, reducing with its concentration. On the contrary, water absorption capacity increased with the increase of the drug content. The film can be applied in the form of a ring around the tooth that is to be extracted after a week. The application in the form of a disposable toothpick allows the dentist to access untouched areas of deep pockets and fistulas after mechanical debridement.

Multilayer films made by the solvent evaporation technique were obtained using surface erodible polymers, particularly cellulose acetate phthalate (CAP) and Pluronic F-127. They were loaded with metronidazole, ketoprofen, doxycycline, or simvastine [64]. This bio-erodible system releases the drug sequentially over time, depending on the pathogenesis of the disease, acting on its various stages and thus ensuring a more appropriate treatment. Another multiple layer film was developed by double casting followed by a compression method starting from CMCNa (which constitutes the core layer), sodium alginate (intermediate layer that loads the drug), and thiolated sodium alginate, which constitutes the outer drug free layers to enhance mucoadhesion and allow sustained release of the drug—metformin hydrochloride—for 12 h. Clinical results indicated improvement of all clinical parameters six months post treatment. The results suggested that local application of the mucoadhesive multiple layer films loaded with metformin hydrochloride showed great potential in the nonsurgical management of moderate chronic periodontitis [65].

Alginate in combination with bacterial cellulose and gelatin can generate multifunctional composites capable of sustainably loading and releasing curcumin. [66]. Hydrated films adhered firmly onto the skin, tests were performed on pig skin in artificial saliva, revealing an adhesion time of up to 6 h. Curcumin-bearing films had substantial antibacterial activity against *Escherichia coli* and *Staphylococcus aureus,* which are not cytotoxic to human keratinocytes and human gingival fibroblasts but exhibited potent anticancer activity in oral cancer cells. The versatile character of the films allows the achievement of the desired characteristics depending on the application, some of them being the realization of patches for skin care or the treatment of periodontitis and cancer of the oral cavity. Films based on alginate and gelatin incorporating nanoparticles of hydroxyapatite and an antibiotic were prepared for treating infrabony periodontal defects. It was found that there is the possibility of tailoring drug release (of tetracycline) from the nanocomposite film by varying the drug loading method; the release period can be extended up to 10 days [67].

HPMC-based bucoadhesive buccal films for local release of *Lactobacillus brevis* CD2 were prepared. They show interesting anti-inflammatory properties due to their high levels of arginine deiminase [68]. The film is able to deliver lactobacilli inside the buccal cavity or towards the mucosa by just changing the application side, according to the desired local treatment, and proves their ability to be used in treating bacterial plaque in the oral cavity. The same polysaccharide, in combination with PVA, was used to obtain films by solvent casting and their physical–mechanical characteristics, morphology, mechanical properties, and disintegrating time, were evaluated. The successful incorporation of povidone-iodine and the ability of the system to deliver the drug consistently recommend it for applications in the treatment of periodontitis [69].

Metronidazole-loaded porous matrices based on gelatin and CMCNa or HEC were synthesized by whipping and lyophilization methods [70]. The matrices in the form of film were analyzed firstly from a morphological point of view and then swelling, degradation, and mechanical properties were also investigated. The applied matrices showed a high antibacterial efficacy against anaerobic *Bacteroides* sp. *Bacteria* and a moderate cytotoxicity in vitro on fibroblast and osteoblast cell cultures. The drug is released, in vitro, without any "burst effect", the release efficacy varying between 60% and 80% after 120 min, depending on the composition of the film (Figure 3).

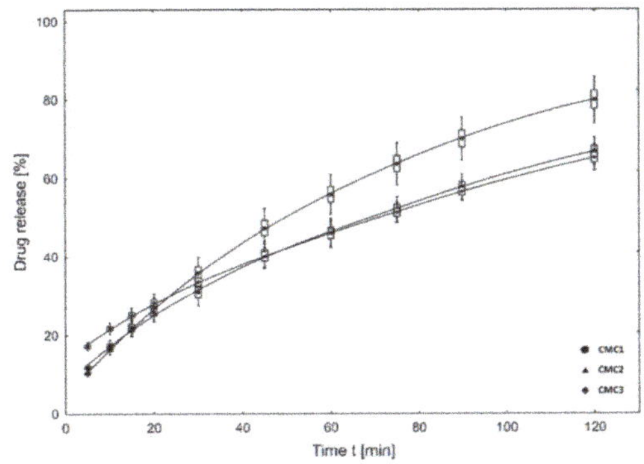

Figure 3. In vitro cumulative release profile of metronidazole from matrices based on CMCNa. Reprinted from ref. [70].

No adverse effects were showed by the metronidazole-loaded matrix based on hydroxyethyl cellulose after the clinical application. The decrease in periodontal pockets' depth and bleeding was observed 1 month after a single application.

In conclusion of this section, the characteristics of polymer–drug systems in the form of films, mentioned above, are complemented by their mechanical properties that are superior to gels and hydrogels, which ensure a longer resistance at the site of application and, consequently, a longer period of the loaded drug release. Their disadvantage, compared to gels and hydrogels, is their lower drug loading efficiency, compensated by their release with a less pronounced "burst effect" but with a constant rate that ensures a constant concentration of the drug at the site of the disease, at the therapeutic level.

2.4. Fibers

Electrospun fibers can be produced from various polymeric materials, which can be biopolymers, synthetic, or a combination of both and have several practical applications. They can serve as drug delivery systems (DDS) or carriers of cells for tissue engineering. In this section, only the studies concerning the preparation of drug-loaded electrospun fibers, fabricated from biopolymers, for the treatment of periodontitis were taken into account. Generally, these fibers are placed in the periodontal pocket around the tooth with the help of an applicator and sometimes fixed with a cyanoacrylate adhesive [71]. The encapsulated drug, especially in the lumen fibers, is thus released directly into the periodontal pocket. A disadvantage, however, is that the introduction of fibers into the bag is a time-consuming operation, and some patients reported discomfort during their placement and a slight inflammation of the gums manifested by redness [72]. An example of such a system reports tetracycline-loaded collagen-based fibers, effective in treating chronic periodontitis for 3 months, which is even commercially available [73]. In another study, the authors obtained fibers from mixtures of alginate and glycerin cross-linked by ionotropic gelling (in the presence of Ba^{2+} cations) and loaded with ciprofloxacin and diclofenac sodium salt. This system has proven a good drug release and it is capable of inhibiting the growth of microbial cultures of *Escherichia coli*, *Escherichia fecalis*, and *Staphylococcus mutans* for over 10 days [74]. Other systems based on PCL were loaded with gentamicin sulfate. In this case, the system effectively suppressed the development of *Stapyhlococcus epidermidis* for 2 weeks [75].

Coaxial electrospinning was used for the preparation of electrospun fibers based on PLGA, gum tragacanth (GT), and tetracycline hydrochloride (TCH) as a hydrophilic model drug [76]. Drug release studies showed that both the fraction of GT and the core–shell structure can effectively control TCH release rate for 75 days with only 19% of burst release within the first 2 h. These membranes might be strong enough to be easily inserted into the periodontal pocket and sustainably release the incorporated drug, while affording patient compliance with low rigidity/stiffness of the membrane during the treatment.

Two antibacterial agents, ampicillin (AMP) and metronidazole (MNZ), were loaded into a fiber mat system obtained from PLA [77]. This combination of drugs successfully suppressed *A. actinomycetemcomitans* in addition to the other pathogens, *F. nucleatum*, *P. gingivalis*, and *E. faecalis*.

Meloxicam (MX)-immobilized biodegradable CS-/PVA-/hydroxyapatite-based electrospun fibers were prepared by Yar et al., [78]. Smoother fibers were obtained at the highest drug concentration.

A tinidazole (TNZ)-loaded CS/PCL mucoadhesive hybrid nanofiber membrane (TNZ-PCHNF) was prepared in order to alleviate the existing shortcomings in the treatment of periodontitis [79]. Antibacterial activity of this membrane (at TNZ concentration of 0%, 10%, 20%, and 30%, *w/w*) was tested against *S. aureus* (MTCC1303), and it appeared that the inhibition zone increased, as expected, with the increase in the drug concentration.

2.5. Microparticles

Such systems are solid polymeric particles, especially spherical in shape, with a diameter between 1 and 1000 μm and contain the drug dispersed throughout their volume. There may be physical, less often chemical (ionic bonds), interactions between the loaded bioactive agent and the constituent polymeric matrix, or the drug may only be adsorbed on the surface of the carrier. They are powdery materials obtained from both natural and synthetic polymers. The category of polymers usable for this purpose includes biodegradable ones, such as PLGA, polylactide, or poly(hydroxy alkanoates), or polysaccharides, such as CS, pectin, hyaluronic acid, etc. The following techniques are used to obtain microparticles: emulsification, solvent evaporation, coacervation, spray drying, electrospray, etc. [16]. The administration can be done in the form of suspensions, toothpaste, or by direct injection into the periodontal pocket [80]. The advantages of using these type of systems include the following: ensuring a controlled release, increasing patient compliance, and achieving a sustained therapeutic effect.

Doxycycline-hiclate-loaded microparticles, having a sustained release for 47 days following a Fickian mechanism, have been shown to be effective in inhibiting microbial cultures of *Staphylococcus aureus*, *Porphyromonas gingivalis*, and *Staphylococcus mutans* [81]. Doxycycline hyclate and ornidazole were loaded in CS-vanillin cross-linked microspheres dispersed in situ gel (MLIG) implants [82]. In contrast to the previous study, the in vitro dissolution study demonstrated a non-Fickian type of drug release mechanism for 12 days. In addition, these formulations exhibited significant antimicrobial activity against *Staphylococcus aureus*, *Escherichia coli*, and *Enterococcus faecalis* and were found to be biocompatible and biodegradable during preclinical studies. Metronidazole-loaded CS particles, obtained by cross-linking in emulsion, showed a longer release compared to mixtures of the same microparticles and drug [83]. Microspheres of CS loaded with ornidazole (ORDZ) were prepared by the emulsification ionotropic gelation method [84]. These microspheres showed drug encapsulation in the range of 11.02 ± 0.98–$32.45 \pm 0.62\%$ and sustained the release up to 5 days. The incorporation into a Pluronic (F-127) gel increased the release up to 7 days. The antimicrobial study indicated an inhibition of growth of *Staphylococcus aureus* at all drug concentrations. In another interesting study, microspheres, based on cationic CS and anionic xanthan gum, were prepared by coacervation and were further transformed into a gel that was loaded with chlorhexidine (CHX) [85]. CHX-containing gels exhibited selective antibacterial effects against the growth of *P. gingivalis*. CHX was also loaded into alginate-based microparticles and the authors have demonstrated that a higher amount of CHX was released from these microparticles in comparison with PLGA microparticles having the same particle size [86].

An antimicrobial decapeptide, KSL-W (KKVVFWVKFK-CONH2), which could maintain stable antimicrobial activity in saliva was loaded into PLGA/CS composite microspheres (Figure 4), prepared by electrospraying and combined cross-linking–emulsion methods [87].

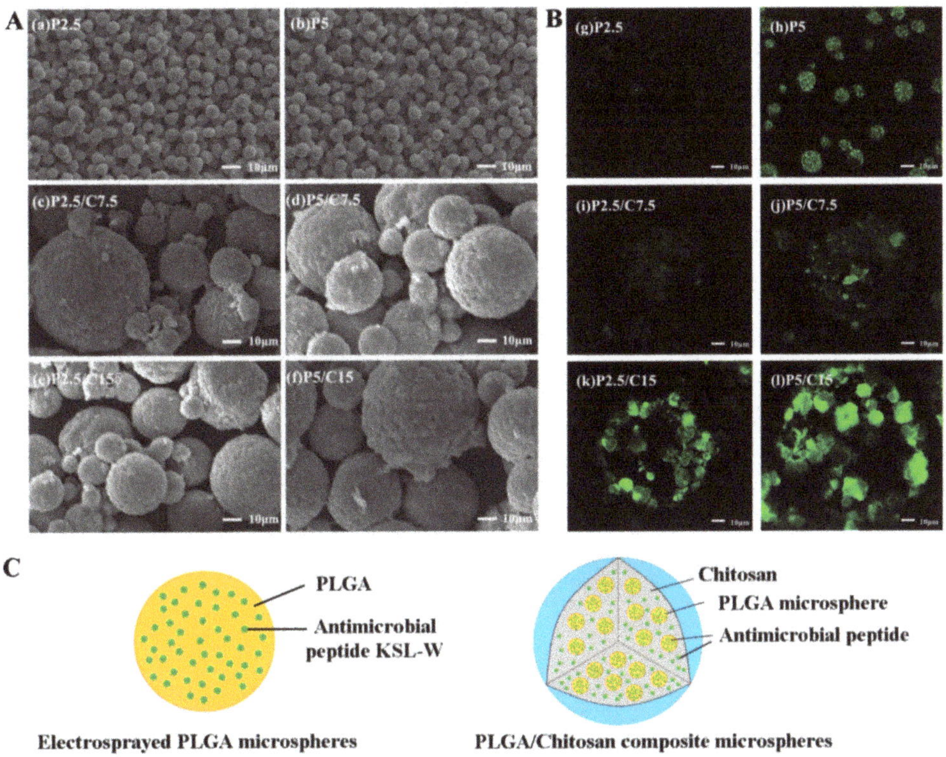

Figure 4. (**A**) Typical scanning electron microscope (SEM) morphologies of KSL-W-loaded PLGA and PLGA/CS composite microspheres; (**B**) visualized distributions of FITC-conjugated KSL-W in PLGA and PLGA/CS microspheres under laser scanning confocal microscope (LSCM); (**C**) a schematic diagram of a KSL-loaded PLGA microsphere and a KSL-loaded PLGA/CS microsphere. PLGA, poly(lactide-co-glycolide); CS, chitosan; FITC, fluorescein isothiocyanate. Reprinted from ref. [87].

Antibacterial experiments demonstrated the prolonged antimicrobial and inhibitory effects of these KSL/PLGA/CS microspheres on oral bacteria.

The combination of polysaccharides with bio-based polymers was also used in a study where interleukin 1 receptor antagonist (IL-1ra) was loaded into dextran/PLGA microspheres in order to evaluate the physicochemical characteristics and anti-inflammatory properties [88,89]. It was demonstrated that these microparticles blocked the IL-1β-induced production of pro-inflammatory cytokines in vitro.

Microparticulate polymer systems, according to experts in the field, have great potential for development in the near future as commercial products after performing several clinical tests.

2.6. Nanoparticles

Nanoparticles are the most promising strategy in the treatment of periodontal disease, because, due to their small diameter (between 10 and 1000 nm), they can penetrate regions that cannot be reached by other drug delivery systems. This advantage leads to a reduction in the frequency of administration and also ensures a more even distribution of the drug. Submicron polymeric particles with a diameter below 500 nm were prepared by many techniques, using in particular natural polymers (polysaccharides, proteins), biocompatible synthetic polymers, or mixtures thereof.

Table 2 briefly presents a series of results reported in the abundant literature of the last 3–4 years that illustrate different nanoparticle systems carrying drugs or other active principles with applicability in the treatment of periodontal disease.

Table 2. Examples of nanoparticulate systems reported in the literature of recent years.

Polymer-Based Nanoparticles	Encapsulated Bioactive Compound	Preparation Method	Antibacterial Action Against	Application	Ref.
CS	Platelet-rich plasma	Ionic gelation method	*Staphylococcus mutans*	Complex/chronic wound healing and soft/hard tissue regeneration following periodontitis treatment or tooth extraction that needs prolonged growth factor release	[90]
	-	Precipitation from acetic acid solution with NaOH	*Enterococcus faecalis, Staphylococcus aureus, Bacillus subtilis*	Eliminating plasmid mediated resistance acquired by periodontal pathogens	[91]
	Amoxicillin and clavulanic acid	Ionotropic gelation with tri-polyphosphate	dentobacterial plaque	Higher efficacy by killing the pathogen bacteria in a sustained manner while reducing the cellular toxicity to non-bacterial cells	[92]
	Antimicrobial peptide	Ionotropic gelation with tri-polyphosphate and coating with the peptide	*Fusobacterium nucleatum, Porphyromonas gingivalis, Staphylococcus gordonii*	Treat root caries restorations to inhibit periodontitis related pathogens in periodontitis care	[93]
	Indocyanine green as a photosensitizer for antimicrobial photodynamic therapy	Ionotropic gelation with tripoly-phosphate	*Aggregatibacter Actino-mycetemcomitans*	Potential implications for the treatment of *A. action-mycetemcomitans* infections in periodontitis and peri-implantitis in vivo	[94,95]
	Asiaticoside containing sulfobutyl ether-β-cyclodextrine complex	Ionotropic gelation		Carrier to deliver asiaticoside for periodontal tissue regeneration	[96]
CS-carboxymethyl CS	Doxycycline	Polyelectrolyte complexation and ionic gelation	*Porphyromonas gingivalis*	New option for the rational administration of doxycycline in the clinical treatment of periodontal disease	[97]
Lecitin-based liposomes coated with quaternary ammonium *N,N,N* trimethyl CS	Doxycycline	Electrostatic adsorption of CS derivative on liposome surface	*Porphyromonas gingivalis, Prevotella intermedia*	Potential applications in the clinical treatment of periodontal disease by extensive and efficient antibacterial activity	[98]
CS in collagen membrane	Chlorhexidine	Ionic gelation with sodium tri-polyphosphate	*Enterococcus faecalis*	Endodontic failure improves regenerative procedures in periapical surgery	[99]
Alginate coated CS core–shell nanoparticles	Transforming growth factor (TGF)-β1 and dexamethasone	Ionic gelation and polyelectrolyte complexation		Achieving healthy connective tissue ingrowth into the apical portion of the root canal space and subsequently a biologically based healing in root canal treatment	[100]
CS–sodium alginate polyelectrolyte complexes	Dimocarpus longan leaves extract	Polyelectrolyte complexation	*Staphylococcus aureus*	High antibacterial potential against bacteria that triggers periodontitis	[101]

Table 2. *Cont.*

Polymer-Based Nanoparticles	Encapsulated Bioactive Compound	Preparation Method	Antibacterial Action Against	Application	Ref.
Sodium alginate	Metronidazole	Emulsion–solvent evaporation method: single emulsion and double emulsification		Delivery of MNZ in periodontitis up to 24 h	[102]
CS coated poly(D,L-lactide-*co*-glycolide)	Lovastatin tetracycline	Double emulsion–solvent evaporation	*Aggregatibacter actinomycetemcomitans, Prevotella nigrescens*	Adjunctive treatment in periodontitis, promoted new bone formation in three-walled defects in beagle dogs	[103]
	Metronidazol *N*-phenacyl-thiazolium	Oil-in-water single emulsion–solvent evaporation		Reduced inflammation of experimental periodontitis, greater potential to resist further periodontal breakdown	[104]
	Simvastatin doxycycline	Double emulsion technology	*Porphyromonas gingivalis Staphylococcus sanguinis*	Loaded with SIM-DOX synergistically promoted the repair of the periodontium	[105]
Cellulose acetate phthalate	Chlorhexidine	Emulsion–solvent diffusion technique		Reduced the dentobacterial plaque index by 65.78%,	[106]
Hydroxyl propyl methyl cellulose, methyl cellulose, Carbopol 934	Coenzyme Q10	Nanoprecipitation, solvent evaporation, lyophilization		Management of chronic periodontitis	[107]

CS is one of the most used polysaccharides for the preparation of nanoparticles, having the advantage of an intrinsic antibacterial activity [91]. It is often used in combination with alginates [100,101] or with PLGA [103–105], associated with the collagen matrix [97], in the form of derivatives [97,98] or biocomposites.

Cellulose derivatives (CAP, HPMC, MC) [105,106] or other polysaccharides (sodium alginate) [102] are reported in recent literature.

The brief presentation in this review highlights that polymer-based controlled release systems developed to date due to the benefits of nanotechnology are viable alternatives for the prevention and treatment of periodontal disease but require in-depth in vivo testing on animals and ultimately on human patients.

3. Conclusions

Periodontal disease is a worldwide health problem affecting people's lifestyles. The best way to fight it is to prevent it, and the best way to do this is to remove its causes, particularly to prevent or remove the bacterial plaque. Classical methods of treating the disease are proving ineffective, which required finding a novel approach, with the potential use of nanotechnology being the most appropriate. The trend of recent years in the treatment of this disease is the local release of antibiotics, antimicrobials, and anti-inflammatory drugs, and the most appropriate method to achieve the desired effect is the local administration of drug delivery systems based on polymers. The advantages of this treatment are multiple because they directly target the affected area, maintain a relatively constant level of drug concentration, and minimize systemic side effects.

Formulations based on drug-loaded polymer systems are diverse, each with advantages and disadvantages.

Nanofiber-based films and membranes containing encapsulated drugs are easy to apply either on the gingival mucosa or by insertion into the periodontal pocket, but the amount of encapsulated drug is lower and its release faster. In addition, their removal must often be done by surgery, which reduces the patient's compliance.

Gels and hydrogels administered by injection have the advantage of forming directly in the periodontal pocket, especially under the influence of external factors. The release of the drug is slower, especially from hydrogels, maintaining the constant concentration of the active principle for a longer time and thus increasing the effectiveness of treatment.

The advances registered by nanotechnology have allowed the realization of micro- and nanoparticle carrying drugs, which have a great potential for development in the near future as commercial products. This type of release system has been found to be the most viable alternative to date for the prevention and treatment of periodontal disease, but their implementation in current treatment requires in-depth in vivo testing on animals and ultimately on human patients.

Author Contributions: Conceptualization, N.B., M.P., L.I.A. and D.L.I.; methodology, N.B., M.P., L.I.A. and D.L.I.; writing—original draft preparation, N.B., M.P., L.I.A. and D.L.I.; writing—review and editing, M.P. and L.I.A.; supervision, M.P. and L.I.A. All authors have read and agreed to the published version of the manuscript.

Funding: This research received no external funding.

Institutional Review Board Statement: Not applicable.

Informed Consent Statement: Not applicable.

Data Availability Statement: Not applicable.

Conflicts of Interest: The authors declare no conflict of interest.

References

1. DeRossi, D. *Polymer Gels*; Plenum Press: New York, NY, USA, 1991.
2. Dunn, R.L.; Ottenbrite, R.M. *Polymeric Drugs and Drug Delivery Systems*; American Chemical Society: Washington, DC, USA, 1991.
3. Kononen, E.; Gursoy, M.; Gursoy, U.K. Periodontitis: A multifaceted disease of tooth-supporting tissues. *J. Clin. Med.* **2019**, *8*, 1135. [CrossRef] [PubMed]
4. Zieba, M.; Chaber, P.; Duale, K.; Maksymiak, M.M.; Basczok, M.; Kowalczuk, M.; Adamus, G. Polymeric carriers for delivery systems in the treatment of chronic periodontal disease. *Polymers* **2020**, *12*, 1574. [CrossRef]
5. Socransky, S.S.; Haffajee, A.D.; Cugini, M.A.; Smith, C.; Kent, R.L.J. Microbial complexes in subgingival plaque. *J. Clin. Periodontol.* **1998**, *25*, 134–144. [CrossRef]
6. Lamont, R.J.; Koo, H.; Hajishengallis, G. The oral microbiota: Dynamic communities and host interactions. *Nat. Rev. Microbiol.* **2018**, *16*, 745–759. [CrossRef] [PubMed]
7. Rams, T.E.; Degener, J.E.; van Winkelhoff, A.J. Antibiotic resistance in human chronic periodontitis microbiota. *J. Periodontol.* **2014**, *85*, 160–169. [CrossRef] [PubMed]
8. Wade, G.W. The oral microbiome in health and disease. *Pharmacol. Res.* **2013**, *69*, 137–143. [CrossRef]
9. Pflughoeft, K.J.; Versalovic, J. Human microbiome in health and disease. *Annu. Rev. Pathol.* **2012**, *7*, 99–122. [CrossRef]
10. Konig, J.; Schwahn, C.; Fanghanel, J.; Plotz, J.; Hoffmann, T.; Kocher, T. Repeated scaling versus surgery in young adults with generalized advanced periodontitis. *J. Periodontol.* **2008**, *79*, 1006–1013. [CrossRef]
11. Lundgren, D.; Asklöw, B.; Thorstensson, H.; Härefeldt, A.M. Success rates in periodontal treatment as related to choice of evaluation criteria. Presentation of an evaluation criteria staircase for costbenefit use. *J. Clin. Periodontol.* **2001**, *28*, 23–30. [CrossRef] [PubMed]
12. Darveau, R.P. Periodontitis: A polymicrobial disruption of host homeostasis. *Nat. Rev. Microbiol.* **2010**, *8*, 481–490. [CrossRef]
13. Holpuch, A.; Desai, K.G.; Schwendeman, S.; Mallery, S. Optimizing therapeutic efficacy of chemopreventive agents. A critical review of delivery strategies in oral cancer chemoprevention clinical trials. *J. Carcinog.* **2011**, *10*, 23.
14. Sankar, V.; Hearnden, V.; Hull, K.; Juras, D.V.; Greenberg, M.S.; Kerr, A.R.; Lockhart, P.B.; Patton, L.L.; Porter, S.; Thornhill, M. Local drug delivery for oral mucosal diseases. Challenges and opportunities. *Oral Dis.* **2011**, *17*, 73–84. [CrossRef]
15. Rata, D.M.; Cadinoiu, A.N.; Burlui, V.; Atanase, L.I. Polysaccharide-based orodental delivery systems. In *Polysaccharide Carriers for Drug Delivery*; Maiti, S., Jana, S., Eds.; Woodhead Publishing: Sawston, UK, 2019; pp. 685–711.
16. Rajeshwari, H.R.; Dhamecha, D.; Jagwani, S.; Rao, M.; Jadhav, K.; Shaikh, S.; Puzhankara, L.; Jalalpure, S. Local drug delivery systems in the management of periodontitis: A scientific review. *J. Control. Rel.* **2019**, *307*, 393–409.
17. Slomkowski, S.; Alemán, J.V.; Gilbert, R.G.; Hess, M.; Horie, K.; Jones, R.G.; Przemyslaw, K.; Meisel, I.; Mormann, W.; Penczek, S.; et al. Terminology of polymers and polymerization processes in dispersed systems (IUPAC Recommendations 2011). *Pure Appl. Chem.* **2011**, *83*, 12, 2229–2259. [CrossRef]
18. Yadav, R.; Kanwar, I.L.; Haider, T.; Pandey, V.; Gour, V.; Soni, V. In situ gel drug delivery system for periodontitis: An insight review. *Future J. Pharm. Sci.* **2020**, *6*, 1–13. [CrossRef]
19. Pragati, S.; Ashok, S.; Kuldeep, S. Recent advances in periodontal drug delivery systems. *Int. J. Drug Deliv.* **2009**, *1*, 1–14.
20. Popa, L.; Ghica, M.V.; Dinu-Pîrvu, C.E. Periodontal chitosan-gels designed for improved local intra-pocket drug delivery. *Farmacia* **2013**, *661*, 240–250.

21. Özdogan, A.I.; Akca, G.; Senel, S. Development and in vitro evaluation of chitosan based system for local delivery of atorvastatin for treatment of periodontitis. *Eur. J. Pharm. Sci.* **2018**, *124*, 208–216. [CrossRef] [PubMed]
22. Özdogan, A.I.; Ilarslan, Y.D.; Kösemehmetoglu, K.; Akca, G.; Comerdov, E.; Iskit, A.B.; Senel, S.; Kutlu, H.B. In vivo evaluation of chitosan based local delivery systems for atorvastatinin treatment of periodontitis. *Int. J. Pharm.* **2018**, *550*, 470–476. [CrossRef] [PubMed]
23. Özdogan, A.I.; Akca, G.; Sevda, S. Development and in vitro evaluation of gel formulation of atorvastatin solid dispersions. *J. Drug Deliv. Sci. Technol.* **2021**, *61*, 102199. [CrossRef]
24. Aminu, N.; Chan, S.Y.; Yam, M.F.; Toh, S.M. A dual-action chitosan-based nanogel system of triclosan and flurbiprofen for localised treatment of periodontitis. *Int. J. Pharm.* **2019**, *570*, 118659. [CrossRef]
25. Alves, M.C.C.; Chaves, D.S.A.; Benevenuto, B.R.; De Farias, B.O.; Coelho, S.M.O.; Ferreira, T.P.; Pereira, G.A.; Dos Santos, G.C.M.; Moreira, L.O.; Freitas, J.P.; et al. Chitosan gels for buccal delivery of *Schinus molle L.* essential oil in dogs: Characterization and antimicrobial activity in vitro. *An. Acad. Bras. Cienc.* **2020**, *92*, e20200562. [CrossRef] [PubMed]
26. Sheshala, R.; Quah, S.Y.; Sahu, P.S.; Tan, G.C.; Meka, V.S.; Jnanendrappa, N. Investigation on solution-to-gel characteristic of thermosensitive and mucoadhesive biopolymers for the development of moxifloxacin-loaded sustained release periodontal in situ gels. *Drug Deliv. Transl. Res.* **2019**, *9*, 434–443. [CrossRef]
27. Bansal, M.; Mittal, N.; Yadav, S.K.; Khan, G.; Gupta, P.; Mishra, B.; Gopal, N. Periodontal thermoresponsive, mucoadhesive dual antimicrobial loaded in-situ gel for the treatment of periodontal disease: Preparation, in-vitro characterization and antimicrobial study. *J. Oral Biol. Craniofac. Res.* **2018**, *8*, 126–133. [CrossRef]
28. Bansal, M.; Mittal, N.; Yadav, S.K.; Khan, G.; Mishra, B.; Nath, G. Clinical evaluation of thermoresponsive and mucoadhesive chitosan in situ gel containing levofloxacin and metronidazole in the treatment of periodontal pockets–a split-mouth, clinical study. *J. Pierre Fauchard Acad.* **2016**, *30*, 6–14. [CrossRef]
29. Aslani, A.; Malekpour, N. Design, formulation, and physicochemical evaluation of periodontal propolis mucoadhesive gel. *Dent. Res. J.* **2016**, *13*, 484–493. [CrossRef]
30. Nasri, L.; De Rosa, A.; Donnarumma, G.; De Gregorio, V.; Grassia, V. A new controlled-release material containing metronidazole and doxycycline for the treatment of periodontal and peri-implant diseases: Formulation and in vitro testing. *Int. J. Dent.* **2019**, 9374607.
31. Rajendran, S.; Kumar, K.S.; Ramesh, S.; Rao, S.R. Thermoreversible in situ gel for subgingival delivery of simvastatin for treatment of periodontal disease. *Int. J. Pharm. Investig.* **2017**, *7*, 101–106. [PubMed]
32. Swain, G.P.; Patel, S.; Gandhi, J.; Shah, P. Development of Moxifloxacin hydrochloride loaded in-situ gel for the treatment of periodontitis: In-vitro drug release study and antibacterial activity. *J. Oral. Biol. Craniofac. Res.* **2019**, *9*, 190–200. [CrossRef] [PubMed]
33. Mihalache, C.; Rata, D.M.; Cadinoiu, A.N.; Patras, X.; Sindilar, E.V.; Bacaita, S.E.; Popa, M.; Atanase, L.I.; Daraba, O.M. Bupivacaine-loaded chitosan hydrogels for topical anesthesia in dentistry. *Polym. Int.* **2020**, *69*, 1152–1160. [CrossRef]
34. Merlusca, I.P.; Ibanescu, C.; Tuchilus, C.; Danu, M.; Atanase, L.I.; Popa, M.I. Characterization of neomycin-loaded xanthan-chitosan hydrogels for topical applications. *Cellulose Chem. Technol.* **2019**, *53*, 709–719. [CrossRef]
35. Huang, J.; Wang, Z.; Krishna, S.; Hu, Q.; Xuan, M.; Xie, H. Environment-sensitive hydrogels as potential drug delivery systems for the treatment of periodontitis. *Mater. Express* **2020**, *10*, 975–985. [CrossRef]
36. Xu, X.; Gu, Z.; Xi, C.; Shi, C.; Liu, C.; Liu, M.; Wang, L.; Zhang, K.; Liu, Q.; Shen, Y.; et al. An injectable and thermosensitive hydrogel: Promoting periodontal regeneration by controlled-release of aspirin and erythropoietin. *Acta Biomater.* **2019**, *86*, 235–246. [CrossRef] [PubMed]
37. Petit, C.; Batool, F.; Stutz, C.; Anton, N.; Klymchenko, A.; Vandamme, T.; Benkirane-Jessel, N.; Huck, O. Development of a thermosensitive statin loaded chitosan-based hydrogel promoting bone healing. *Int. J. Pharm.* **2020**, *586*, 119534. [CrossRef] [PubMed]
38. Shen, Z.; Kuang, S.; Zhang, Y.; Yang, M.; Qin, W.; Lin, Z.; Shi, X. Chitosan hydrogel incorporated with dental pulp stem cell-derived exosomes alleviates periodontitis in mice via a macrophage-dependent mechanism. *Bioactive Mater.* **2020**, *5*, 1113–1126. [CrossRef] [PubMed]
39. Ji, Q.X.; Zhao, Q.S.; Deng, J.; Lü, R. A novel injectable chlorhexidine thermosensitive hydrogel for periodontal application: Preparation, antibacterial activity and toxicity evaluation. *J. Mater. Sci. Mater. Med.* **2010**, *21*, 2435–2442. [CrossRef]
40. Chang, P.C.; Chao, Y.C.; Hsiao, M.S.; Chou, H.S.; Jheng, Y.H.; Yu, X.H.; Lee, N.; Yang, C.; Liu, D.M. Inhibition of periodontitis induction using a stimuli responsive hydrogel carrying Naringin. *J. Periodontol.* **2017**, *88*, 190–196. [CrossRef]
41. Yu, M.C.; Chang, C.Y.; Chao, Y.C.; Jheng, Y.H.; Yang, C.; Lee, N.; Yu, S.H.; Yu, X.H.; Liu, D.M.; Chang, P.C. pH-responsive hydrogel with an anti-glycation agent for modulating experimental periodontitis. *J. Periodontol.* **2016**, *87*, 742–748. [CrossRef]
42. Xu, L.; Bai, X.; Yang, J.; Li, J.; Yuan, H.; Xie, J.; Li, J. Preparation and characterization of a gellan gum-based hydrogel enabling osteogenesis and inhibiting Enterococcus faecalis. *Int. J. Biol. Macromol.* **2020**, *165*, 2964–2973. [CrossRef]
43. Tong, X.; Qi, X.; Mao, R.; Pan, W.; Zhang, M.; Chan, G.; Shen, J.; Deng, H.; Hu, R.; Wu, X. Construction of functional curdlan hydrogels with bio-inspired polydopamine for synergistic periodontal antibacterial therapeutics. *Carbohydr. Polym.* **2020**, *245*, 116585. [CrossRef]
44. Qasim, S.S.B.; Nogueira, L.P.; Fawzy, A.S.; Daood, U. The effect of cross-linking efficiency of drug-loaded novel freeze gelated chitosan templates for periodontal tissue regeneration. *AAPS Pharm. Sci. Tech.* **2020**, *2*, 173. [CrossRef]

45. Liao, Y.; Li, H.; Shu, R.; Chen, H.; Zhao, L.; Song, Z.; Zhou, W. Mesoporous Hydroxyapatite/Chitosan loaded with recombinant-human amelogenin could enhance antibacterial effect and promote periodontal regeneration. *Front. Cell. Infect. Microbiol.* **2020**, *10*, 180. [CrossRef] [PubMed]
46. Xiao, Y.; Gong, T.; Jiang, Y.; Wang, Y.; Wen, Z.T.; Zhou, S.; Bao, C.; Xu, X. Fabrication and characterization of a glucose-sensitive antibacterial Chitosan-polyethylene oxide hydrogel. *Polymer* **2016**, *82*, 1–10. [CrossRef]
47. Johnson, A.; Kong, F.; Miao, S.; Lin, H.T.V.; Thomas, S.; Huang, Y.C.; Kong, Z.L. Therapeutic effects of antibiotics loaded cellulose nanofiber and κ-carrageenan oligosaccharide composite hydrogels for periodontitis treatment. *Sci. Rep.* **2020**, *10*, 18037. [CrossRef] [PubMed]
48. Hirun, N.; Tantishaiyakul, V.; Sangfai, T.; Ouiyangkul, P.; Li, L. In situ mucoadhesive hydrogel based on methylcellulose/xyloglucan for periodontitis. *J. Sol-Gel Sci. Tech.* **2019**, *89*, 531–542. [CrossRef]
49. Liu, J.; Xiao, Y.; Wang, X.; Huang, L.; Chen, Y.; Bao, C. Glucose-sensitive delivery of metronidazole by using a photo-crosslinked chitosan hydrogel film to inhibit Porphyromonas gingivalis proliferation. *Int. J. Biol. Macromol.* **2019**, *122*, 19–28. [CrossRef]
50. Wei, Y.; Chang, Y.H.; Liu, C.J.; Chung, R.J. Integrated oxidized-hyaluronic acid/collagen hydrogel with β-TCP using proanthocyanidins as a crosslinker for drug delivery. *Pharmaceutics* **2018**, *10*, 37. [CrossRef] [PubMed]
51. Dong, Z.; Sun, Y.; Chen, Y.; Liu, Y.; Tang, C.; Qu, Z. Injectable adhesive hydrogel through a microcapsule cross-link for periodontitis treatment. *ACS Appl. Bio Mater.* **2019**, *2*, 5985–5994. [CrossRef]
52. Li, H.; Ji, Q.; Chen, X.; Sun, Y.; Xu, Q.; Deng, P.; Hu, F.; Yang, J. Accelerated bony defect healing based on chitosan thermosensitive hydrogel scaffolds embedded with chitosan nanoparticles for the delivery of BMP2 plasmid DNA. *J. Biomed. Mater. Res. A* **2017**, *105*, 265–273. [CrossRef]
53. Chichiricco, P.M.; Riva, R.; Thomassin, J.M.; Lesoeur, J.; Struillou, X.; Le Visage, C.; Jérôme, C.; Weiss, P. In situ photochemical crosslinking of hydrogel membrane for guided tissue regeneration. *Dent. Mater.* **2018**, *34*, 1769–1782. [CrossRef]
54. Jain, N.; Jain, G.K.; Javed, S.; Iqbal, Z.; Talegaonkar, S.; Ahmad, F.J.; Khar, R.K. Recent approaches for the treatment of periodontitis. *Drug Discov.* **2008**, *13*, 932–943. [CrossRef]
55. Ganjoo, R.; Sonil, S.; Ram, V.; Verma, A. Medium molecular weight chitosan as a carrier for delivery of lincomycin hydrochloride from intra-pocket dental film: Design, development, in vitro and ex vivo characterization. *J. Appl. Pharm. Sci.* **2016**, *6*, 8–19. [CrossRef]
56. Ikinci, G.; Senel, S.; Akincibay, H.; Kaş, S.; Erciş, S.; Wilson, C.G.; Hincal, A.A. Effect of chitosan on a periodontal pathogen Porphyromonas gingivalis. *Int. J. Pharm.* **2002**, *235*, 121–127. [CrossRef]
57. Khan, G.; Yadav, S.K.; Patel, R.R.; Nath, G.; Bansal, M.; Mishra, B. Development and evaluation of biodegradable chitosan films of metronidazole and levofloxacin for the management of periodontitis. *AAPS Pharmscitech.* **2015**, *17*, 1–14. [CrossRef]
58. Abouhussein, D.; El Nabarawi, M.A.; Shalaby, S.H.; El-Bary, A.A. Cetylpyridinium chloride chitosan blended mucoadhesive buccal films for treatment of pediatric oral diseases. *J. Drug Deliv. Sci. Technol.* **2020**, *57*, 101676. [CrossRef]
59. Labib, G.S.; Aldawsari, H.M.; Badr-Eldin, S.M. Metronidazole and pentoxifylline films for the local treatment of chronic periodontal pockets: Preparation, in vitro evaluation and clinical assessment. *Expert Opin. Drug Deliv.* **2014**, *11*, 855–865. [CrossRef]
60. Ashri, L.Y.; El Sayeh, F.A.; El Ela, A.; Ibrahim, M.A.; Alshora, D.H.; Naguib, M. Optimization and evaluation of chitosan buccal films containing tenoxicam for treating chronic periodontitis: In vitro and in vivo studies. *J. Drug Deliv. Sci. Technol.* **2020**, *57*, 101720. [CrossRef]
61. El-Kamel, A.H.; Ashri, L.Y.; Alsarra, I.A. Micromatricial metronidazole benzoate film as a local mucoadhesive delivery system for treatment of periodontal diseases. *AAPS Pharmscitech.* **2007**, *8*, E184–E194. [CrossRef]
62. Wu, W.; Chen, W.; Jin, Q. Oral mucoadhesive buccal film of ciprofloxacin for periodontitis: Preparation and characterization. *Trop. J. Pharm. Res.* **2016**, *15*, 447. [CrossRef]
63. Ghavami-Lahiji, M.; Shafiei, F.; Najafi, F.; Erfan, M. Drug-loaded polymeric films as a promising tool for the treatment of periodontitis. *J. Drug Deliv. Sci. Technol.* **2019**, *52*, 122–129. [CrossRef]
64. Sundararaj, S.C.; Thomas, M.V.; Peyyala, R.; Dziubla, T.D.; Puleo, D.A. Design of a multiple drug delivery system directed at periodontitis. *Biomaterials* **2013**, *34*, 8835–8842. [CrossRef]
65. Kassem, A.A.; Issa, D.A.E.; Kotry, G.S.; Farid, R.M. Thiolated alginate-based multiple layer mucoadhesive films of metformin for intra-pocket local delivery: In vitro characterization and clinical assessment. *Drug Dev. Ind. Pharm.* **2017**, *43*, 120–131. [CrossRef]
66. Chiaoprakobkij, N.; Suwanmajo, T.; Sanchavanakit, N.; Phisalaphong, M. Composite film curcumin-loaded bacterial cellulose/alginate/gelatin as a multifunctional biopolymer. *Molecules* **2020**, *25*, 3800. [CrossRef]
67. Madhumathi, M.; Rekha, L.J.; Sampath Kumar, T.S. Tailoring antibiotic release for the treatment of periodontal infrabony defects using bioactive gelatin-alginate/apatite nanocomposite films. *J. Drug Deliv. Sci. Technol.* **2018**, *43*, 57–64. [CrossRef]
68. Abruzzo, A.; Vitali, B.; Lombardi, F.; Guerrini, L.; Cinque, B.; Parolin, C.; Bigucci, F.; Cerchiara, T.; Arbizzani, C.; Gallucci, M.C.; et al. Mucoadhesive buccal films for local delivery of lactobacillus brevis. *Pharmaceutics* **2020**, *12*, 241. [CrossRef]
69. Kida, D.; Gładysz, O.; Szulc, M.; Zborowski, J.; Junka, A.; Janeczek, M.; Lipińska, A.; Skalec, A.; Karolewicz, B. Development and evaluation of a polyvinylalcohol-cellulose derivative-based film with povidone-iodine predicted for wound treatment. *Polymers* **2020**, *12*, 1271. [CrossRef] [PubMed]
70. Kida, D.; Karolewicz, B.; Junka, A.; Sender-Janeczek, A.B.; Duś, I.; Marciniak, D.; Szulc, M. Metronidazole-loaded porous matrices for local periodontitis treatment: In vitro evaluation and in vivo pilot study. *Appl. Sci.* **2019**, *9*, 4545. [CrossRef]

71. Joshi, D.; Garg, T.; Goyal, A.K.; Rath, G. Advanced drug delivery approaches against periodontitis. *Drug Deliv.* **2016**, *23*, 363–377. [CrossRef]
72. Garg, T.; Goyal, A.K. Biomaterial-based scaffolds-current status and future directions. *Expert Opin. Drug Deliv.* **2014**, *11*, 767–789. [CrossRef] [PubMed]
73. Sachdeva, S.; Agarwal, V. Evaluation of commercially available biodegradable tetracycline fiber therapy in chronic periodontitis. *J. Indian Soc. Periodontol.* **2011**, *15*, 130–134. [PubMed]
74. Johnston, D.; Choonara, Y.E.; Kumar, P.; Lisa, C.; van Vuuren, S.; Pillay, V. Prolonged delivery of ciprofloxacin and diclofenac sodium from a polymeric fiber device for the treatment of peridontal disease. *Biomed. Res. Int.* **2013**, 460936. [CrossRef]
75. Chang, H.I.; Lau, Y.C.; Yan, C.; Coombes, A. Controlled release of an antibiotic, gentamicin sulphate, from gravity spun polycaprolactone fibers. *J. Biomed. Mater. Res.* **2008**, *84*, 230–237. [CrossRef]
76. Ranjbar-Mohammadi, M.; Zamani, M.; Prabhakaran, M.P.; Bahrami, S.H.; Ramakrishna, S. Electrospinning of PLGA/gum tragacanth nanofibers containing tetracycline hydrochloride for periodontal regeneration. *Mater. Sci. Eng. C* **2016**, *58*, 521–531. [CrossRef]
77. Schkarpetkin, D.; Reise, M.; Wyrwa, R.; Volpel, A.; Berg, A.; Schweder, M.; Schnabelrauch, M.; Watts, D.C.; Sigusch, B.W. Development of novel electrospun dual-drug fibermats loaded with a combination of ampicillin andmetronidazole. *Dent. Mater.* **2016**, *32*, 951–960. [CrossRef]
78. Yar, M.; Farooq, A.; Shahzadi, L.; Khan, A.S.; Mahmood, N.; Rauf, A.; Chaudhry, A.A.; Ur Rehman, I. Novel meloxicam releasing electrospun polymer/ceramic reinforced biodegradable membranes for periodontal regeneration applications. *Mater. Sci. Eng. C* **2016**, *64*, 148–156. [CrossRef] [PubMed]
79. Khan, G.; Yadav, S.K.; Patel, R.R.; Kumar, N.; Bansal, M.; Mishra, B. Tinidazole functionalized homogeneous electrospun chitosan/poly(ε-caprolactone) hybrid nanofiber membrane: Development, optimization and its clinical implications. *Int. J. Biol. Macromol.* **2017**, *103*, 1311–1326. [CrossRef] [PubMed]
80. Sholapurkar, A.; Sharma, D.; Glass, B.; Miller, C.; Nimmo, A.; Jennings, E. Professionally delivered local antimicrobials in the treatment of patients with periodontitis-a narrative review. *Dent. J.* **2021**, *9*, 2. [CrossRef]
81. Phaechamud, T.; Chanyaboonsub, N.; Setthajindalert, O. Doxycycline hyclate-loaded bleached shellac in situ forming microparticle for intraperiodontal pocket local delivery. *Eur. J. Pharm. Sci.* **2016**, *93*, 360–370. [CrossRef] [PubMed]
82. Yadav, S.K.; Khan, G.; Bansal, M.; Thokala, S.; Bonde, G.V.; Upadhyay, M.; Mishra, B. Multiparticulate based thermosensitive intrapocket forming implants for better treatment of bacterial infections in periodontitis. *Int. J. Biol. Macromol.* **2018**, *116*, 394–408. [CrossRef]
83. Pichayakorn, W.; Boonme, P. Evaluation of cross-linked chitosan microparticles containing metronidazole for periodontitis treatment. *Mater. Sci. Eng. C* **2013**, *33*, 1197–1202. [CrossRef]
84. Dias, R.J.; Daulatrao Havaldar, V.; Ghorpade, V.S.; Mali, K.K.; Kondibahu Gaikwad, V.; Madhukar Kumbhar, D. Development and evaluation of in-situ gel containing ornidazole loaded microspheres for treatment of periodontitis. *J. Appl. Pharm. Sci.* **2016**, *6*, 200–209. [CrossRef]
85. Kim, J.; Hwang, J.; Seo, Y.; Jo, Y.; Son, J.; Choi, J. Engineered chitosan–xanthan gum biopolymers effectively adhere to cells and readily release incorporated antiseptic molecules in a sustained manner. *J. Ind. Eng. Chem.* **2017**, *46*, 68–79. [CrossRef]
86. Scholz, M.; Reske, T.; Bohmer, F.; Hornung, A.; Grabow, N.; Lang, H. In vitro chlorhexidine release from alginate based microbeads for periodontal therapy. *PLoS ONE* **2017**, *12*, e0185562. [CrossRef]
87. Li, Y.; Na, R.; Wang, X.; Liu, H.; Zhao, L.; Sun, X.; Ma, G.; Cui, F. Fabrication of antimicrobial peptide-loaded PLGA/Chitosan composite microspheres for long-acting bacterial resistance. *Molecules* **2017**, *22*, 1637. [CrossRef]
88. Lu, J.; Ren, B.; Wang, L.; Li, M.; Liu, Y. Preparation and evaluation of IL-1ra-loaded Dextran/PLGA microspheres for inhibiting periodontal inflammation in vitro. *Inflammation* **2020**, *43*, 168–178. [CrossRef]
89. Ren, B.; Lu, J.; Li, M.; Zou, X.; Liu, Y.; Wang, C.; Wang, L. Anti-inflammatory effect of IL-1ra-loaded dextran/PLGA microspheres on Porphyromonas gingivalis lipopolysaccharide-stimulated macrophages in vitro and in vivo in a rat model of periodontitis. *Biomed. Pharmacother.* **2021**, *134*, 111171. [CrossRef] [PubMed]
90. Barreras, U.S.; Méndez, F.T.; Martínez Martínez, R.E.; Valencia, C.S.; Martinez Rodríguez, P.R.; Loyola Rodríguez, J.P. Chitosan nanoparticles enhance the antibacterial activity of chlorhexidine in collagen membranes used for periapical guided tissue regeneration. *Mater. Sci. Eng. C* **2016**, *58*, 1182–1187. [CrossRef]
91. Subbiah, U.; Elayaperumal, G.; Elango, S.; Ramanathan, A.; Gita, B.; Subramani, K. Effect of Chitosan, Chitosan nanoparticle, Anacyclus pyrethrum and Cyperus rotundus in combating plasmid mediated resistance in periodontitis. *AntiInfective Agents* **2019**, *17*, 1. [CrossRef]
92. Zegan, G.; Anistoroaei, D.; Carausu, E.M.; Cernei, E.R.; Golovcencu, L. Amoxicilin and clavulanic acid intercaled nanostructures for dentistry uses. *Mater. Plast.* **2019**, *56*, 396–398. [CrossRef]
93. Hu, Y.; Chen, Y.; Lin, L.; Zhang, J.; Lan, R.; Wu, B. Studies on antimicrobial peptide-loaded nanomaterial for root caries restorations to inhibit periodontitis related pathogens in periodontitis care. *J. Microencapsul.* **2021**, *38*, 89–99. [CrossRef] [PubMed]
94. Pourhajibagher, M.; Rokn, A.R.; Rostami-Rad, M.; Barikan, H.R.; Bahador, A. Monitoring of virulence factors and metabolic activity in aggregatibacter actinomycetemcomitans cells surviving antimicrobial photodynamic therapy via nano-chitosan encapsulated indocyanine green. *Front. Phys.* **2018**, *6*, 124. [CrossRef]

95. Pourhajibaghera, M.; Rokna, A.R.; Barikania, H.R.; Bahado, A. Photo-sonodynamic antimicrobial chemotherapy via chitosan nanoparticlesindocyanine green against polymicrobial periopathogenic biofilms: Ex vivo study on dental implants. *Photodiagnosis Photodyn. Ther.* **2020**, *31*, 101834. [CrossRef]
96. Soea, H.M.S.H.; Luckanagula, J.A.; Pavasant, P.; Jansooka, P. Development of in situ gel containing asiaticoside/cyclodextrin complexes. Evaluation in culture human periodontal ligament cells (HPLDCs). *Int. J. Pharm.* **2020**, *586*, 119589. [CrossRef] [PubMed]
97. Xu, S.; Zhou, Q.; Jiang, Z.; Wang, Y.; Yang, K.; Qiu, X.; Ji, Q. The effect of doxycycline-containing chitosan/carboxymethyl chitosan nanoparticles on NLRP3 inflammasome in periodontal disease. *Carbohydr. Polym.* **2020**, *237*, 116163. [CrossRef]
98. Hu, F.; Zhou, Z.; Xu, Q.; Fan, C.; Wang, L.; Ren, H.; Ji, Q.; Chen, X.; Xu, S. A novel pH-responsive quaternary ammonium chitosan-liposome nanoparticles for periodontal treatment. *Int. J. Biol. Macromol.* **2019**, *129*, 1113–1119. [CrossRef] [PubMed]
99. Ikono, R.; Mardliyati, E.; Agustin, I.T.; Ulfi, M.M.F.; Andrianto, D.; Hasanah, U.; Bachtiar, B.M.; Mardianingsih, N.; Bachtiar, E.W.; Maulana, N.N.; et al. Chitosan—PRP nanosphere as a growth factors slow releasing device with superior antibacterial capability. *Biomed. Phys. Eng. Express* **2018**, *4*, 045026. [CrossRef]
100. Shrestha, S.; Kishen, A. Temporal-controlled bioactive molecules releasing core-shell nano-system for tissue engineering strategies in endodontics. *Nanomedicine* **2019**, *18*, 11–20. [CrossRef] [PubMed]
101. Fuloria, N.K.; Ko, M.Y.; Rui, C.S.; Hang, C.Z.; Karupiah, S.; Paliwal, N.; Kumari, U.; Gupta, K.; Sathasivam, K.; Fuloria, S. Green synthesis and evaluation of dimocarpus longan leaves extract based chitosan nanoparticles against periodontitis triggering bacteria. *Asian J. Chem.* **2020**, *32*, 1660–1666. [CrossRef]
102. Singh, M.; Jain, G. Development of sodium alginate coated nanoparticles of metronidazole for the treatment of periodontitis. *Int. J. App. Pharm.* **2019**, *11*, 105–110. [CrossRef]
103. Lee, B.S.; Lee, C.C.; Wang, Y.P.; Chen, H.J.; Lai, C.H.; Hsieh, W.L.; Chen, Y.W. Controlled-release of tetracycline and lovastatin by poly(d,l-lactide-co-glycolide acid)-chitosan nanoparticles enhances periodontal regeneration in dogs. *Int. J. Nanomed.* **2016**, *11*, 285–297.
104. Lin, J.H.; Feng, F.; Yu, M.C.; Wang, C.H.; Chang, P.C. Modulation of periodontitis progression using pH-responsive nanosphere encapsulating metronidazole or *N*-phenacylthialzolium bromide. *J. Periodontal Res.* **2018**, *53*, 22–28. [CrossRef] [PubMed]
105. Chang, P.C.; Tai, W.C.; Luo, H.T.; Lai, C.H.; Lin, H.H.; Lin, Z.J.; Chang, Y.C.; Lee, B.S. Core-shell poly-(D,L-Lactide-co-Glycolide)-chitosan nanospheres with simvastatin-doxycycline for periodontal and osseous repai. *Int. J. Biol. Macromol.* **2020**, *158*, 627–635. [CrossRef] [PubMed]
106. Vidal-Romero, G.; Zambrano-Zaragoza, M.L.; Martínez-Acevedo, L.; Leyva-Gómez, G.; Mendoza-Elvira, S.E.; Quintanar-Guerrero, D. Design and evaluation of pH-dependent nanosystems based on cellulose acetate phthalate, nanoparticles loaded with chlorhexidine for periodontal treatment. *Pharmaceutics* **2019**, *11*, 604. [CrossRef] [PubMed]
107. Shaheen, M.A.; Elmeadawy, S.E.; Bazeed, F.B.; Anees, M.M.; Saleh, N.M. Innovative coenzyme Q10-loaded nanoformulation as an adjunct approach for the management of moderate periodontitis: Preparation, evaluation, and clinical study. *Drug Deliv. Transl. Res.* **2020**, *10*, 2. [CrossRef] [PubMed]

molecules

MDPI

Article

Hydrogel Films Based on Chitosan and Oxidized Carboxymethylcellulose Optimized for the Controlled Release of Curcumin with Applications in Treating Dermatological Conditions

Mohamed Dellali [1,2,3], Camelia Elena Iurciuc (Tincu) [3,4,*], Corina Lenuța Savin [3], Nawel Spahis [2], M'hamed Djennad [1] and Marcel Popa [3,5,*]

1 Laboratory of Structure, Elaboration, and Application of Molecular Materials, Abdelhamid Ibn Badis University of Mostaganem, Mostaganem 27000, Algeria; m.dellali@univ-chlef.dz (M.D.); mhamed.djennad@univ-mosta.dz (M.D.)
2 Faculty of Technology, Hassiba Benbouali University of Chlef, BP 151, Chlef 02000, Algeria; n.spahis@univ-chlef.dz
3 Department of Natural and Synthetic Polymers, Gheorghe Asachi Technical University of Iasi, Mangeron Blvd. no. 73, 700050 Iasi, Romania; savincorina@yahoo.com
4 Department of Pharmaceutical Technology, Faculty of Pharmacy, "Grigore T. Popa" University of Medicine and Pharmacy, University Street, no. 16, 700115 Iaşi, Romania
5 Academy of Romanian Scientists, Splaiul Independentei Street, No 54, 050094 Bucharest, Romania
* Correspondence: camelia_tincu83@yahoo.com (C.E.I.); marpopa2001@yahoo.fr (M.P.)

Citation: Dellali, M.; Iurciuc (Tincu), C.E.; Savin, C.L.; Spahis, N.; Djennad, M.; Popa, M. Hydrogel Films Based on Chitosan and Oxidized Carboxymethylcellulose Optimized for the Controlled Release of Curcumin with Applications in Treating Dermatological Conditions. *Molecules* **2021**, *26*, 2185. https://doi.org/10.3390/molecules26082185

Academic Editors: Ivan Gitsov and Angelina Angelova

Received: 6 February 2021
Accepted: 6 April 2021
Published: 10 April 2021

Publisher's Note: MDPI stays neutral with regard to jurisdictional claims in published maps and institutional affiliations.

Abstract: Cross-linked chitosan (CS) films with aldehyde groups obtained by oxidation of carboxymethyl cellulose (CMC) with $NaIO_4$ were prepared using different molar ratios between the CHO groups from oxidized carboxymethyl cellulose (CMCOx) and NH_2 groups from CS (from 0.25:1 to 2:1). Fourier-transform infrared (FTIR) and nuclear magnetic resonance (NMR) spectroscopy demonstrated the aldehyde groups' presence in the CMCOx. The maximum oxidation degree was 22.9%. In the hydrogel, the amino groups' conversion index value increased when the -CHO/-NH_2 molar ratio, cross-linking temperature, and time increased, while the swelling degree values decreased. The hydrogel films were characterized by scanning electron microscopy (SEM) and FTIR analysis. The curcumin encapsulation efficiency decreases from 56.74% to 16.88% when the cross-linking degree increases. The immobilized curcumin release efficiency (R_{Ef}%) and skin membrane permeability were evaluated *in vitro* in two different pH solutions using a Franz diffusion cell, and it was found to decrease when the molar ratio -CH=O/NH_2 increases. The curcumin R_{Ef}% in the receptor compartment was higher at pH = 7.4 (18%- for the sample with a molar ratio of 0.25:1) than at pH = 5.5 (16.5%). The curcumin absorption in the skin membrane at pH = 5.5 (47%) was more intense than at pH = 7.4 (8.6%). The curcumin-loaded films' antioxidant activity was improved due to the CS presence.

Keywords: chitosan; oxidized carboxymethyl cellulose; hydrogel film; curcumin immobilization; delivery system

1. Introduction

Dermatology is a constantly evolving medical field that deals with skin, nail, hair diseases, and systemic diseases treatment, especially for those whose symptoms can be observed primarily on the skin. The treatments performed aim to improve the skin's external appearance by treating various diseases and pathological conditions. Although skin disease incidence is lower than other diseases, they significantly impact life quality, and skin cancer or severe infections are life-threatening [1]. The most common skin diseases reported during a patient's lifetime were warts (41.3%), acne (19.2%), and contact dermatitis (15.0%), followed by other forms of eczema or atopic dermatitis and urticaria [2].

Hydrogels are three-dimensional polymeric structures that can absorb large amounts of water or biological fluids. One of the main advantages of the hydrogels used in the local treatment of skin diseases is their easy applicability and the possibility of reducing the immobilized drugs' side effects. During topical (dermal) administration, the drug initially penetrates the stratum corneum, then diffuses into the deeper epidermis until it reaches the dermis, being absorbed into the systemic circulation through dermal microcirculation [3].

The drug's local administration depends on many factors, including the skin's barrier properties, the incorporated therapeutic compound's physicochemical properties, and the delivery system used [4]. The immobilized drug interaction with the polymer matrix is essential for drug effectiveness in topical applications. The application of these delivery systems with immobilized drugs results in the therapeutic concentration achievement in the individual layers of the skin located in the application area. At the same time, the serum concentration is low, which causes a decrease in the drug's side effects [5,6].

Natural polymer-based (proteins and polysaccharides) hydrogels are most often used in biomedical applications due to their biocompatibility, non-toxicity, biodegradability, the similarities of their physical properties to natural tissues, low immunogenicity, and their functional groups could be modified to obtain release systems with improved properties [1].

Polysaccharides offer a wide range of versatile functionalities and structural diversity due to their variable molecular weight and numerous and varied functional groups (amino, carboxyl, carbonyl, and hydroxyl) on the main chain [7]. These hydrophilic groups determine an increased solubility in water and strengthen their bioadhesive and biorecognition characteristics (electrostatic interactions between biological tissues and polysaccharides) [8]. For example, chitosan, the only polysaccharide with a positive charge (polycation), can be attached to negatively charged mucosal layers through electrostatic interactions [9,10].

Chitosan (CS) is a natural, cationic polysaccharide obtained by chitin's alkaline deacetylation process [11]. It is a semicrystalline, semisynthetic polymer with a linear structure, composed of (1→4)-2-acetamido-2-deoxy-β-D-glucan (N-acetyl-D-glucosamine) and (1→4)-2-amino-2-deoxy-β-D-glucan (D-glucosamine) units [12]. It has excellent properties such as biodegradability, biocompatibility, non-toxicity, and mucoadhesivity [13,14]. The CS has high reactivity and can be easily functionalized due to the hydroxyl and amine groups found in large numbers along its chain and allow the reaction with different cross-linking agents. Also, chitosan itself possesses biological activity. The hydrogels obtained are used as delivery systems for the controlled release of drugs and genes [15,16], in biosensors obtaining [17], and in medical imaging [18,19].

The hydrogels' properties and the release of the drug from these delivery systems depend on the type of cross-linking agent used and its concentration. The morphology of such hydrogels is porous and depends on the cross-linking agent used [20].

Aldehydes, epoxy compounds, esters are chemical cross-linking agents commonly used to prepare CS-based hydrogels, and sodium tripolyphosphate or sodium sulfate have been used as ionic cross-linking agents [21–23]. In an aqueous solution, the aldehydes form an imine bond with the amine group in the CS to give a biocompatible hydrogel. Glutaric aldehyde acts as a cross-linking agent, and it has been shown that the morphology of the obtained hydrogel is porous, and the pores are evenly distributed in the formed matrix [24]. Glutaric aldehyde often causes neurological and other cytotoxicities [25]. Macromolecules containing aldehydes can be used to prepare CS hydrogels [26]. Thus, poly(ethylene glycol) dialdehyde was used in the CS hydrogel and has been shown to improve the polymer matrix mechanical properties as the cross-linking agent's concentration increases [27,28]. Natural carbonyl compounds, including partially oxidized polysaccharides, are a preferred alternative as cross-linking agents, thus increasing the application area for the hydrogels obtained [29]. Research has shown that amino groups in CS react with aldehyde groups in oxidized cellulose to form Schiff bases [30].

This paper reports the results obtained after preparing a biocompatible hydrogel based on CS cross-linked with oxidized carboxymethyl cellulose (CMCOx) in which curcumin was incorporated. Previous in vivo research and clinical trials have shown that

CS-based hydrogel films with topical applications can cause rapid wound healing, stop bleeding, reduce healing time, faster vascularization, and colonization with fibroblasts were observed [31]. Curcumin was chosen as an active principle model used to treat skin diseases due to its remarkable therapeutic properties and strong antioxidant character [32,33]. The main disadvantage of curcumin is its insolubility in water and low bioavailability in cells. Curcumin can degrade rapidly under the influence of several factors such as natural light, alkaline pH. It is a good metal chelator, but previous studies have shown that curcumin-based metal complexes cause a decrease in their pharmacological action compared with free curcumin [34]. Curcumin can be protected from degrading factors by its encapsulation in various pharmaceutical formulations such as release systems based on natural and synthetic polymers, liposomes, micro/nanoemulsions, solid lipid nanoparticles [35]. Lipids associated with curcumin indeed increase bioavailability, but particles of this kind do not ensure a continuous, sustained release, even if the amount of immobilized curcumin is higher compared with biopolymers nanoparticles [36,37]. We choose CS to obtain the hydrogel because it has antimicrobial activity and antioxidant activity, which is an advantage over other polymers, including polysaccharides.

Oxidized derivatives (CMCOx) were obtained by converting 1,2-dihydroxyl groups into dialdehyde groups at C2 and C3 without significant side reactions under the action of $NaIO_4$ [38]. Fourier-transform infrared (FTIR) and nuclear magnetic resonance (NMR) spectroscopy identified the modified polysaccharide structure. The reverse titration reaction with $Na_2S_2O_3$ was used to determine the aldehyde groups' content and the polysaccharide's oxidation degree. The CMCOx molecular weight was determined by the viscosimetric method. Several hydrogels were obtained by varying the molar ratio between the $-NH_2$ groups from CS and -CHO groups from CMCOx.

In order to obtain the hydrogels, the reaction temperature, the molar ratio between CS and CMCOx (practically, between the number of moles of -CH=O and $-NH_2$ groups), and the cross-linking time were the parameters whose influence was studied. The different classical characterization methods make it possible to define between the hydrogel films obtained, those with the best water retention capacity, and the best porosity for the active principle's encapsulation and controlled release. The structural characterization and study of hydrogel films' morphology are detailed in this article.

This article's main objective and point of originality were to optimize the reaction conditions to obtain CS-based hydrogel films cross-linked with CMCOx with the best physicochemical characteristics and use them as delivery systems for controlled and sustained release of curcumin. The hydrogels obtained are sensitive to pH and have the ability to encapsulate and release the active principle. The release kinetics and permeability of curcumin from the obtained hydrogels were studied on chicken skin used as a membrane in two different pH environments, at pH = 5.5 and pH = 7.4 at 37 °C up to equilibrium using a Franz diffusion cell. The antioxidant activity, expressed by IC50 (the inhibition concentration at which the DPPH radicals were scavenged by 50%) calculated from the interpolation of the linear regression analysis for curcumin and hydrogel films with or without encapsulated curcumin was determined. The influencing factors leading to the intensification of antioxidant activity in hydrogel films with immobilized curcumin were studied.

2. Results and Discussions

2.1. Preparation and Characterization of CMCOx

Sodium periodate has been used to oxidize the C2-C3 vicinal hydroxyl groups in the carboxymethyl cellulose (CMC) glucoside cycle, but glycosidic bonds' cleavage is also inevitable leading to polymer degradation [39,40]. The oxidation reaction did not form toxic compounds [41], and it took place in the dark to avoid the advanced oxidation, in double-distilled water pH = 6.5 and a temperature of 30 °C. Figure S1 (Supplementary information) schematically shows CMC's oxidation reaction.

2.1.1. FTIR Spectroscopy of CMC and CMCOx

FTIR spectra were recorded to demonstrate CMC oxidation by the presence of aldehyde groups in the modified polysaccharide. Figure 1 shows the FTIR spectra of CMC and CMCOx for 6 h.

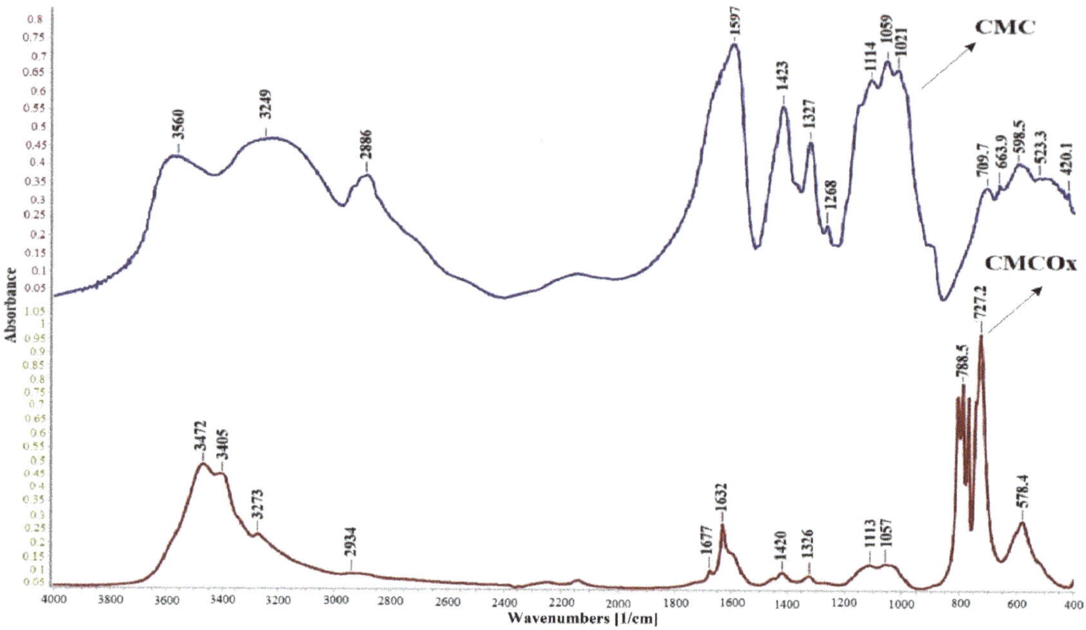

Figure 1. Fourier-transform infrared (FTIR) spectra for carboxymethyl cellulose (CMC) and oxidized carboxymethyl cellulose (CMCOx)- the time used for the oxidation was 6 h.

The absorption peak from 1599 cm^{-1} in the CMC spectrum appears shifted in the CMCOx spectra to 1631 cm^{-1}. Absorption at 1599 cm^{-1} can be assigned to carboxylate anion (νc = o) asymmetric stretch band from the glucuronic acid in the CMC spectrum [42], and the absorption bands' appearance at a lower frequency from 1631 cm^{-1}, which is characteristic for conjugated C=O, from carboxylate anion and could be determined by CMCOx water absorption and degradation [43]. The appearance of a new absorption peak of a lower intensity from 1677 cm^{-1} attributed to unsaturated ketones groups is a clear indication that partial oxidation of CMC has occurred [44,45]. The absorption peak from 1725–1740 cm^{-1}, specific to the aldehyde group [44], did not occur because there is the possibility that aldehydes groups were found in the hemiacetals forms in CMCOx; the appearance of the peak could be masking it. For example, in the case of dextran and inulin oxidation, the authors could not identify in the FTIR spectrum the absorption band specific to the aldehyde group [46,47]. Absorption bands from about 800 cm^{-1} can be attributed to semiacetals forming between aldehyde groups and vicinal hydroxyl groups [48,49]. The absorption band from 3593 cm^{-1} correspondingly to the OH-group's elongation vibration in the CMC spectrum becomes narrower in the CMCOx spectra. The absorption band moves to approximately 3472 cm^{-1}, which probably indicates that the hydroxyl groups number has decreased due to the aldehyde group formation after the oxidation reaction. The absorption bands from 1424 cm^{-1} to 1328 cm^{-1}, also attributed to the carboxylate anion in the CMC spectrum, are found in the CMCOx spectrum but appear slightly shifted, indicating structural changes following the oxidation reaction.

2.1.2. NMR Spectra of CMC and CMCOx

^{1}H NMR and ^{13}C NMR spectra were recorded for both CMC and CMCOx. The spectra are shown in Figure 2 (^{1}H NMR) and Figure 3 (^{13}C NMR).

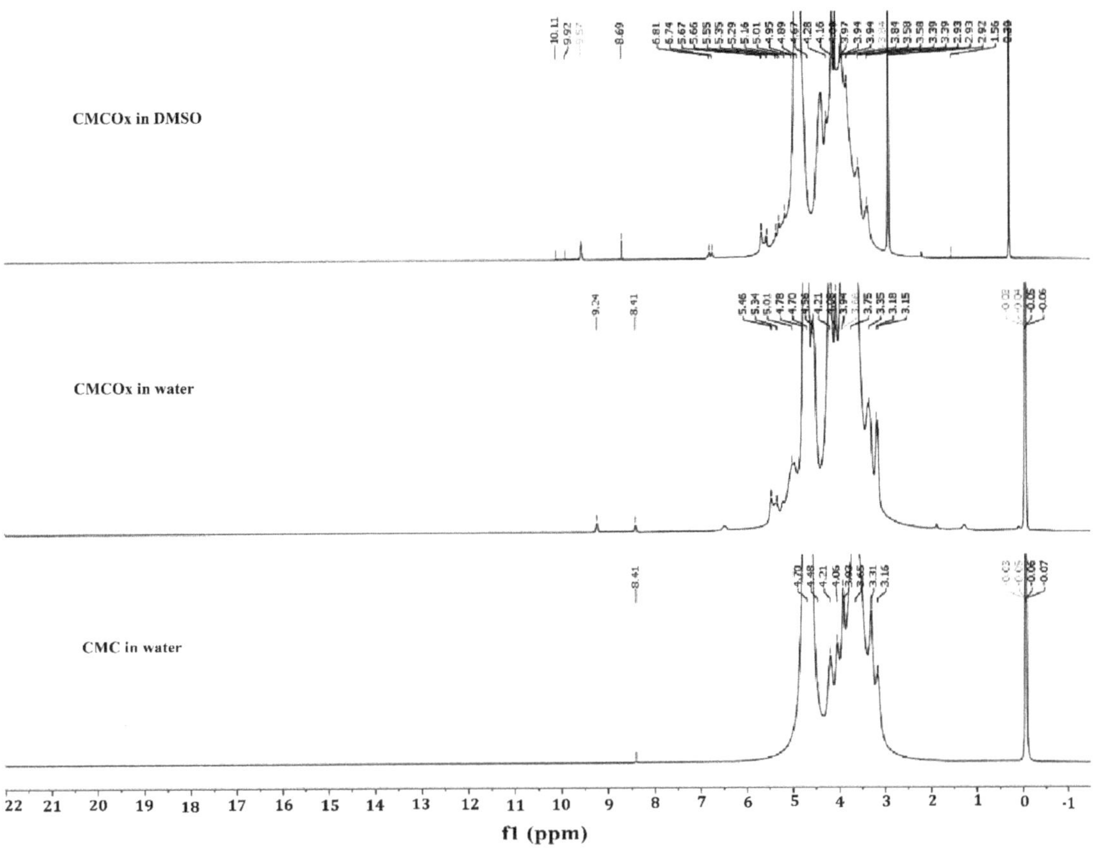

Figure 2. Proton nuclear magnetic resonance (^{1}H NMR) spectra for CMC dissolved in water, CMCOx dissolved in water, and CMCOx dissolved in DMSO.

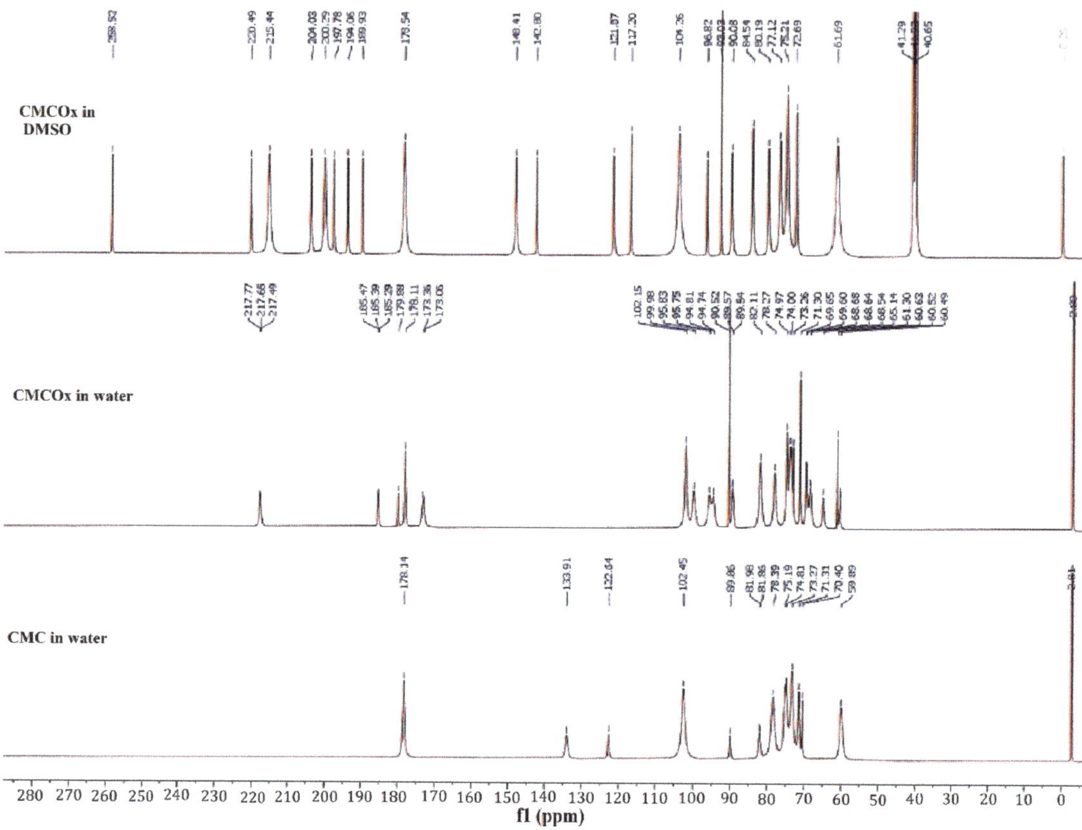

Figure 3. Carbon-13 nuclear magnetic resonance (^{13}C NMR) spectra for CMC dissolved in water, CMCOx dissolved in water, and CMCOx dissolved in DMSO.

^{1}H NMR and ^{13}C NMR spectroscopy were used to assess whether the CMC oxidation had occurred and identified if the signals corresponding to the aldehyde groups were found in the CMCOx spectra. Figure 2 shows that the ^{1}H NMR spectra of the CMCOx sample (the oxidation reaction time for this sample was 6 h) dissolved in deuterated water or DMSO present characteristic peaks similar to those in the CMC spectrum. The formation of aldehyde groups in the structural unit of CMC is highlighted in the ^{1}H NMR spectrum of CMCOx dissolved in water by the presence of the singlet signal at 9.24 ppm and in the CMCOx spectrum dissolved in DMSO at 10.11 ppm, 9.92 ppm, or 9.58 ppm, these peaks are characteristic for the proton from the aldehyde groups. Several signals in the region of 3.14–4.7 ppm could be attributed to the CH or OH groups. ^{13}C NMR spectroscopy (Figure 3) shows that in the CMCOx spectrum dissolved in deuterated water, there are no signals in the aldehyde groups region between 190 ppm and 200 ppm. Still, a characteristic signal of carbonyl groups in ketones appears at 217.77 ppm. There is a possibility that the hydrated form of the aldehyde groups may appear in an aqueous solution [50,51], but the characteristic signals were not found in the spectrum of CMCOx dissolved in deuterated water. In the spectrum of CMCOx dissolved in DMSO, there are signals of the aldehyde groups at 204.03 ppm, 200.29 ppm, 197.78 ppm, and 194.06 ppm.

In the CMC spectrum dissolved in deuterated water, the signal at 102.45 ppm corresponds to the C1 atom in the anhydroglucose unit. The signal from 71.315 ppm is attributed to the C7 methylene group. The carboxylic group signal appears at 178.146 ppm and can be attributed to the carboxylated ion. The signal from 59.88 ppm can be assigned to the C6 atom from the native CMC anhydroglucose unit, while the signal corresponding to the C6 atom in the CMC was shifted to 70.4 ppm [52]. In the CMCOx dissolved in the DMSO ^{13}C NMR spectrum, we observe that the signals belonging to the C2 and C3 atoms from 74.96 ppm and 73.28 ppm, respectively, from the CMC spectrum, no longer occur, being an evident proof of CMC oxidation. In the ^{13}C NMR spectrum of CMC dissolved in water, a slight reduction of the signal intensity from these two carbon atoms was observed. Literature mentions that when the oxidation degree is higher, these peaks' intensity is lower. CMC oxidation determines numerous signals in the region of 80–105 ppm. The signals are attributed to carbon atoms from C-O-C groups of hemiacetals [50]. The results obtained are consistent with those reported by other studies [50–52].

2.1.3. Oxidation Reaction Kinetic and Quantitative Determination of the Aldehyde Groups Obtained in CMCOx

The variation in time of the oxidation degree of CMC was monitored by determining the consumption of $NaIO_4$ during the reaction. Figure 4 shows the results obtained.

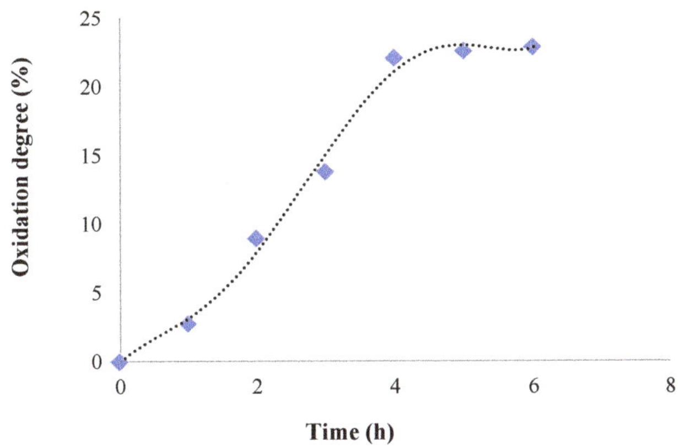

Figure 4. Variation in time of the oxidation degree of CMC.

For the oxidation degree determination, a volume of 100 mL of CMC solution of 3% concentration (w/v) was prepared, and the working protocol was described in the Section 3.4. It was found that the oxidation degree gradually increased up to 4 h when it reaches the value of 22.11%, after which it was practically stabilized, reaching the value of 22.9% after 6 h. Consequently, partial oxidation was achieved, which was our demand, because advanced oxidation can result in oxidation of carbonyl groups to carboxylic groups and, possibly, the polysaccharide chain degradation with a drastic reduction of the molecular weight. The number of moles of aldehyde groups varies, and it was between 0.71×10^{-3} and 5.86×10^{-3} moles CHO/g of CMCOx. Similar results were found and reported in the literature, where it was shown that the aldehyde groups' number increases over time in the oxidation reaction [31,53].

The first effect of CMC oxidation is the increase in its water solubility. On the one hand, this effect is determined by the increase of the carbonyl group number, which are less involved in the hydrogen bonds formed between the polymer chains than the -OH groups and, on the other hand, by the decrease in the CMCOx molecular weight. The oxidation process is completed in about 6 h in the dark. In some papers, it is estimated that the oxidation process retardation can be caused by the hydrogen bond that is formed between the carboxyl group of one polymer chain and the hydroxyl group from the C3 atom of neighboring chains. The oxidation degree can also be adjusted by changing the ratio between the reactants [50].

2.1.4. Molecular Weight Determination

The oxidation reaction can also determine the polysaccharide degradation, reducing its molecular weight. It was considered necessary to monitor the variation of this characteristic for CMCOx obtained after different oxidation times using the viscometric method. Figure 5 shows CMC's average viscosimetric molecular weight variation with the oxidation reaction time.

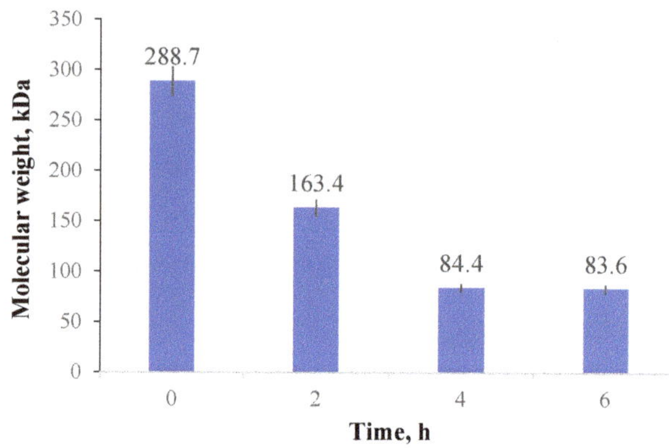

Figure 5. Variation of the average CMC molecular weight with the oxidation process duration determined by the viscosimetric method.

The oxidation process consequences on the CMC average molecular weight were evaluated. Three samples were obtained by varying the oxidation reaction time (between 2 and 6 h). The reactants' amounts were maintained constant (1 g of $NaIO_4$ and 1 g of CMC) in each synthesis. There was a significant reduction in CMC's molecular weight with increasing oxidation time due to polysaccharide degradation. The results obtained are consistent with the results obtained by Eremeeva and Bykova [54].

The average viscosimetric molecular weight decrease from 288.7 kDa for the initial CMC to about 83.6 kDa for CMCOx (after 6 h of oxidation), when the oxidation degree was maximum, respectively 22.9%). The reduction of the polysaccharide's molecular weight took place by hydrolysis of the glycosidic bonds during the oxidation process, decreasing when the oxidation degree increases. It can be noted that the average viscosimetric molecular weight variation of CMCOx was consistent with the oxidation degree. After four hours of CMC oxidation, the average molecular weight value was almost constant.

2.2. Obtaining and Characterization of the CS Films with CMCOx

Hydrogel films based on CS with CMCOx were obtained at pH = 2.4 and a temperature of 55 °C. At a higher pH value, CS precipitates due to the reduction of its solubility. The reaction between the protonated amino groups from CS and the aldehyde groups from CMCOx determines the obtaining of imine groups or Schiff bases. Some of the unreacted amine groups form physical bonds (mainly electrostatic) with the CMCOx carboxylic groups, as mentioned in Section 2.2.3. The structure proposed by us for the obtained film is schematized in Figure S2 presented in Supplementary information. The cross-linking reaction is influenced by the cross-linking time or temperature, and the amount of CMCOx used (the molar ratio between the amino groups in CS and the aldehyde groups in CMC). The influence of these factors on cross-linking was studied in the paper.

Curcumin was incorporated by diffusion and alcohol evaporation after the films were dried (described in Section 3.4.9). The results of the encapsulation efficiency are given in Section 2.2.6.

2.2.1. FTIR Spectroscopy of the Hydrogels Obtained

Figure 6 shows the FTIR spectra of CS, CMCOx for 6 h, and the P2 sample without curcumin in which the molar ratio -CHO/-NH_2 was 0.375:1. The spectra were restricted only to the wavenumber range between 1800–600 cm^{-1}, where the differences were expected.

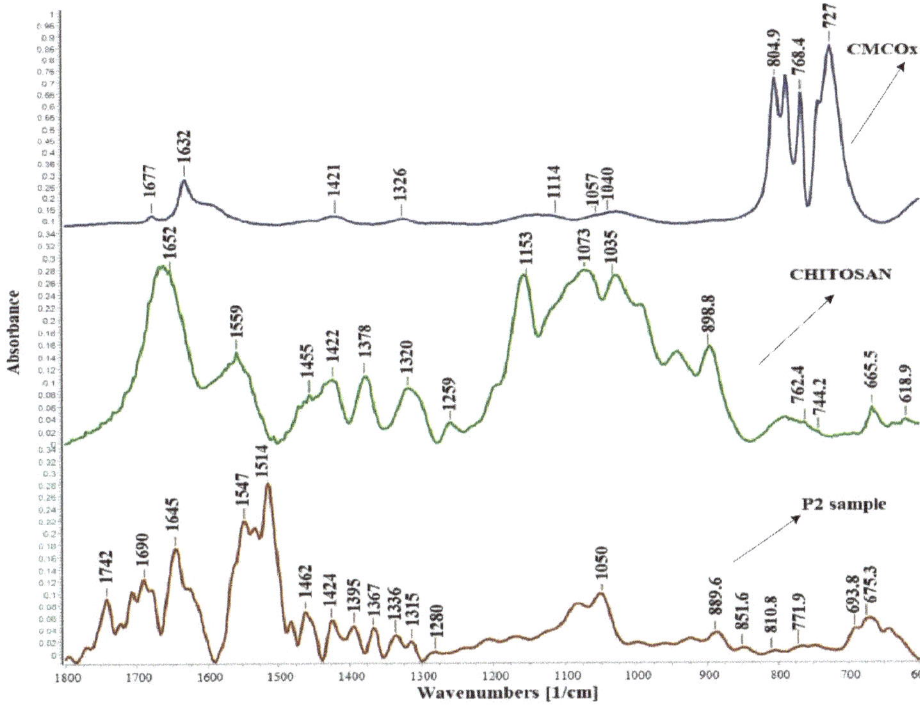

Figure 6. FTIR spectra of CS, CMCOx, and P2 films (with a molar ratio NH_2/CHO = 1:0.375 mol/mol).

In the P2 sample's spectrum, the absorption peaks characteristic of the constituents polymers are found. The imine bond (-N=CH-), commonly called the Schiff base, is the most common covalent bond used to create CS-based hydrogels by reaction with the carbonyl group. Figure 6 shows in the hydrogel spectrum the appearance of new absorption peaks at 1645 cm^{-1}, 1690 cm^{-1}, or 1742 cm^{-1}. The peak from 1320 cm^{-1} from the chitosan spectrum was shifted at 1336 cm^{-1} and split into many other lower intensity peaks than the CS spectrum. The literature states that the imine group's (CH=N) characteristic absorption band (1660–1630 cm^{-1}) cannot be observed clearly, probably because it overlaps with the chitosan amide I absorption band from the chitosan spectrum [48]. The aldehyde groups' peak characteristic from 1677 cm^{-1} in the CMCOx spectrum appears shifted in the spectrum of hydrogel films to about 1690 cm^{-1}, indicating possible interactions between the amine and aldehyde groups with the possibility of Schiff bases formation. The characteristic chitosan band from 1559 cm^{-1} shifts in the spectrum of the P2 sample and several peaks of lower intensities are observed, which may indicate some intermolecular interactions of amine groups [55]. The absorption peak from 1514 cm^{-1} can be attributed to the stretching vibration C=C or C=N [44,56,57].

The absorption band from 1742 cm^{-1} indicates the electrolytic dissociation of the -CH $2COO^{-}Na^{+}$ group and occurs, especially in polyelectrolyte complexes. The literature states that this absorption band is stronger if amino groups' consumption intensifies [58].

Visually, the change in the hydrogel films' yellow color is due to the imine bond formed by the two types of functional groups' condensation [59].

2.2.2. Scanning Electron Microscopy

The film's surface morphology was highlighted by scanning electron microscopy. Figure 7 shows the SEM photos at the surface of the two obtained films. To highlight the cross-linking degree's influence on the films' morphology, we selected the P1 and P3 samples. The molar ratios $CH{=}O/\text{-}NH_2$ used for the film obtaining are quite different 0.25:1 for the P1 sample and 0.5:1 for the P3 sample.

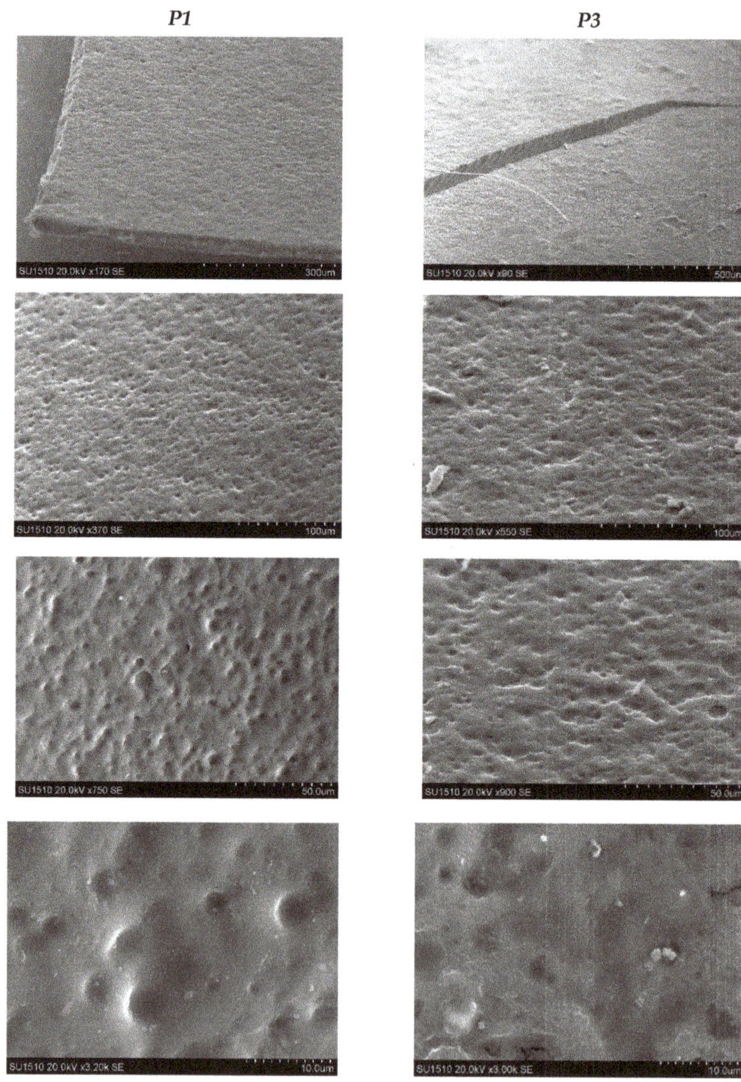

Figure 7. Scanning electron microscopy photographs on the P1 and P3 films surface.

There was a slight difference in the morphology between these two samples. It was found that both are relatively compact and homogeneous, with the surface revealing a slight porosity and roughness. The porosity decreases when the CMCOx amount increases—an effect that can be explained by the films cross-linking density improvement determined by a higher number of aldehyde groups employed in reaction with the amine ones from CS.

2.2.3. Amino Groups Conversion Index Determination

The free amino groups determination from CS and cross-linked CS-based hydrogel films was essential because it provides information on the amino groups' conversion degree from CS to Schiff bases in the presence of aldehyde groups from CMCOx, necessary to optimize the parameters that could influence the cross-linking reaction. The ninhydrin test was used for this purpose. The conversion index (CI%) was analyzed for several films differing by the initial molar ratio between the carbonyl and amine groups, compared to the amino groups from free CS.

Based on the CS calibration curve, the number of moles of free amino groups in the obtained hydrogel films (N_a) was determined. The number of moles of amino groups that react was established by the difference between the number of initial free amino groups in CS (determined based on the initial CS amount added in the cross-linking reaction) (Nb) and those determined in the films. The CI was calculated using equation (3). Several factors that can influence the CI values were studied: the molar ratio (-CHO/-NH_2), cross-linking reaction time, and temperature. Figure 8 presents the results obtained after the amino group conversion index determination by studying the factors influencing the cross-linking reaction mentioned previously.

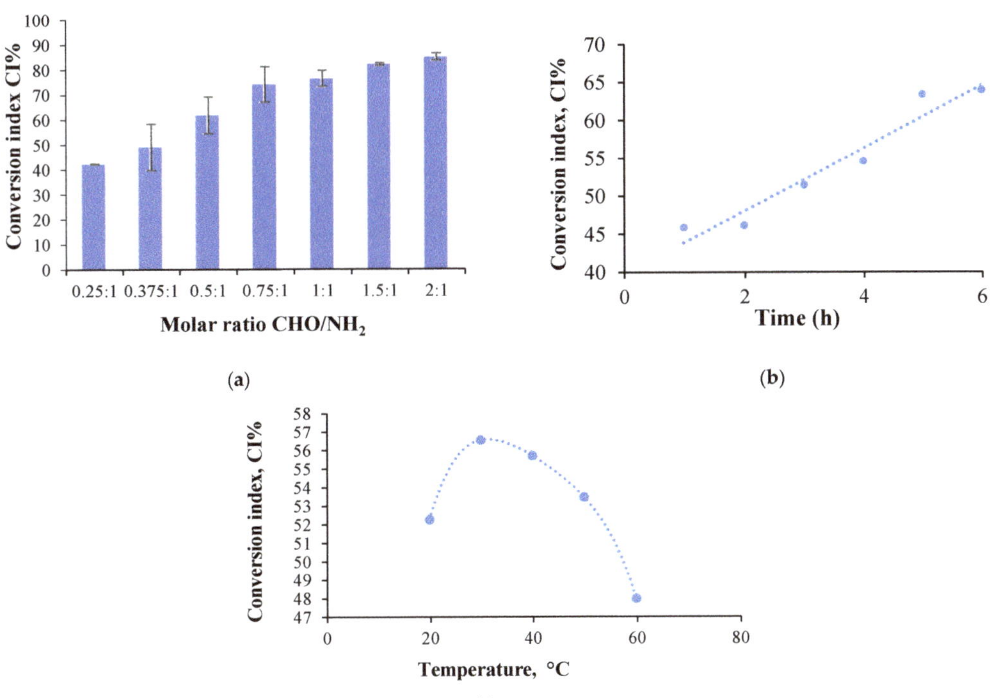

Figure 8. The variation of the amino groups' conversion index values by studying the cross-linking reaction influencing factors such as (**a**) the molar ratios influence ($t_{reaction}$ = 2.5 h; $T_{reaction}$ = 55 °C, pH = 2.4—it was determined for all the samples (**b**) variation of CI values as a function of cross-linking time for P2 sample ($T_{reaction}$ = 55 °C, molar ratio CHO: NH_2 = 0.375:1, pH = 2.4); (**c**) variation of CI as a function of cross-linking temperature for P2 sample ($T_{reaction}$ = 55 °C, molar ratio CHO: NH_2 = 0.375:1, pH = 2.4).

- **The molar ratio CHO/NH_2 influence the CI values**

The films were obtained using different molar ratios between the amino groups from CS and the aldehydes from CMCOx. The number of moles of amino groups from CS was maintained constant for all samples, respectively 1.75×10^{-3} moles, varying the number of moles of carbonyl groups. The amino groups' CI was determined for all the samples, without active principle immobilized, and Figure 8a presents the obtained results. The CI values were found to increase when the CMCOx amount in the samples was higher, reaching 85% for the 2:1 molar ratio. It was observed that the CI value for the P1 sample, for example, in which the molar ratio was 0.25:1, was 42.27% and not 25% as expected theoretically. The unexpectedly high value of CI could be explained only by the existence of strong intermolecular interactions such as hydrogen bonds or by the complexation of carboxylic groups with amine groups, which could capture the $-NH_2$ groups not involved in Schiff-base bonds formation leading to an erroneous conclusion about the CI value. The effect is maintained, less and less intense, up to the molar ratio CHO/NH_2 of 0.5:1 (mol/mol).

The intermolecular interactions mentioned before decrease with the increase of the CMCOx amount by amino group consumption in the condensation reaction. Other authors also indicate the possibility of the formation of polyelectrolytic complexes at pH = 2.4 or 2.5. A research study states that CMC can form polyelectrolytic complexes at pH = 2.4 with gelatin at the gelatin: CMC ratio of 1.7: 1. The number of amino groups that could interact electrostatically with the carboxylic groups of carboxymethyl cellulose is lower at pH = 2 compared to pH = 3 [60]. Also, at pH = 2.5, polyelectrolytic complexes based on CMC and CS were prepared, and it was showed that the efficiency of obtaining them depends on the ratio between CMC and CS [61].

The possibility of obtaining polyelectrolytic complexes at pH = 2.4 was tested to elucidate whether the electrostatic interactions between the amino groups from chitosan and the carboxylic groups from CMCOx can influence the CI value mentioned above. Three films were prepared based on CS and CMC to test this hypothesis, keeping the same polymers amounts as those corresponding to CS mixtures with CMCOx obtained at different molar ratios CHO/NH_2 (0.25:1-P1; 0.375:1-P2, and 0.5:1-P3). After drying the films, the amino groups were determined by dosing them with ninhydrin. The obtained results are presented in Table S1 (in Supplementary information), keeping the samples' initial codes, with the mention that for the CI (%) were introduced the following notations: $CI_{\text{chemical cross-linking and physical interactions}}$ (%), which comprise the total amino groups percentage involved in chemical bonds and physical interactions (column 3); $CI_{\text{physical interactions}}$ (%) which express the amino groups' percentage involved only in physical interactions (column 4); $CI_{\text{chemical cross-linking and physical interactions}} - CI_{\text{physical interactions}} = CI_{\text{chemical cross-linking (Shiff base)}}$ (%) which expresses the of amino groups percentage involved only in the chemical reaction of Shiff base formation (column 5).

A first finding is that, indeed, it was found that in non-chemically cross-linked samples, some of the amino groups from CS did not participate in the reaction with ninhydrin, obviously because amine groups were involved in stable interactions with carboxylic groups of CMC. It is assumed that the same effect occurs when cross-linked samples are obtained, i.e., some of the amino groups of CS react with the CMCOx carbonyl groups, and others interact with the carboxylic group of CMXOx. These interactions could explain the high CI values corresponding to these samples, which do not correctly express the CI of amino groups into Schiff bases.

Thus, for the P1 sample, $CI_{\text{chemical cross-linking and physical interactions}} - CI_{\text{physical interactions}} = CI_{\text{chemical cross-linking (Shiff base)}}$ (%) difference was 25.8%. The maximum theoretical value of the corresponding CI being 25%, according to the molar ratio of the functional groups involved in t e hP1 sample cross-linking reaction (0.25 moles of $-CH{=}O$ for 1 mol of $-NH_2$). Similarly, for the P2 sample, the difference was 31.08%, given that the molar ratio between these two types of functional groups was 0.375:1. For sample P3, corresponding to a molar ratio of 0.5 moles of -CHO groups for 1 mol of -NH_2 groups, was 37.94% (compared to the maximum theoretical value of 50%).

The high amino groups consumption once with the increase of the molar ratio between the functional groups involved in the chemical cross-linking reaction and as the cross-linking degree increase could lead us to the conclusion that the interaction between the unreacted -NH_2 groups from CS with -COOH from CMCOx could be diminished when the molar ratio increase. Therefore, at a molar ratio value higher than 0.75:1 (CHO: NH_2), the CI value accurately reflects the reaction between amino groups from CS with the aldehyde groups from CMCOx; as a result, the Schiff bases formation. Following the data presented in Figure 8a, it was found that for the P4 sample, the CI values approached the theoretical one and continue to increase slightly with the molar ratio value for samples P5–P7, without reaching the maximum of 100%, even if the carbonyl groups content from samples in some cases was in excess. In principle, this effect is not possible; the drastic reduction of the chain fragments mobility between the obtained cross-linked network nodes and the possibility of the functional groups' interaction in the cross-linking reaction was reduced.

It should be noted that intermolecular bonds (physical or complexed) are reversible and can be readily cleaved compared to irreversible chemical bonds [62]. Thus, the Schiff bases can stabilize the hydrogel based on CS and CMCOx, but in the case of these bonds, the possibility can exist of hydrolytic degradation in aqueous solutions. Lü et al. have shown that hydrogels' degradation after 24 h decreases in intensity as the cross-linking degree increases because water molecules cannot enter into the cross-linked polymer network, and thus the degradation is diminished [63].

- **The cross-linking time and temperature influence on the amine group CI**

To determine the influence of cross-linking time and temperature on CI (more precisely, on $CI_{\text{chemical cross-linking and physical interactions}}$), sample P2 was selected because the CI has a medium value, and the influence of these parameters could be studied. Six hydrogel films were prepared to evaluate the cross-linking time influence in which all reaction conditions were kept constant, and only the cross-linking time was varied between 1 to 6 h. The amino groups' CI variation as a function of the cross-linking time for the P2 sample is shown in Figure 8b.

Six other films were obtained to evaluate the cross-linking temperature influence on the CI at different cross-linking temperatures between 20 °C and 60 °C, maintaining the other conditions constant as it was described in Materials and Methods. Figure 8c shows the CI variation as a function of the cross-linking temperature for the films obtained.

Figure 8b shows that the CI increases with increasing cross-linking time, reaching a maximum value of 64% after 6 h. Given the -CH O/-NH_2 molar ratio's initial value, only 37.50% of the total amino groups could react with the aldehyde ones in CMCOx. It results in the involvement of 26.5% of the total amino groups in CS in intermolecular interactions, such as hydrogen or electrostatic bonds, with their number being slightly higher when the cross-linking time increases.

Of course, the increase of CI with the cross-linking time can be caused both to the increase in the number of imine-type chemical bonds between chains and the number of intermolecular interactions of physical nature. Also, the pKa value of the carboxylic groups in CMCO is 3.5: at a lower pH value, hydrogen bonds and electrolyte interactions can be formed that blocks the -NH_2 groups, preventing them from reacting with ninhydrin. In our case, the pH of the cross-linking medium was 2.4. Their number may increase as the polymers' contact time increases at a temperature of 55 °C. The results are consistent with those obtained when analyzing the molar ratio influence on the CI using the ninhydrin test.

Figure 8c shows that the CI value increases until a temperature of 30 °C when it reaches a maximum value, after which its value begins to decrease. The range of CI variation on this temperature range was not very high, with a maximum of 8%. Still, the values remain superior to the theoretical one, with a maximum value of 37.5%. The CI would be expected to increase continuously with the temperature (which positively influences the chemical reactions). On the other hand, the increase of this parameter has the effect of cleaving the physical or hydrogen bonds between the functional groups within polymers. As a result, ninhydrin will react with an increasing number of amino groups released from their physical interactions, and the CI value will be reduced. However, it should be noted that even if the temperature reaches 60 °C, not all physical connections were cleaved, as the apparent value of CI remained above the theoretical value of 37.5%. The previously presented influencing factors study allowed us to obtain important information, qualitatively, on the most favorable conditions for obtaining hydrogel films with the dermal application.

2.2.4. Hydrogel Films Ability to Absorb Aqueous Solutions

The hydrogel films swelling behavior in physiological fluids is essential for their biomedical applications. On the one hand, this behavior determines the ability of the films to be loaded with drugs and, on the other hand, influences the transport of these drugs through the polymer network, which will influence the release kinetics of the active ingredient. The swelling degree was determined gravimetrically for P1, P2, and P3 samples at 37 °C to equilibrium in a 0.1 M phosphate buffer solution at pH = 7.4 and a 0.1 M acetate buffer solution pH = 5.5. The influence of the CHO/NH_2 molar ratio, cross-linking time, and cross-linking temperature on the swelling degree values was studied to establish the optimal parameters for obtaining biocompatible hydrogels for controlled and sustained release of active principles with transdermal applications.

- **The molar ratio CHO/NH_2 influence on the swelling degree value**

In order to study the influence of this parameter, the P1, P2, P3 samples obtained at three -CHO/-NH_2 molar ratios values were selected because the amino groups' CI value was medium and were representative for this determination. Figure 9 presents the swelling degree variation in time for P1, P2, and P3 samples in two mediums of different pH values.

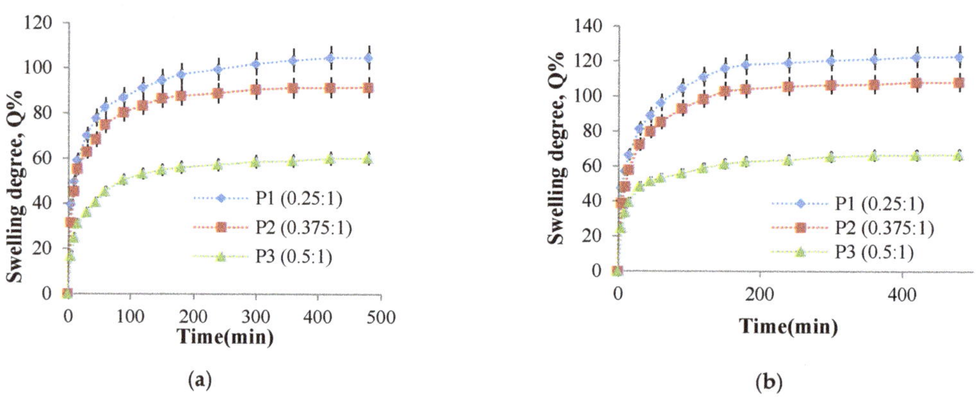

Figure 9. The variation in time of the swelling degree for P1, P2, and P3 samples using different molar ratios in two solutions of different pH: (**a**) pH = 5.5 and (**b**) pH = 7.4.

Figure 9 shows that the value of the swelling degree decreases when the CMCOx amount increases, regardless of the solution pH value in which the hydrogel films were immersed, an effect absolutely consistent with the CI of the amino groups' evolution. The weakly alkaline environment causes an increase in the swelling degree due to the CMC composition's carboxylate groups. It should be noted that the electrostatic repulsions between the polymer network chain segments are intense and are influenced by the carboxylate groups' number from CMCOx and determines the absorption of a higher buffer solution amount.

- **The influence of the cross-linking temperature on the hydrogel films swelling degree**

The influence of temperature on the cross-linking density and the swelling degree in aqueous media with different pH values was studied. As mentioned in Section 3.4.7, the films were weighed and immersed in an acetate buffer solution at pH = 5.5, respectively, in a phosphate buffer solution at pH = 7.4. The swelling degree was determined gravimetrically at well-established time intervals up to equilibrium. The P2 sample was considered of reference, and it was selected for this study (the P2 sample was chosen for the study of the cross-linking temperature influence on amino groups CI value). Four hydrogel films were prepared using different cross-linking temperatures between 30 °C and 60 °C. As mentioned in the Section 3.4, all other parameters were maintained at constant values. The four hydrogel films' swelling degree kinetics in buffer solutions were studied. The kinetic curves obtained are presented in Figure 10.

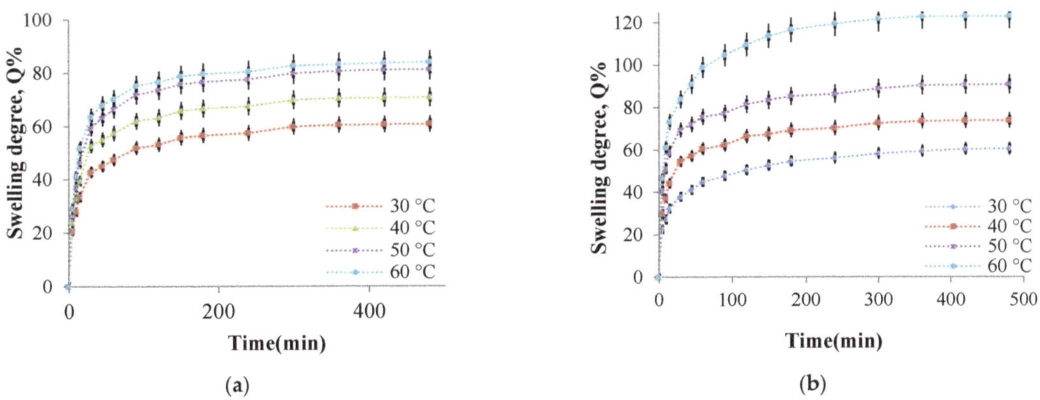

Figure 10. The variation in time of the swelling degree for the P2 sample prepared at different cross-linking temperatures in acetate buffer solution, pH = 5.5 (**a**) and in phosphate buffer solution at pH = 7.4 (**b**).

The analysis of Figure 10a allows us to ascertain that the films can absorb water at pH = 5.5, and the swelling degree values depend on the cross-linking temperature. The swelling was instantaneous when the obtained samples were immersed in aqueous solutions, and the polymers' strong hydrophilic character determined it. At the same time, the presence of amine groups that were protonated in a weak acid medium leads to electrostatic repulsions between the ammonium cations formed, the effect being the distance of the chain segments between the cross-linking nodes and, consequently, the diffusion of an increased water amount into the polymer network. The swelling degree increases rapidly in the first hour and then slower in the time interval range of 1–3 h with a tendency to reach equilibrium after 8 h. The swelling degree depends on the cross-linking temperature at which the hydrogels are obtained. In agreement with the CI evolution, the cross-linking temperature increase after 30 °C determines the decrease of CI values and a higher swelling degree value (Figure 10). Because there are strong physical interactions between the CS chains (responsible for the amino groups' partial blocking that can no longer react with ninhydrin), it could be concluded that these physical interactions act as an additional cross-linking. These interactions decrease in intensity as the cross-linking temperature increase (see Section 2.2.3), and the swelling degree values in the aqueous solutions increases. For the samples prepared at different cross-linking temperatures, the same swelling degree evolution in time was recorded in the films immersed in weakly basic aqueous solution at pH = 7.4 (Figure 10b). Still, the swelling degree values obtained at this pH were superior to those obtained at pH = 5.5.

At temperature values above 40 °C, the weakly alkaline medium's swelling degree values are higher than in the acidic medium. However, at pH = 7.4, the -NH_2 groups' protonation from CS is no longer possible. The possible explanation of this behavior could be that the hydrogels obtained contain -COOH groups from CMCOx. In the alkaline medium, the electrostatic repulsions between the formed -COO- (carboxylate) groups become very strong and determine the polymer network relaxation, expressed by an intensified absorption of the aqueous solution; therefore, the swelling degree values increase [23]. At higher temperatures, many -COOH groups are released from their interactions with the $-NH_2$ groups, becoming carboxylated groups, and the electrostatic repulsions between the chain segments of the network were intensified.

Based on these results, it can be concluded that the CS/CMCOx hydrogels are pH-sensitive. Therefore they can be included in the category of smart hydrogels.

- **The influence of the cross-linking time on the hydrogel films swelling degree**

It was previously stated that the cross-linking reaction time causes higher amino groups' CI values, suggesting that the cross-linking degree increase even if significant intermolecular interactions such as hydrogen bonds and electrostatic interactions occur at the same time. Therefore, the cross-linking time influence on the swelling degree values in buffer solutions at pH = 7.4 and pH = 5.5 of the P2 sample (with the molar ratio 1:0.375) was studied. Six hydrogel films were obtained using different cross-linking times (between 1 and 6 h). The kinetic study of the swelling process in time in buffer solutions at pH = 5.5 and pH = 7.4 is presented in Figure 11.

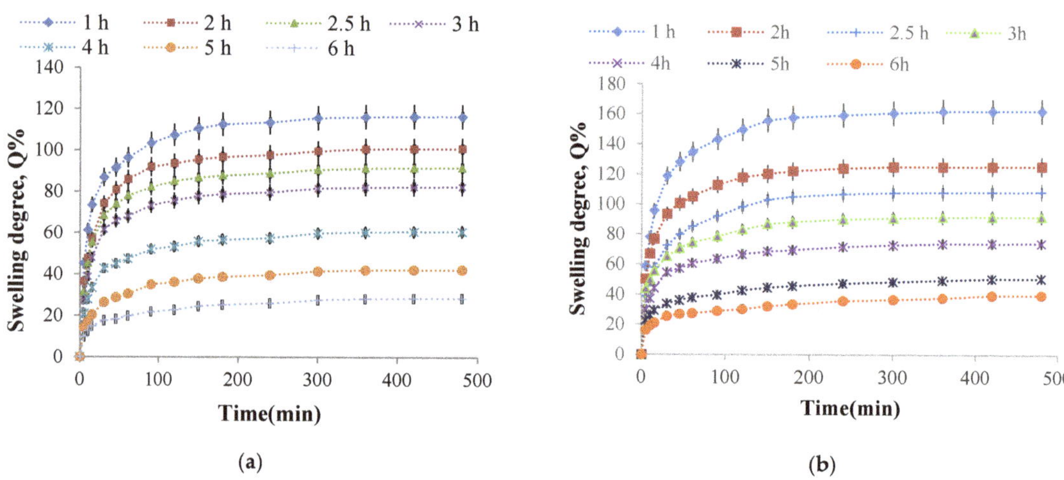

Figure 11. Variation in time of the swelling degree for P2 sample obtained at different cross-linking times in different pH solutions: (**a**) pH =5.5 and (**b**) pH = 7.4.

The first finding was that the swelling was faster in the first 30 min, followed by a slower increase in swelling degree value, and finally, there was a tendency to stabilize it after about 3 h. The cross-linking reaction duration, obviously accompanied by the network density increase, reduces the swelling degree values in aqueous buffer solutions. Such swelling degree evolution with the cross-linking time was expected given that the CI of the amino groups into Schiff bases increase, and a higher aldehyde groups' number from CMCOx were involved in the reaction with the amine groups from CS, causing the formation of an increasingly dense polymer network. The hydrogels' pH-sensitive nature was highlighted again. It is observed that, regardless of the cross-linking time, the swelling degree values were higher in the weak alkaline medium (pH = 7.4). The explanation was given in the previous paragraph. It is verified that if the cross-linking time increase, reduces the swelling degree maximum value, reached after about 8 h.

2.2.5. Encapsulation Efficiency

Curcumin was incorporated into the samples according to the protocol described in Section 3.4.9. The results obtained for the curcumin encapsulation efficiency in the analyzed samples are presented in Figure 12.

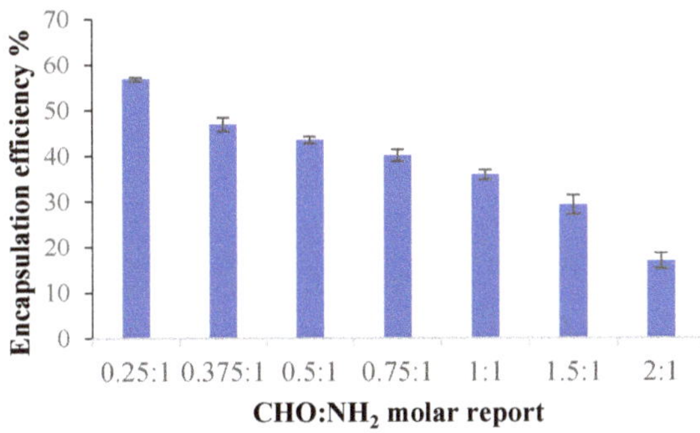

Figure 12. Curcumin encapsulation efficiency in CS/CMCOx films obtained at different CHO: NH_2 molar ratios.

It was found that the curcumin encapsulation efficiency decreases as the CMCOx amount in the films increases, consequently as the cross-linking degree increases. This effect was determined because the polymer network meshes became smaller when the cross-linking density increased, and the curcumin amount that could diffuse into the cross-linked polymeric matrix was lower.

2.2.6. *In Vitro* Release Kinetics of Curcumin from Hydrogel Films

The curcumin release kinetics from the obtained films were studied on a Franz cell using chicken skin as a membrane. The method has been described in the Section 3.4. Figure 13 shows the curcumin release kinetic curves in 0.1 M acetate buffer solution at pH = 5.5 containing 1% Tween 80 and in a 0.1 M phosphate buffer solution containing 1% Tween 80 at pH = 7.4 for P1C, P3C, P5C. These samples were selected to highlight very clearly the influence of the cross-linking degree, determined by the molar ratio -CHO/NH_2, being consistent with the conversion index of the amino groups, on the process of curcumin releasing from the films. Table 1 shows the release efficiency and permeability values of immobilized curcumin through the membrane in the Franz cell's receptor compartment. The exponential factor n calculated from the Ritger–Peppas [64] equation is also presented in Table 1.

Table 1. The release efficiency and permeability of curcumin-loaded in hydrogel films through the skin membrane.

Sample	Ratio	Efabs% on Skin Membrane, %		Ef% in the Receptor, %		Total Ef, %		P (μg/cm^2) in the Receptor (after 48 h)		P (μg/cm^2) in Skin Membrane, 48 h		Total P (μg/cm^2), 48 h		n		R2	
		pH 5.5	pH 7.4	pH 5.5	pH 7.4	pH 5.5	pH 7.4	pH 5.5	pH 7.4	pH 5.5	pH 7.4	pH 5.5	pH 7.4	pH 5.5	pH 7.4	pH 5.5	pH 7.4
P1C	0.25:1	47	8.6	16.5	18	64	26	41.1	71	117	35.1	158	106	0.63	0.43	0.92	0.97
P3C	0.5:1	31.6	3.6	9.5	12	41	15	33.9	66.3	113	20.4	147	87	0.48	0.53	0.81	0.91
P5C	1:01	23.3	3.2	7.4	11	31	14	28.5	49.5	90	14.9	118	64	0.53	0.31	0.97	0.99

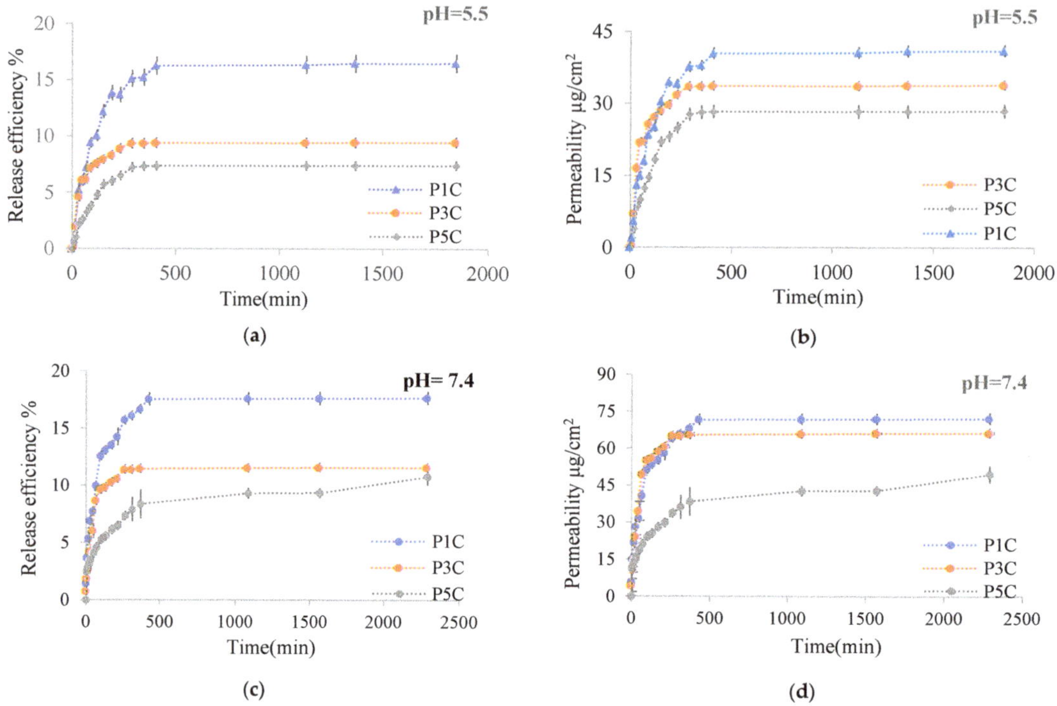

Figure 13. The curcumin release kinetics in time in acetate buffer solution at pH = 5.5 expressed as release efficiency (**a**) or in terms of permeability-µg/cm^2 (**b**) and phosphate buffer solution at pH = 7.4 expressed as release efficiency (**c**) or in terms of permeability-µg/cm^2 (**d**), for samples P1C, P3C, and P5C.

Figure 13a,c shows the kinetic curves of curcumin release in the receptor compartment after approximately 30 h at pH = 5.5 and after 48 h at pH = 7.4. Table 1 shows that the total curcumin released efficiency (the curcumin release efficiency from receptor compartment cumulated with the release efficiency of curcumin calculated based on the curcumin amount retained within skin membrane) was higher at pH = 5.5 than at pH = 7.4. The total release efficiency at pH = 5.5 was 63.5% compared with release efficiency of 26% at pH = 7.4 for films with a low cross-linking density (P1 sample). The release efficiency increases when the cross-linking degree decreases, being affected by the polymer network meshes size, which was higher when the cross-linking degree was lower. The amount of curcumin retained in the skin membrane (based on which the release efficiency within the skin was calculated) was lower at pH = 7.4 than at pH = 5.5 (Table 1). The pKa value of CS varies between 6.17 and 6.51 [65,66]. At a pH lower than the pKa value, the free amino groups could be protonated and interact with different skin proteins leading to an increased curcumin absorption efficiency. Regardless of the pH used, the skin retention efficiency of the curcumin released from the hydrogel films increases when the cross-linking degree decreases. It can be observed that the curcumin release efficiency in the receptor compartment at pH = 7.4 has a higher value compared to the curcumin release efficiency at pH = 5.5 (Figure 13), but the retention of curcumin released from hydrogel films at pH = 7.4 in the skin membrane was lower than at pH = 5.5 (Table 1).

In the weakly alkaline medium, curcumin was found in the phenolate form and became more soluble; therefore, its release efficiency in the Franz cell receptor compartment was intensified. The curcumin released from films through the skin membrane in the receptor compartment at pH = 7.4 depends on the carboxylate groups found in the polymer matrix that lead to an increase of the swelling degree, which intensifies the active principle diffusion. At this pH, the free amino groups are not protonated, and the hydrogel film in which curcumin was included does not interact with the skin proteins.

Curcumin-loaded films were used to study in vitro transdermal permeability through the Franz diffusion cell.

The curcumin release kinetics were also evaluated by determining the permeability of curcumin released through the skin membrane in time, expressed as μg curcumin/cm^2. Regarding the curcumin permeability through the skin, it was found that the results obtained are comparable to those obtained at the curcumin release efficiency determination. Thus, for the encapsulated curcumin, the total permeability (the curcumin permeability from receptor compartment cumulated with the permeability of curcumin retained in the skin membrane) was higher at pH = 5.5 compared to pH = 7.4 (Table 1) due to the protonated amino groups that could interact with proteins found in the skin membrane and, therefore, the retention of curcumin released from hydrogels into the skin was improved.

The permeability variation curve s in time through the skin membrane in the receptor compartment is presented in Figure 13b,d for P1C, P3C, P5C samples, and the results are consistent with those obtained when the release efficiency was determined. Thus, a better permeability of curcumin released from hydrogels through the skin in the receptor compartment of the Franz diffusion cell at pH = 7.4 compared to pH = 5.5 was observed, this effect being determined by the slightly increased solubility of curcumin (which partially passes into phenolate form) in the alkaline medium at pH = 7.4. The curves have a typical appearance for polymer/drug systems from which the drug was released in a controlled and sustained manner, with a "burst effect" in the first 300 min. After this period, the curcumin release was slower, with a tendency to equilibrium until approximately 2000 min. The curcumin release depends on the film's cross-linking degree because the active principle diffusion through the hydrogel becomes more and more slowly as the polymer network's cross-linking degree increases.

The exponential factor n value from the Ritger Peppas equation calculated based on the curcumin release curves shows that the diffusion mechanism was practically Fickian for most of the samples used to study the release kinetics. For the curcumin release from the P1C sample, the exponential factor n value was 0.63; thus, the diffusion mechanism was of the non-Fickian type and disturbed by certain factors that have not been studied. Also, at pH = 7.4, the exponential factor n value for curcumin release from the P5C sample was 0.31, indicating a Fickian type diffusion mechanism, but probably disturbed by interactions between curcumin and the constituents polymers functional groups from hydrogels. It can be observed from the permeability and efficiency curves that the curcumin release from the P5C sample did not reach an equilibrium after 48 h. There is a possibility that these interactions between curcumin and the polymer matrix determine the active principle diffusion disruption.

Agrawal et al. showed that the curcumin permeability in a based hydrogel based on Carbopol 940F through mouse skin was 0.67 ± 0.01 μg/cm^2/h and the free curcumin permeability in aqueous solutions is 0.46 ± 0.02 μg/cm^2/h [67]. Ternullo et al. also obtained a curcumin permeability of 1.5 μg/cm^2/h on human skin from a CS hydrogel with immobilized curcumin [68].

In our case, the total curcumin permeability after its diffusion from the film into the chicken skin at pH = 5.5 was between 2.27 μg/cm^2/h and 3.3 μg/cm^2/h, and at pH = 7.4, it was between 1.33 μg/cm^2/h and 2.21 μg/cm^2/h. The CS bioadhesive properties could determine this improved curcumin permeability from the films through the chicken skin, and the CMCOx contributes to the increase of the film's hydrophilic character due to the free carboxylic groups that could interact at pH = 5.5 with skin proteins that contains lysine residues.

2.2.7. Antioxidant Activity

The antioxidant activity expressed by IC50 was determined for P1C, P3C, P5C films, for CS, CMCOx, for P5 films without curcumin immobilized, for free curcumin, and as well as for three mixtures obtained using the identical amounts of CMCOx and curcumin as in P1C, P3C, and P5C samples. Ascorbic acid was used as a standard. The curcumin mixtures with CMC were dissolved in 50 mL of ethanol/water solution, were coded with M1, M3, and M5, and different curcumin concentrations solution expressed in µmoles/mL were used to determine the inhibition percent of the free radicals from DPPH and based on these results to evaluate the antioxidant activity expressed by IC50 for the mixtures mentioned before. The antioxidant activity evaluation of the immobilized curcumin within hydrogel films allows the study of the influence of the constituents polymers and the molar ratio between polymers on this active principle feature. In principle, when the IC50 value is lower, the antioxidant activity is higher. As mentioned in Section 3.4.10, the IC50 value was determined based on the graphical representation between the inhibition percentage versus concentration expressed in µmoles/mL. The results obtained are presented in Table 2 and Figure S3 (Supplementary Information).

Table 2. Antioxidant activity expressed by IC50 for the analyzed samples.

Sample	IC50, µmoles/mL
Ascorbic acid	0.031 ± 0.00053
Curcumin	0.051 ± 0.00033
P1C	0.054 ± 0.00039
P3C	0.046 ± 0.00039
P5C	0.039 ± 0.00012
M1	0.082 ± 0.0032
M3	0.081 ± 0.00041
M5	0.092 ± 0.00846
P5 + M5	0.035 ± 0.00012
CS	0.4 ± 0.0103
P5 (without curcumin)	10.02 ± 0.28

It was found that the IC50 values for free curcumin and curcumin included in the P1C film are close, but the IC50 value decreases when the CMCOx amount in the films increases. The curcumin-loaded hydrogel film's antioxidant activity increases as the hydrogel film's cross-linking degree increases. In order to determine whether CS or CMCOx influences the curcumin antioxidant activity, the IC50 value was determined for CMCOx, CS, P5 sample without curcumin and for curcumin/CMCO mixtures coded with M1, M3, M5. The P5 hydrogel film influence on the antioxidant activity was evaluated and was used the same film amounts to determine the inhibition percentage of DPPH free radicals as in the case of curcumin-loaded P5C films inhibition percentage determination.

As previously mentioned, the curcumin encapsulation efficiency in films decreases with the cross-linking degree increasing. Therefore, to obtain the same curcumin concentration in the ethanol solution used to determine the DPPH free radicals inhibition percentage for P1C, P3C, and P5C hydrogel films, the film amount used for this determination, was higher as the films cross-linking degree was improved. The obtained results showed that CMCOx did not have antioxidant activity. CS had antioxidant activity, but the IC50 value was much higher than that of curcumin. It was also found that the IC50 value was significantly higher for the P5 film compared to that obtained for CS. However, the CS film's concentration used to determine the inhibition percentage was higher than the free CS concentration used. There is a possibility that the CS antioxidant activity decreases as the free amino groups are fewer and fewer as they react with the aldehyde groups from CMCOx. The amino group's CI in Schiff bases in hydrogel films with a molar ratio of 1:1 was 76.5%, which means that only 23.5% of the total amino groups are free.

The IC50 values for curcumin from the M1 or M3 mixtures were close, but the IC50 value increased for sample M5 compared to the other two samples. The M1, M3, and M5 samples' antioxidant activities were higher than that of free curcumin. We notice that the antioxidant activity decreased with the increase of the CMCOx amount in the curcumin/CMCO mixture.

The free curcumin could interact with the aldehyde groups from CMCOx, and new compounds could be formed. The recorded UV spectra show two peaks that are not characteristic of curcumin (425 nm) at wavelengths of $\lambda = 292$ nm and $\lambda = 360$ nm. In a future study, we will try to elucidate this aspect. For the P5C film, the IC50 value was 0.039 µmoles/ml, and the antioxidant activity was higher than that of free curcumin. However, the IC50 value increases, being 0.092 µmoles/mL for the M5 mixture, in which the same curcumin and CMCOx amounts were used as in P5C hydrogel film. In order to determine whether the IC50 value obtained of 0.039 µmol could be influenced by the CS antioxidant activity, the inhibition percentages obtained from the analysis of each amount of P5 film without curcumin were summed with the inhibition percentage obtained for each curcumin concentration from the M5 mixture. Then, based on the curve obtained from the graph I% vs. the concentration of curcumin (µmoles/mL), IC50 was determined. The IC50 value obtained was 0.035 µmoles/mL, being very close to the IC50 value obtained for curcumin-loaded in the P5C film, explaining the curcumin-loaded films' antioxidant activity increase. The obtained results show that CS influences the curcumin IC50 values and determines curcumin-loaded films with improved antioxidant activity. The obtained CS films were cross-linked with aldehyde groups from CMCOx and dried before curcumin encapsulation, and the possibility of interactions between curcumin and CMCOx was very low. Thus, the wavelength where curcumin included in the film had a maximum absorption does not change compared to free curcumin's characteristic wavelength (425 nm). Therefore, the CS films influence the curcumin antioxidant activity, and the IC50 value decreases.

The obtained results allowed the optimization of the cross-linking reaction between CS with CMCOx to obtain hydrogel films with biomedical applications. An optimal film must be flexible, mechanically stable, and have a high capacity to include and controlled release active principles. A high cross-linking degree causes incorporation of a small amount of active principle and a slow diffusion of it from the polymer matrix in quantities that may not reach the therapeutic concentration but with high mechanical stability. The optimal parameters for obtaining the film must be identified to ensure good mechanical stability and a high capacity for inclusion and controlled release of the active ingredient. Considering the results presented above, the optimal parameters established for obtaining the films used for dermal applications were: the cross-linking temperature of 55 °C, the cross-linking time of 2.5 h, and the molar ratio -CH O/NH_2 that range between 0.25:1 and 0.5:1 (0.375:1 is preferable).

3. Materials and Methods

3.1. Materials

Medium molecular weight chitosan (CS) with a degree of deacetylation of 75% (M = 190–310 kDa and viscosity of 200–800 cPs), carboxymethyl cellulose (CMC) with a degree of substitution of 70%, powder curcumin (extracted from Curcuma Longa), DPPH (1,1-diphenyl-2-picrylhydrazyl), Tween 80, ninhydrin, sodium meta periodate were purchase from Sigma Aldrich, potassium iodide, sodium thiosulfate, ethyl alcohol, acetic acid, disodium phosphate, monosodium phosphate, sodium chloride were purchase from Chemical Company.

3.2. Preparation of Oxidized Carboxymethylcellulose

The method was adapted from A. Kulikowska [69] with some modifications. Briefly, 3 g of CMC was dissolved in 100 mL of distilled water at 80 °C, and the solution thus prepared was added into a 250 mL flask and allowed to cool at 30 °C. A stoichiometrically calculated amount of 3 g of $NaIO_4$ was dissolved in bi-distilled water (20 mL) at room temperature and then added dropwise over the CMC solution at 30 °C. The reaction takes place in the dark, under stirring (500 rpm) for 2, 4, and 6 h, respectively. The reaction product was precipitated in ethanol (at 4 °C), and the precipitate was filtered, washed three times with ethanol, and dried at room temperature.

3.3. Obtaining Hydrogel Films Based on CS and CMCOx

Three grams of CS were dispersed in 200 mL of an acetic acid solution at pH = 2.4, and the solution was maintained under stirring at room temperature for 24 h. The solution at pH = 2.4 was prepared using 10 mM NaCl and glacial acetic acid to adjust the pH. According to Caneret al. [70], acetic acid is the best solvent for obtaining CS-based hydrogel films with optimal mechanical and barrier properties. Then by centrifugation, it was obtained a clear CS solution. The undissolved CS resulting from the centrifugation was dried at 100 °C, weighed, and the solution concentration was adjusted to 1.5%. Different molar ratios between the amino groups from CS and the aldehyde groups from CMCOx were used, maintaining the CS amount constant and varying the CMCOx amount. The number of moles/g chitosan was determined by calculations knowing that the structural unit's molecular weight was 171.5 g/mol; this value was calculated knowing that the chitosan deacetylation degree was 75%, and in each structural unit, there is only one amino group. The number of moles of aldehyde/g was calculated using titration with $Na_2S_2O_3$, as described in Section 3.4.3 from Materials and Methods.

The molar ratios of -CHO/-NH_2 used were: 0.25:1, 0.375:1, 0.5:1, 1:1, 1.5:1, 2:1 (mol/mol). In the second step, 20 mL of the previously prepared 1.5% CS solution was heated to 55 °C, and then, under stirring, the required amount of CMCOx dissolved in 10 mL solution of pH = 2.4 was added dropwise. The solution thus prepared was stirred for 2.5 h. 0.3 g glycerin (which ensures a weight ratio of 1 g glycerin for 100 g solution of polymers mixture) was added to obtain a flexible, fragility-free hydrogel that can be applied to the skin, ensuring good contact with it. The polymer solution was poured into 9 cm diameter Petri dishes and dried at room temperature. After drying, the films are detached from the Petri dishes and stored in the refrigerator at 4 °C until further characterizations. Table 3 shows the experimental program for obtaining hydrogel films.

Table 3. The experimental program used to obtain covalently cross-linked hydrogel films based on CS and CMCOx.

Samples Code *	The Molar Ratio (-CH=O/-NH_2)	Moles of Aldehyde Groups from CMCOx ($\times 10^3$)
P1	0.25:1	0.4375
P2	0.375:1	0.656
P3	0.5:1	0.875
P4	0.75:1	1.3125
P5	1:1	1.75
P6	1.5:1	2.625
P7	2:1	3.5

* The number of moles of -NH_2 groups in the CS of 1.75×10^{-3} moles was kept constant. The CS solution volume was 20 mL, and the CMCOx solution volume was 10 mL. 0.3 g of glycerin was added in each synthesis to avoid the obtaining of fragile and brittle films. The films containing curcumin and containing the same molar ratios were coded with P1C, P2C, P3C, P4C, P5C, P6C, and P7C.

3.4. Characterisation Methods

3.4.1. FTIR Spectroscopy

FTIR spectra were obtained for CMCOx, and films based on CS cross-linked with CMCOx without curcumin were recorded using the KBr pellet method. The spectra were recorded on a Bruker Vertex FTIR spectrophotometer (Billerica, MA, USA) over a frequency range of 4.000–400 cm^{-1} at 4 cm^{-1} resolution-32 scans (discussions were made in Section 2.1.1 for CMC or CMCOx, and in Section 2.2.1 for films based on CS cross-linked with CMCOx).

3.4.2. ^{1}H NMR and ^{13}C NMR Spectra of CMC and CMCOx

Nuclear magnetic resonance spectroscopy (Bruker NEO 1-400, Billerica, MA, USA) was employed to investigate the product's molecular structure before and after oxidation. The samples were prepared by dissolving CMC in deuterated water at 80 °C, and the

CMCOx was dissolved in deuterated water or DMSO, after which they were analyzed by ^{1}H NMR and ^{13}C NMR technique.

3.4.3. Oxidation Reaction Kinetics and Quantitative Determination of Aldehyde Groups Obtained in CMCOx

- **Aldehyde groups determination**

The working protocol was adapted with some modifications after Teotia [31] and McSweeny et al. [71]. The aldehyde group content in CMCOx was determined indirectly by dosing the residual sodium periodate in a reaction mixture by iodometric titration. Briefly, 1 mL of CMCOx solution was added over 1 mL of 20% KI solution mixed with 1 mL of 37% HCL. I_2 was titrated with 0.05 N $Na_2S_2O_3$ until the solution color turns light yellow. Afterward, to visualize I_2, 1 mL of starch solution with a concentration of 1% was added, and the solution's color was turned blue. It was titrated until the color becomes transparent, and I_2 no longer existed in the reaction. Lange previously described the iodometric titration reactions in 1961 [72].

Based on the reactions involved, the amount of sodium periodate that did not react in the oxidation reaction was determined. The $NaIO_4$ amount that reacts was calculated as the difference from the initial amount of $NaIO_4$ added in the oxidation reaction. The number of moles of aldehyde groups could be calculated, considering that 1 mole of $NaIO_4$ reacted was required to obtain 2 moles of aldehyde groups in CMCOx.

- Oxidation kinetics studies

Three grams of CMC was dissolved at 80 °C in 100 mL of bi-distilled water. After the solution temperature decreases at 30 °C, 2 g of $NaIO_4$ dissolved in 20 mL of water was added dropwise over the polymer solution, under stirring. The reaction occurs in a glass flask, on a water bath, in the dark at 30 °C, under stirring to prevent air and light influence on the oxidation reaction. After every hour for up to 6 h, 1 mL of sample was taken from the reaction medium and titrated with $Na_2S_2O_3$. The degree of oxidation was calculated with Equation (1).

$$OD\% = \frac{Q_R}{Q_T} \times 100 \tag{1}$$

where Q_R is the $NaIO_4$ amount that reacted, and Q_T is the total amount of $NaIO_4$ introduced into the reaction.

3.4.4. Molecular WEIGHT determination of CMC and CMCOx by the Viscometric Method

The CMC and CMCOx intrinsic viscosities were determined by working at different polymer concentrations (0.1%, 0.2%, 0.3%; 0.35%, 0.4% and 0.5% *w/v*) in NaCl solution in a weight ratio polymer: NaCl = 1:1 (for example for a polymer solution preparation with 0.2% concentration, 0.2 g polymer were dissolved in 100 mL solution of 0.2% NaCl concentration), using an Ubellohde viscometer (K = 0.00999) at 25 °C. The average viscosimetric molecular weight (M_η) was calculated using the Kuhn–Mark–Houwink Equation (2):

$$[\eta] = K \times M_\eta{}^{a} \tag{2}$$

where $K = 1.23 \times 10^{-3}$ and a = 0.91 (for both initial CMC and CMCOx) [54], [η] represent the intrinsic viscosity, and M_η is the average molecular weight determined by the viscosimetric method.

[η]—was deduced from the equation of the graphical representation ηred as a function of the solution concentration.

3.4.5. Scanning Electron Microscopy

Hydrogel films were characterized by scanning electron microscopy (SEM) to determine their surface morphology. They were dried, metalized with gold using a spray deposition device, and analyzed using a HITACHI SU 1510 electron microscope. (Hitachi SU-1510, Hitachi Company, Chiyoda City, Tokyo, Japan).

3.4.6. The Amino Groups' CI% Determination into Shiff Bases in Hydrogel Films

The CS used (average molecular weight) has a degree of deacetylation of 75%, and the molecular weight of the structural unit is 171.5 g/mol. A CS stock solution of 0.1% (*g*/*v*) concentration was initially prepared to plot the calibration curve using the ninhydrin test by dissolving the CS amount in acetic acid 0.1%. Different volumes were taken to prepare six different CS concentrations between 0.1 and 0.4 mg/mL and were brought to sign using acetate buffer solution pH = 5.6, 0.1 M in volumetric flasks of 10 mL. The ninhydrin test protocol used to plot the calibration curve was as follows: in the test tubes to 1 mL of CS solution of different concentrations prepared before was added 2 mL of 2% ninhydrin solution in ethanol. Then the reaction mixture was heated at 95 °C for 30 min, and the color of the solutions became dark blue. After the solutions were cooled, 8 mL of ethanol/distilled water solution (1:1 volume ratio) was added. The spectrophotometer's calibration was done with a blank prepared, as we mentioned before but without chitosan. Instead of chitosan, 1 mL of acetate buffer of pH = 5.6 was used. The solutions' absorbances were recorded with a UV spectrophotometer at a wavelength of 570 nm. The concentrations expressed first in mg/mL were converted to molar concentrations (number of moles of amino groups'/mL). The CS calibration curve was plotted using the ninhydrin assay, and it is presented in Figure S4 (absorbance as a function of the number of moles of amino groups from the CS). The CS calibration curve has the equation y = 0.056x.

It was considered necessary to determine the number of free amino groups from the obtained films using the ninhydrine test. The working method was similar to the one presented above used to plot the chitosan calibration curve. The small film amounts previously weighted were added in the test tubes, and 1 mL of buffer acetate 0.1 M, pH = 5.6 with 2 mL of 2% ninhydrine solution was added, then heated at 95 °C for 30 min. After the solution was cooled, 8 mL of ethanol/distilled water solution (1:1 volume ratio) was added, and the absorbance was measured at 570 nm. Based on the chitosan calibration curve previously plotted, the free amino groups within hydrogel films were determined. The total number of amino groups involved in the formation of Schiff bases, but also in strong physical interactions with the carboxylic groups in CMCOx, can be calculated by the difference between the total number of moles of free amino groups initially added in the reaction (calculated based on the initial chitosan amount). The free amino groups within the hydrogel were determined using the ninhydrin test. The conversion index was calculated with the relation (3) and expressed:

$$CI\% = \frac{N_b - N_a}{N_b} \times 100 \quad (3)$$

where N_b—number of moles of free amino groups before cross-linking; N_a—number of moles of free amino groups after cross-linking. $N_b - N_a$—represent the bonded amino groups within hydrogel film.

The influence of the cross-linking temperature, cross-linking time, and the molar ratio on CI value was studied. The CI of the amino group determination was performed in triplicate. The standard deviation was within ±3%.

3.4.7. Hydrogel Films Ability to Absorb Aqueous Solutions

The obtained films have hydrogel character, so it was considered useful to determine their ability to retain water—usually quantified by the degree of swelling (Qt,%). This feature is significant because it causes the more or less intense diffusion of the active ingredient from the hydrogel matrix.

For the obtained films, Qt was determined gravimetrically. The buffer solutions simulating physiological fluids used were phosphate buffer at pH = 7.4 and an acetate buffer simulating skin pH at pH = 5.5.

The films were dried to constant weight, and a quantity of precisely weighed film (M_{dry}) was immersed in a 5 mL solution of different pH values at 37 °C. At various intervals, the sample was removed from the liquid medium, and its surface was buffered with filter paper to remove the excess water. The mass of the swelling films ($M_{swelling\ films}$) was determined by weighing. The solution absorbed by the films ($M_{solution}$) represents the difference between the swollen film's weight ($M_{swelling\ films}$) and the weight of the dry film (M_{dry}). After weighing, the samples were reintroduced into the solution, the operation being repeated after well-established time intervals until equilibrium was reached. The swelling degree was expressed as the ratio between the amount of solution retained in the films at each time interval measured and the amount of completely dry film (relation (4)). The influence of the cross-linking temperature, cross-linking time, and the molar report on swelling degree value was studied. The swelling degree was performed in triplicate. The standard deviation was within ±5%.

$$Q\% = \frac{M_{solution}}{M_{dry}} \times 100 \quad (4)$$

3.4.8. Encapsulation Efficiency

The whole film was used for curcumin encapsulation by diffusion and alcohol evaporation method. The procedure was as follows: a quantity of 20 mg of curcumin was dissolved in 15 mL of ethanol, and the obtained films (see the Methods section) were immersed in the curcumin solution and maintained for 24 h at 45 °C for ethanol evaporation. After 24 h, the dry films were placed for 30 s in 25 mL of phosphate buffer solution at pH = 7.4 to remove the curcumin found on the film's surface that was not immobilized. In order to determine the curcumin quantity that was not immobilized, 20 mL of ethanol was added to the containers in which curcumin encapsulation took place. The curcumin amount was determined spectrophotometrically at λ = 425 nm based on the curcumin calibration curves in ethanol or phosphate buffer solution previously plotted. The difference between the initial curcumin amount used for encapsulation and the curcumin amount determined spectrophotometrically, it was determined the encapsulated curcumin amount in the hydrogel films. The encapsulation efficiency was determined with the relation (5):

$$Ef\% = (\text{curcumin amount encapsulated})/(\text{initial curcumin amount}) \times 100 \quad (5)$$

-initial curcumin amount = 20 mg

3.4.9. *In Vitro* Release Kinetics of Curcumin from Hydrogel Films

The in vitro release kinetics of curcumin from the obtained hydrogel films were studied in two different pH media: 0.1 M acetate buffer solution at pH = 5.5 (similar with skin pH) and 0.1 M phosphate buffer solution at pH = 7.4 (similar with blood pH), at 37 °C, until equilibrium.

In order to simulate the dermal application, curcumin release was studied using a Franz cell (2 cm diameter, 15 mL receptor volume), the cell compartments being separated by a chicken skin membrane. The skin was degreased with 96% ethyl alcohol using sterile dressings, and only skin without defects and open-pore was used for the experiment. After preparation, the skin was stored in 10% glycerin solution for 24 h before use. The membrane thus prepared was fixed between the donor and the receiving compartment. The films with immobilized curcumin (size 2.5 × 2.5) were placed in the Frantz cell's donor compartment on the skin membrane's surface. Acetate buffer solution at pH = 5.5 or phosphate buffer solution at pH = 7.4 with 1% Tween 80 were used as a release medium. 1 mL solution of pH = 5.5 or pH = 7.4 was placed in the donor container. The curcumin release kinetics took place in the dark (the Franz cell was covered with aluminum folium), the temperature in the receiving compartment was maintained at 37 °C and the stirring at 150 rpm. 0.5 mL of the receptor compartment medium were taken after different time intervals for quantitative measurement of released curcumin and replaced with a fresh medium. The release profiles were obtained by measuring the curcumin concentration using a nanodrop UV-vis spectrophotometer at 425 nm. The permeability of curcumin (P) from films was also calculated through the skin membrane expressed as μg curcumin/cm^2 with the relation (6), and the release efficiency (%) was calculated with relation (7):

$$P = \frac{M_r + M_s}{S},\ \mu g/cm^2 \tag{6}$$

$$R_{Ef}\% = \frac{M_r + M_s}{M_i} \times 100 \tag{7}$$

where M_r represents the curcumin amount released in the receptor; M_s represents the curcumin amount found in the skin membrane; S represents the surface of the film (12.56 cm^2); M_i represents the total curcumin amount immobilized in the film disc used for release.

The curcumin calibration curves shown in Figure S5 (Supplementary Information) have the following equations: y = 0.0167x (calibration curve in ethanol); y = 0.0079x (calibration curve in phosphate buffer at pH = 7.4), respectively, y = 0.0104x (calibration curve in acetate buffer pH = 5.5). All the determinations were performed in triplicate. The standard deviation was within ±5%.

3.4.10. Antioxidant Activity

With some modifications, the work method was described before by Choi et al. [73]. The curcumin stock solution was prepared by dissolving 5 mg of curcumin in 50 mL of ethanol. Several dilutions were made to test the antioxidant activity, and the final concentrations of curcumin solutions were between 10 and 50 μg/mL. 2 mL of each solution concentration was added in the test tubes, over which 2 mL of 0.1 mM DPPH solution (in ethanol) was added. The samples prepared were vortexed for 20–30 s and maintained in the dark at a temperature of 37 °C for 1 h. The samples' absorbance was measured after 60 min using a UV spectrophotometer at a wavelength of 517 nm. Ascorbic acid was used as a standard. The absorbance values were converted into antioxidant activity percentage (free radicals inhibition percentage in DPPH-I%) using the following relation (8):

$$I\% = 100 - \left[(A_s - A_b) \times \frac{100}{A_c}\right] \tag{8}$$

IC50 (expressed as μmoles/ml) was calculated from the graphical representation I% vs. concentration and represented the sample's concentration that can capture 50% of the free radicals in DPPH. Ethanol was used to calibrate the spectrophotometer—the A_s represents the absorbance value of different concentrations of solutions. As a blank (A_b), it was used a solution prepared from 2 mL of ethanol and 2 mL of curcumin solutions of different concentrations was used (the blank's absorbance was measured for each concentration). The control solution was prepared using 2 mL of DPPH solution and 2 mL of ethanol. P1C, P3C, and P5C films were selected to determine the immobilized curcumin antioxidant activity. Different film amounts containing between 10 and 50 μg of curcumin were used, and the percentage of free radical inhibition percentage (I%) was determined as described above. It was also determined the antioxidant activity for CS solution, hydrogel films without curcumin, and curcumin mixed with CMCOx dissolved in an ethanol/water mixture. It was prepared mixtures of curcumin with CMCO by adding identical amounts of curcumin and CMCO as in the films P1C, P3C, P5C, and the samples were marked with M1, M3, and M5. The antioxidant activity of M1, M3, M5 samples was also determined. All the determinations were performed in triplicate. The standard deviation was within ±5%.

4. Conclusions

CMC was oxidized with sodium meta-periodate, and the aldehyde group formation was demonstrated by FTIR and NMR spectroscopy. The polysaccharide oxidation degree increases in time, its maximum value being 22.9%. CMC was degraded during the oxidation reaction; its molecular mass decrease when the oxidation time increases oxidation, up to 1186 g/mol after 6 h of oxidation. Hydrogel films were obtained using different molar ratios between the amino groups from CS and the aldehyde groups from CMCOx. FTIR spectroscopy demonstrated the Schiff bases' formation, and analysis of the films' surfaces by SEM showed that the hydrogels' porosity/rugosity decreases with the increasing of the CMCOx amount in their composition. The amino groups' CI into Schiff bases depended on several influencing factors such as the CMCOx concentration in the film composition, the intermolecular interactions between the constituent polysaccharides, the cross-linking time, and the cross-linking temperature. It was found that some of the amino groups from CS are involved in strong intermolecular interactions with carboxylic groups of CMCOx, their reaction with ninhydrin was restricted, and the values of the amino groups' CI were distorted. The swelling degree determined in two different pH environments (pH = 5.5 and pH = 7.4) increases when the CMCOx amount decreases in the film and when the amino groups' CI value decreases. The cross-linking temperature and the cross-linking time for the films prepared significantly influence the swelling degree, obtaining a maximum value at the temperature of 60 °C and a short cross-linking time (one hour). The CI values decrease with the temperature and cross-linking time increasing. Higher values of the swelling degree at pH = 7.4 compared to those at pH = 5.5 were obtained due to the presence in the film of free carboxylic groups from CMCOx, which causes strong electrostatic repulsions and the absorption of a higher amount of aqueous solutions.

The thermal stability of the films depends on the cross-linking degree. Thermogravimetric analysis showed that the maximum weight loss is lower for the samples with a higher crosslinking degree (P3). Curcumin was included in the hydrogel films by diffusion method and evaporation of alcohol, and it was observed that the encapsulation efficiency decreases with increasing the cross-linking degree. The kinetics of curcumin release from the film and permeability were studied in buffer solutions of pH = 5.5 and pH = 7.4 through a chicken skin membrane, using a Franz diffusion cell. The results showed that the curcumin release from films intensified at pH = 7.4. Still, at pH = 5.5, there is a higher absorption of curcumin amount in the skin membrane, making the total release efficiency higher at pH = 5.5. Curcumin-loaded in the hydrogel films retains its antioxidant activity, and CS contributes to its increase. Obtaining CS-based hydrogel films cross-linked with aldehyde groups from CMCOx can be optimized for various applications such as biomedical ones.

Supplementary Materials: The following are available online, Figure S1: Schematic representation of CMC's oxidation reaction under the sodium periodate action, Figure S2: The structure of the hydrogel films based on chitosan and oxidized carboxymethyl cellulose—schematic presentation, Figure S3: Antioxidant activity determination expressed by IC50 values for the analyzed samples using the DPPH assay, Figure S4: The CS calibration curve determined with ninhydrine test, Figure S5: Calibration curves of curcumin in ethanol (**a**), phosphate buffer at pH = 7.4 (**b**), and acetate buffer at pH = 5.5 (**c**), Table S1: Table SI. The CI values (%) for samples obtained by chemical cross-linking and physical interaction between CS and CMCOx, respectively, by the CS amino groups' interaction with CMC's carboxylic groups.

Author Contributions: Conceptualization, M.P., C.E.I. and M.D. (Mohammed Dellali); methodology, C.E.I., C.L.S. and M.P.; software, C.L.S., M.D. (M'hamed Djennad) and N.S.; validation, M.D. (M'hamed Djennad) and N.S.; formal analysis, M.D. (Mohammed Dellali) and C.E.I.; investigation, M.D. (Mohammed Dellali), C.E.I.; resources, C.E.I. and M.P.; data curation, M.D.(Mohammed Dellali), M.D. (M'hamed Djennad), N.S.; writing—original draft preparation, C.E.I., M.D. (Mohammed Dellali), M.P.; writing—review and editing, C.E.I., M.D. (Mohammed Dellali), M.P supervision, C.E.I. and M.P. All authors have read and agreed to the published version of the manuscript.

Funding: This research received no external funding.

Institutional Review Board Statement: Not applicable.

Informed Consent Statement: Not applicable.

Data Availability Statement: Not available.

Acknowledgments: Dellali Mohamed wants to thank the ERASMUS+ International Dimension Programme for the studentship at the University of Iasi, Romania.

Conflicts of Interest: The authors declare no conflict of interest.

Sample Availability: The data presented in this study are available on request from the corresponding author.

References

1. Zagórska-Dziok, M.; Sobczak, M. Hydrogel-Based Active Substance Release Systems for Cosmetology and Dermatology Application: A Review. *Pharmaceutics* **2020**, *12*, 396. [CrossRef]
2. Svensson, A.; Ofenloch, R.; Bruze, M.; Naldi, L.; Cazzaniga, S.; Elsner, P.; Goncalo, M.; Schuttelaar, M.-L.; Diepgen, T. Prevalence of skin disease in a population-based sample of adults from five European countries. *Br. J. Dermatol.* **2018**, *178*, 1111–1118. [CrossRef]
3. Donnelly, R.F.; Singh, T.R.R.; Morrow, D.I.J.; Woolfson, A.D. Microneedle-Mediated Intradermal Delivery. In *Microneedle-Mediated Transdermal and Intradermal Drug Delivery*; Donnelly, R.F., Singh, T.R.R., Morrow, D.I., Woolfson, A.D., Eds.; John Wiley & Sons, Ltd.: Hoboken, NJ, USA, 2012; pp. 113–151.
4. Larrañeta, E.; Stewart, S.; Ervine, M.; Al-Kasasbeh, R.; Donnelly, R.F. Hydrogels for Hydrophobic Drug Delivery. Classification, Synthesis and Applications. *J. Funct. Biomater.* **2018**, *9*, 13. [CrossRef]
5. Benson, H.A.; Grice, J.E.; Mohammed, Y.; Namjoshi, S.; Roberts, M.S. Topical and Transdermal Drug Delivery: From Simple Potions to Smart Technologies. *Curr. Drug Deliv.* **2019**, *16*, 444–460. [CrossRef] [PubMed]
6. Ruela, A.L.M.; Perissinato, A.G.; Lino, M.E.D.S.; Mudrik, P.S.; Pereira, G.R. Evaluation of skin absorption of drugs from topical and transdermal formulations. *Braz. J. Pharm. Sci.* **2016**, *52*, 527–544. [CrossRef]
7. Miao, T.; Wang, J.; Zeng, Y.; Liu, G.; Chen, X. Polysaccharide-Based Controlled Release Systems for Therapeutics Delivery and Tissue Engineering: From Bench to Bedside. *Adv. Sci.* **2018**, *5*, 1700513. [CrossRef] [PubMed]
8. Lee, J.W.; Park, J.H.; Robinson, J.R. Bioadhesive-based dosage forms: The next generation. *J. Pharm. Sci.* **2000**, *89*, 850–866. [CrossRef]
9. Sogias, I.A.; Williams, A.C.; Khutoryanskiy, V.V. Why is Chitosan Mucoadhesive? *Biomacromolecules* **2008**, *9*, 1837–1842. [CrossRef] [PubMed]
10. Li, J.; Mooney, D.J. Designing hydrogels for controlled drug delivery. *Nat. Rev. Mater.* **2016**, *1*, 1–17. [CrossRef] [PubMed]
11. Islam, S.; Bhuiyan, M.A.R.; Islam, M.N. Chitin and Chitosan: Structure, Properties and Applications in Biomedical Engineering. *J. Polym. Environ.* **2017**, *25*, 854–866. [CrossRef]
12. Aguilar, A.; Zein, N.; Harmouch, E.; Hafdi, B.; Bornert, F.; Offner, D.; Clauss, F.; Fioretti, F.; Huck, O.; Benkirane-Jessel, N.; et al. Application of Chitosan in Bone and Dental Engineering. *Molecules* **2019**, *24*, 3009. [CrossRef] [PubMed]
13. Zhao, D.; Yu, S.; Sun, B.; Gao, S.; Guo, S.; Zhao, K. Biomedical Applications of Chitosan and Its Derivative Nanoparticles. *Polymers* **2018**, *10*, 462. [CrossRef] [PubMed]
14. Cheung, R.C.F.; Ng, T.B.; Wong, J.H.; Chan, W.Y. Chitosan: An Update on Potential Biomedical and Pharmaceutical Applications. *Mar. Drugs* **2015**, *13*, 5156–5186. [CrossRef] [PubMed]
15. Verma, M.S.; Liu, S.; Chen, Y.Y.; Meerasa, A.; Gu, F.X. Size-tunable nanoparticles composed of dextran-b-poly(D,L-lactide) for drug delivery applications. *Nano Res.* **2011**, *5*, 49–61. [CrossRef]
16. Guo, P.; Martin, C.R.; Zhao, Y.; Ge, J.; Zare, R.N. General Method for Producing Organic Nanoparticles Using Nanoporous Membranes. *Nano Lett.* **2010**, *10*, 2202–2206. [CrossRef] [PubMed]
17. Li, J.; Xie, B.; Xia, K.; Zhao, C.; Li, Y.; Li, D.; Han, J. Facile synthesis and characterization of cross-linked chitosan quaternary ammonium salt membrane for antibacterial coating of piezoelectric sensors. *Int. J. Biol. Macromol.* **2018**, *120*, 745–752. [CrossRef]
18. Sengiz, C.; Congur, G.; Eksin, E.; Erdem, A. Multiwalled Carbon Nanotubes-Chitosan Modified Single-Use Biosensors for Electrochemical Monitoring of Drug-DNA Interactions. *Electroanalysis* **2015**, *27*, 1855–1863. [CrossRef]
19. Wei, P.-R.; Cheng, S.-H.; Liao, W.-N.; Kao, K.-C.; Weng, C.-F.; Lee, C.-H. Synthesis of chitosan-coated near-infrared layered double hydroxide nanoparticles for in vivo optical imaging. *J. Mater. Chem.* **2012**, *22*, 5503–5513. [CrossRef]
20. Martínez-Martínez, M.; Rodríguez-Berna, G.; Bermejo, M.; Gonzalez-Alvarez, I.; Gonzalez-Alvarez, M.; Merino, V. Covalently crosslinked organophosphorous derivatives-chitosan hydrogel as a drug delivery system for oral administration of camptothecin. *Eur. J. Pharm. Biopharm.* **2019**, *136*, 174–183. [CrossRef]
21. Kim, B.S.; Yeo, T.Y.; Yun, Y.H.; Lee, B.K.; Cho, Y.W.; Han, S.S. Facile preparation of biodegradable glycol chitosan hydrogels using divinyladipate as a crosslinker. *Macromol. Res.* **2009**, *17*, 734–738. [CrossRef]
22. Gonçalves, V.L.; Laranjeira, M.C.M.; Fávere, V.T.; Pedrosa, R.C. Effect of crosslinking agents on chitosan microspheres in controlled release of diclofenac sodium. *Polímeros* **2005**, *15*, 6–12. [CrossRef]
23. Iurciuc-Tincu, C.-E.; Atanase, L.I.; Ochiuz, L.; Jérôme, C.; Sol, V.; Martin, P.; Popa, M. Curcumin-loaded polysaccharides-based complex particles obtained by polyelectrolyte complexation and ionic gelation. I-Particles obtaining and characterization. *Int. J. Biol. Macromol.* **2020**, *147*, 629–642. [CrossRef] [PubMed]
24. Hoffmann, B.; Seitz, D.; Mencke, A.; Kokott, A.; Ziegler, G. Glutaraldehyde and oxidised dextran as crosslinker reagents for chitosan-based scaffolds for cartilage tissue engineering. *J. Mater. Sci. Mater. Med.* **2009**, *20*, 1495–1503. [CrossRef] [PubMed]

25. Huang-Lee, L.L.H.; Cheung, D.T.; Nimni, M.E. Biochemical changes and cytotoxicity associated with the degradation of polymeric glutaraldehyde derived crosslinks. *J. Biomed. Mater. Res.* **1990**, *24*, 1185–1201. [CrossRef]
26. Aramwit, P.; Ekasit, S.; Yamdech, R. The development of non-toxic ionic-crosslinked chitosan-based microspheres as carriers for the controlled release of silk sericin. *Biomed. Microdevices* **2015**, *17*. [CrossRef] [PubMed]
27. Sharma, P.K.; Halder, M.; Srivastava, U.; Singh, Y. Antibacterial PEG-Chitosan Hydrogels for Controlled Antibiotic/Protein Delivery. *ACS Appl. Bio Mater.* **2019**, *2*, 5313–5322. [CrossRef]
28. Akakuru, O.U.; Isiuku, B.O. Chitosan Hydrogels and their Glutaraldehyde-Crosslinked Counterparts as Potential Drug Release and Tissue Engineering Systems-Synthesis, Characterization, Swelling Kinetics and Mechanism. *J. Phys. Chem. Biophys.* **2017**, *7*, 100256. Available online: https://www.longdom.org/open-access/chitosan-hydrogels-and-their-glutaraldehydecrosslinked-counterpartsas-potential-drug-release-and-tissue-engineering-systems--synth-2161-0398-1000256.pdf (accessed on 1 November 2020).
29. Baron, R.I.; Culica, M.E.; Biliuta, G.; Bercea, M.; Gherman, S.; Zavastin, D.; Ochiuz, L.; Avadanei, M.; Coseri, S. Physical Hydrogels of Oxidized Polysaccharides and Poly(Vinyl Alcohol) for Wound Dressing Applications. *Materials* **2019**, *12*, 1569. [CrossRef]
30. George, D.; Maheswari, P.U.; Begum, K.M.S. Synergic formulation of onion peel quercetin loaded chitosan-cellulose hydrogel with green zinc oxide nanoparticles towards controlled release, biocompatibility, antimicrobial and anticancer activity. *Int. J. Biol. Macromol.* **2019**, *132*, 784–794. [CrossRef]
31. Patrulea, V.; Ostafe, V.; Borchard, G.; Jordan, O. Chitosan as a starting material for wound healing applications. *Eur. J. Pharm. Biopharm.* **2015**, *97*, 417–426. [CrossRef]
32. Larasati, Y.A.; Yoneda-Kato, N.; Nakamae, I.; Yokoyama, T.; Meiyanto, E.; Kato, J.-Y. Curcumin targets multiple enzymes involved in the ROS metabolic pathway to suppress tumor cell growth. *Sci. Rep.* **2018**, *8*, 1–13. [CrossRef]
33. Iurciuc-Tincu, C.-E.; Cretan, M.S.; Purcar, V.; Popa, M.; Daraba, O.M.; Atanase, L.I.; Ochiuz, L. Drug Delivery System Based on pH-Sensitive Biocompatible Poly(2-vinyl pyridine)-b-poly(ethylene oxide) Nanomicelles Loaded with Curcumin and 5-Fluorouracil. *Polymers* **2020**, *12*, 1450. [CrossRef] [PubMed]
34. (Tincu), C.-E.I.; Atanase, L.; Jérôme, C.; Sol, V.; Martin, P.; Popa, M.; Ochiuz, L. Polysaccharides-Based Complex Particles' Protective Role on the Stability and Bioactivity of Immobilized Curcumin. *Int. J. Mol. Sci.* **2021**, *22*, 3075. [CrossRef] [PubMed]
35. Zheng, B.; McClements, D.J. Formulation of More Efficacious Curcumin Delivery Systems Using Colloid Science: Enhanced Solubility, Stability, and Bioavailability. *Molecules* **2020**, *25*, 2791. [CrossRef]
36. Rakotoarisoa, M.; Angelov, B.; Garamus, V.M.; Angelova, A. Curcumin- and Fish Oil-Loaded Spongosome and Cubosome Nanoparticles with Neuroprotective Potential against H_2O_2-Induced Oxidative Stress in Differentiated Human SH-SY5Y Cells. *ACS Omega* **2019**, *4*, 3061–3073. [CrossRef]
37. Zamarioli, C.M.; Martins, R.M.; Carvalho, E.C.; Freitas, L.A. Nanoparticles containing curcuminoids (*Curcuma longa*): Development of topical delivery formulation. *Rev. Bras. Farm.* **2015**, *25*, 53–60. [CrossRef]
38. Anjali, T. Modification of carboxymethyl cellulose through oxidation. *Carbohydr. Polym.* **2012**, *87*, 457–460. [CrossRef]
39. Hosoya, T.; Bacher, M.; Potthast, A.; Elder, T.; Rosenau, T. Insights into degradation pathways of oxidized anhydroglucose units in cellulose by β-alkoxy-elimination: A combined theoretical and experimental approach. *Cellulose* **2018**, *25*, 3797–3814. [CrossRef]
40. Li, H.; Wu, B.; Mu, C.; Lin, W. Concomitant degradation in periodate oxidation of carboxymethyl cellulose. *Carbohydr. Polym.* **2011**, *84*, 881–886. [CrossRef]
41. Mansouri, L.; Benghanem, S.; Elkolli, M.; Mahmoud, B. Chemical and biological behaviors of hydrogels based on oxidized carboxymethylcellulose coupled to chitosan. *Polym. Bull.* **2019**, *77*, 85–105. [CrossRef]
42. Lan, W.; He, L.; Liu, Y. Preparation and Properties of Sodium Carboxymethyl Cellulose/Sodium Alginate/Chitosan Composite Film. *Coatings* **2018**, *8*, 291. [CrossRef]
43. Peng, Y.; Liu, R.; Cao, J. Characterization of surface chemistry and crystallization behavior of polypropylene composites reinforced with wood flour, cellulose, and lignin during accelerated weathering. *Appl. Surf. Sci.* **2015**, *332*, 253–259. [CrossRef]
44. Nandiyanto, A.B.D.; Oktiani, R.; Ragadhita, R. How to Read and Interpret FTIR Spectroscope of Organic Material. *Indones. J. Sci. Technol.* **2019**, *4*, 97–118. [CrossRef]
45. Venkateshwarlu, G.; Subrahmanyam, B. Conformations of α,β-unsaturated ketones: An IR spectroscopic study. *Proc. Indian Acad. Sci. (Chem. Sci.)* **1990**, *102*, 45–50. [CrossRef]
46. Afinjuomo, F.; Fouladian, P.; Barclay, T.G.; Song, Y.; Petrovsky, N.; Garg, S. Influence of Oxidation Degree on the Physicochemical Properties of Oxidized Inulin. *Polymers* **2020**, *12*, 1025. [CrossRef] [PubMed]
47. Lisman, A.; Butruk, B.; Wasiak, I.; Ciach, T. Dextran/Albumin hydrogel sealant for Dacron(R) vascular prosthesis. *J. Biomater. Appl.* **2014**, *28*, 1386–1396. [CrossRef] [PubMed]
48. Dou, Y.; Zhang, L.; Zhang, B.; He, M.; Shi, W.; Yang, S.; Cui, Y.; Yin, G. Preparation and Characterization of Edible Dialdehyde Carboxymethyl Cellulose Crosslinked Feather Keratin Films for Food Packaging. *Polymers* **2020**, *12*, 158. [CrossRef] [PubMed]
49. Dacrory, S.; Abou-Yousef, H.; Kamel, S.; Abou-Zeid, R.E.; Abdel-Aziz, M.S.; ElBadry, M. Functionalization and cross-linking of carboxymethyl cellulose in aqueous media. *Cellul. Chem. Technol.* **2019**, *53*, 23–33. [CrossRef]
50. Dyatlov, V.; Gumnikova, V.; Grebeneva, T.; Kruppa, I.; Rustamov, I.; Kireev, V.; Maleev, V. Study of the Chemical Structure of Dialdehyde Carboxymethyl Cellulose Produced by Periodate Oxidation under Different Conditions. *Int. Polym. Sci. Technol.* **2015**, *42*, 19–26. [CrossRef]
51. Gu, H.; He, J.; Huang, Y.; Guo, Z. Fabrication of oxidized sodium carboxymethylcellulose from viscose fibers and their viscosity behaviors. *Fibers Polym.* **2013**, *14*, 1266–1270. [CrossRef]

52. Kono, H.; Oshima, K.; Hashimoto, H.; Shimizu, Y.; Tajima, K. NMR characterization of sodium carboxymethyl cellulose: Substituent distribution and mole fraction of monomers in the polymer chains. *Carbohydr. Polym.* **2016**, *146*, 1–9. [CrossRef]
53. Hou, Q.X.; Liu, W.; Liu, Z.H.; Bai, L.L. Characteristics of Wood Cellulose Fibers Treated with Periodate and Bisulfite. *Ind. Eng. Chem. Res.* **2007**, *46*, 7830–7837. [CrossRef]
54. Eremeeva, T.; Bykova, T. SEC of mono-carboxymethyl cellulose (CMC) in a wide range of pH.; Mark–Houwink constants. *Carbohydr. Polym.* **1998**, *36*, 319–326. [CrossRef]
55. Marin, L.; Harabagiu, V.; Van Der Lee, A.; Arvinte, A.; Barboiu, M. Structure-directed functional properties of symmetrical and unsymmetrical Br-substituted Schiff-bases. *J. Mol. Struct.* **2013**, *1049*, 377–385. [CrossRef]
56. Cheaburu-Yilmaz, C.N.; Dumitriu, R.P.; Nistor, M.-T.; Lupusoru, C.; Popa, M.I.; Profire, L.; Vasile, C. Biocompatible and Biodegradable Chitosan/Clay Nanocomposites as New Carriers for Theophylline Controlled Release. *Br. J. Pharm. Res.* **2015**, *6*, 228–254. [CrossRef]
57. Dovbeshko, G.I.; Gridina, N.Y.; Kruglova, E.B.; Pashchuk, O.P. FTIR spectroscopy studies of nucleic acid damage. *Talanta* **2000**, *53*, 233–246. [CrossRef]
58. Fukuda, H. Polyelectrolyte complexes of chitosan with sodium carboxymethylcellulose. *Bull. Chem. Soc. Jpn.* **1980**, *53*, 837–840. [CrossRef]
59. Benghanem, S.; Chetouani, A.; Elkolli, M.; Bounekhel, M.; Benachour, D. Grafting of oxidized carboxymethyl cellulose with hydrogen peroxide in presence of Cu(II) to chitosan and biological elucidation. *Biocybern. Biomed. Eng.* **2017**, *37*, 94–102. [CrossRef]
60. Lee, S.H. Preparation and Evaluation of Crosslinked Polyelectrolyte Complex Membranes. *Polym. J.* **2000**, *32*, 716–721. [CrossRef]
61. Fukuda, H.; Kikuchi, Y. Polyelectrolyte Complexes of Sodium Carboxymethylcellulose with Chitosan. *Makromol. Chem.* **1979**, *180*, 1631–1633. Available online: https://onlinelibrary.wiley.com/doi/pdf/10.1002/macp.1979.021800629 (accessed on 1 March 2021). [CrossRef]
62. Strätz, J.; Liedmann, A.; Trutschel, M.-L.; Mäder, K.; Groth, T.; Fischer, S. Development of hydrogels based on oxidized cellulose sulfates and carboxymethyl chitosan. *Cellulose* **2019**, *26*, 7371–7382. [CrossRef]
63. Lü, S.; Liu, M.; Ni, B. An injectable oxidized carboxymethylcellulose/N-succinyl-chitosan hydrogel system for protein delivery. *Chem. Eng. J.* **2010**, *160*, 779–787. [CrossRef]
64. Ritger, P.L.; Peppas, N.A. A simple equation for description of solute release I. Fickian and non-fickian release from non-swellable devices in the form of slabs, spheres, cylinders or discs. *J. Control. Release* **1987**, *5*, 23–36. [CrossRef]
65. Wang, Q.Z.; Chen, X.G.; Liu, N.; Wang, S.X.; Liu, C.S.; Meng, X.H. Protonation constants of chitosan with different molecular weight and degree of deacetylation. *Carbohydr. Polym.* **2006**, *65*, 194–201. [CrossRef]
66. Szymańska, E.; Winnicka, K. Stability of Chitosan—A Challenge for Pharmaceutical and Biomedical Applications. *Mar. Drugs* **2015**, *13*, 1819–1846. [CrossRef] [PubMed]
67. Agrawal, R.; Sandhu, S.K.; Sharma, I.; Kaur, I.P. Development and Evaluation of Curcumin-loaded Elastic Vesicles as an Effective Topical Anti-inflammatory Formulation. *AAPS PharmSciTech* **2014**, *16*, 364–374. [CrossRef] [PubMed]
68. Ternullo, S.; Werning, L.V.S.; Holsæter, A.M.; Škalko-Basnet, N. Curcumin-In-Deformable Liposomes-In-Chitosan-Hydrogel as a Novel Wound Dressing. *Pharmaceutics* **2019**, *12*, 8. [CrossRef] [PubMed]
69. Kulikowska, A.; Wasiak, I.; Ciach, T. Carboxymethyl Cellulose Oxidation to Form Aldehyde Groups. *Biomed. Eng.* **2013**, *4*, 11–18. Available online: http://yadda.icm.edu.pl/baztech/element/bwmeta1.element.baztech-4e71d122-c444-458c-9083-57e262ced33f (accessed on 1 May 2018).
70. Caner, C.; Vergano, P.J.; Wiles, J.L. Chitosan Film Mechanical and Permeation Properties as Affected by Acid, Plasticizer, and Storage. *J. Food Sci.* **2006**, *63*, 1049–1053. [CrossRef]
71. McSweeny, J.D.; Rowell, R.M.; Chen, G.C.; Eberhardt, T.L.; Min, S.-H. Periodate and hypobromite modification of southern pine wood to improve sorption of copper ion. *Bioresources* **2008**, *3*, 204–216. Available online: https://bioresources.cnr.ncsu.edu//BioRes_03/BioRes_03_1_0204_McSweeny_RCEM_Periodate_Pine_Copper.pdf (accessed on 1 May 2018).
72. Dean, J.A. *Lange's Handbook of Chemistry*, 15th ed.; McGraw-Hill Book Company, Inc.: New York, NY, USA; Toronto, ON, Canada; London, UK, 1999; p. 1292. Available online: http://depa.fquim.unam.mx/amyd/archivero/ManualdeLange_9164.pdf (accessed on 1 May 2018).
73. Choi, C.W.; Kim, S.C.; Hwang, S.S.; Choi, B.K.; Ahn, H.J.; Lee, M.Y.; Park, S.H.; Kim, S.K. Antioxidant activity and free radical scavenging capacity between Korean medicinal plants and flavonoids by assay-guided comparison. *Plant Sci.* **2002**, *163*, 1161–1168. [CrossRef]

molecules

MDPI

Article

A Theoretical Model for Release Dynamics of an Antifungal Agent Covalently Bonded to the Chitosan

Luminita Marin [1], Marcel Popa [2,3], Alexandru Anisiei [1], Stefan-Andrei Irimiciuc [4,*], Maricel Agop [3,5,*], Tudor-Cristian Petrescu [6], Decebal Vasincu [7] and Loredana Himiniuc [8]

1 "Petru Poni" Institute of Macromolecular Chemistry, 41A Gr. Ghica Voda Street, 700487 Iasi, Romania; lmarin@icmpp.ro (L.M.); anisiei.alexandru@icmpp.ro (A.A.)
2 Department of Natural and Synthetic Polymers, "Gheorghe Asachi" Technical University of Iasi, 700050 Iasi, Romania; marpopa@ch.tuiasi.ro
3 Academy of Romanian Scientists, 54 Splaiul Independentei, 050094 Bucharest, Romania
4 National Institute for Laser, Plasma and Radiation Physics, 409 Atomistilor Street, 077125 Bucharest, Romania
5 Department of Physics, "Gh. Asachi" Technical University of Iasi, 700050 Iasi, Romania
6 Department of Structural Mechanics, "Gh. Asachi" Technical University of Iasi, 700050 Iasi, Romania; tudor.petrescu@tuiasi.ro
7 Department of Biophysics, Faculty of Dental Medicine, "Grigore T. Popa" University of Medicine and Pharmacy, 16 University Str., 700115 Iasi, Romania; decebal.vasincu@umfiasi.ro
8 Department of Obstetrics and Gynecology, "Grigore T. Popa" University of Medicine and Pharmacy Iasi, 16 Universitatii Str., 700115 Iasi, Romania; loredanahiminiuc@gmail.com
* Correspondence: stefan.irimiciuc@inflpr.ro (S.-A.I.); magop@tuiasi.ro (M.A.)

Abstract: The aim of the study was to create a mathematical model useful for monitoring the release of bioactive aldehydes covalently bonded to the chitosan by reversible imine linkage, considered as a polymer–drug system. For this purpose, two hydrogels were prepared by the acid condensation reaction of chitosan with the antifungal 2-formyl-phenyl-boronic acid and their particularities; influencing the release of the antifungal aldehyde by shifting the imination equilibrium to the reagents was considered, i.e., the supramolecular nature of the hydrogels was highlighted by polarized light microscopy, while scanning electron microscopy showed their microporous morphology. Furthermore, the in vitro fungicidal activity was investigated on two fungal strains and the in vitro release curves of the antifungal aldehyde triggered by the pH stimulus were drawn. The theoretical model was developed starting from the hypothesis that the imine-chitosan system, both structurally and functionally, can be assimilated, from a mathematical point of view, with a multifractal object, and its dynamics were analyzed in the framework of the Scale Relativity Theory. Thus, through Riccati-type gauges, two synchronous dynamics, one in the scale space, associated with the fungicidal activity, and the other in the usual space, associated with the antifungal aldehyde release, become operational. Their synchronicity, reducible to the isomorphism of two SL(2R)-type groups, implies, by means of its joint invariant functions, bioactive aldehyde compound release dynamics in the form of "kink–antikink pairs" dynamics of a multifractal type. Finally, the theoretical model was validated through the experimental data.

Keywords: mathematical model; chitosan; bioactive aldehyde release; Scale Relativity Theory; multifractal; Riccati gauge; joint invariant function; SL(2R)-type group

Citation: Marin, L.; Popa, M.; Anisiei, A.; Irimiciuc, S.-A.; Agop, M.; Petrescu, T.-C.; Vasincu, D.; Himiniuc, L. A Theoretical Model for Release Dynamics of an Antifungal Agent Covalently Bonded to the Chitosan. *Molecules* **2021**, *26*, 2089. https://doi.org/10.3390/molecules26072089

Academic Editor: Dimitrios Bikiaris

Received: 4 March 2021
Accepted: 1 April 2021
Published: 6 April 2021

1. Introduction

Chitosan is a polysaccharide extensively investigated over the last decades due to its outstanding biologic properties, which make it an excellent candidate for the development of a large variety of biomaterials [1]. However, the intrinsic properties of chitosan are not sufficiently explored, in order to assure curing ability, and they should be improved by further chemical modifications [2,3]. Among various modification routes, the acid condensation reaction of chitosan with various aldehydes demonstrated efficiency towards a large realm of materials such as films, hydrogels and nanoparticles, in view of application in domains of contemporary interest, such as tissue engineering [4], drug delivery [5], soil conditioners [6], cosmetics [7], adsorption of carbon dioxide [8] or other hazardous pollutants [9], heavy metal sensing [10] and so on. However, recent studies demonstrated that the imination reaction of chitosan has a reversibility degree favored by the acidic medium in which chitosan is soluble [3]. This finding is in line with well-documented investigations that demonstrated that the imine linkage has a reversible character that proved beneficial for the construction of dynamic materials with an ability to adapt under the influence of various environment stimuli, such as moisture, temperature, pressure or pH [11–13]. These recent discoveries opened new perspectives for the use of chitosan as a matrix for the controlled release of the reversible bonded aldehydes [3,14]. Under the influence of environmental stimuli, the imination equilibrium is shifted to the reagents releasing the aldehyde, and the aldehyde consumption further triggers the equilibrium shifting towards the reagents and consequently towards the delivery of new aldehyde amounts. This simple mechanism becomes of particular importance when the released aldehyde has bioactive properties and its delivery is controlled by its consumption.

Usually, the models used to describe drug delivery dynamics are based on a combination of basic theories, derived especially from physics and computer simulations [15–18]. In such a conjecture, their description implies both computational simulations based on specific algorithms [18], as well as developments on standard theories. A class of models was developed on spaces with integer dimension—i.e., differentiable models, such as the Zero order model, First order model, Higuchi model, Hixson–Crowell model, Korsmeyer–Peppas model, etc. [19]. Another class of models was developed on spaces with non-integer dimensions and is explicitly written through fractional derivatives [20,21]—i.e., non-differentiable models, with examples including the fractal models [22]. Expanding on the last class of models, new developments have been made, based on Scale Relativity Theory, either in the monofractal dynamics, as in the case of Nottale [23], or in the multifractal dynamics, as in the case of the Multifractal Theory of Motion [24,25]. Our group has recently published in this framework, proving a good match for describing various drug delivery systems [26,27].

Both in the context of Scale Relativity Theory in the sense of Nottale [23], as well as in the one of Multifractal Theory of Motion [24,25], supposing that any polymer–drug system is assimilated both structurally and functionally to a multifractal object, the said dynamics can be described through motions of the polymer–drug structural units, dependent on the chosen scale resolution, on continuous and non-differentiable curves (multifractal curves). Such an assumption may be illustrated by considering the following scenario: between two successive interactions of the polymer–drug structural units, the trajectory of the polymer–drug structural unit is a straight line that becomes non-differentiable in the impact point. Considering that all interaction points form an uncountable set of points, it results that the trajectories of the polymer–drug structural units become continuous and non-differentiable (i.e., fractal curves). Clearly, the reality is much more complicated, taking into account both the diversity of the structural units which compose the polymer–drug system and the various interactions between them in the form of double interactions etc. Extrapolating the previous reasoning for any polymer–drug system, including our imine-chiotsan system, it results that it can be assimilated to a multifractal. The multifractal model is used here to characterize the dynamic of the atoms and molecule projecting their trajectories in the multifractal plane. Details on the fundamental links between the molecular or atomistic characteristics with the Multifractal Scale Relativity Theory can be found in [23–25].

All these considerations imply that in the description of the imine-chitosan dynamics, instead of "working" with a single variable (regardless of its nature, i.e., velocity, density, etc.) described by a strict non-differentiable function, it is possible to "work" only with approximations of this mathematical function, obtained by averaging them on different-scale resolutions. As a consequence, any variable purposed to describe the imino-chitosan dynamics will perform as the limit of a family of mathematical functions, this being non-differentiable for null scale resolutions and differentiable otherwise [23] (from a mathematical point of view, these variables can be explained through multifractal functions, i.e., functions dependent not only on spatial and temporal coordinates, but also on the scale resolution).

Since for a large temporal scale resolution with respect to the inverse of the highest Lyapunov exponent [28,29], the deterministic trajectories of any structural unit belonging to the polymer–drug system can be replaced by a collection of potential ("virtual") trajectories, the concept of definite trajectory can be substituted by the one of probability density.

With all of the above considerations taken into account, the multifractality expressed through stochasticity, in the description of the dynamics of any polymer–drug system, becomes operational in the multifractal paradigm through the Multifractal Theory of Motion.

In this context, the authors' study was directed to the modeling of the release of an antifungal aldehyde covalently bonded to chitosan by imine units. The in vitro release of aldehyde was investigated in an acidic medium of pH 4.2, characteristic to the vagina, which is highly susceptible to fungal infections. A mathematical model was created considering the imine-chitosan system as a multifractal object, and analyzing its dynamics in the framework of Scale Relativity Theory, using various operational procedures (Riccati-type gauges, isomorphisms of groups, joint invariant functions on groups, etc.). Finally, the theoretical model was validated by means of the experimental data.

2. Experiment

2.1. Materials

2-Formyl-phenyl-boronic acid, low molecular weight chitosan, and phosphate buffer solution were purchased from Aldrich (Sigma-Aldrich, Munchen, Germany) and used as received. The reagents used in antifungal measurements were purchased from Sigma-Aldrich (Sigma-Aldrich, Munchen, Germany) and used as received.

2.2. Synthesis of Imino-Chitosan Derivatives

Two chitosan derivatives with different aldehyde content were synthetized by reacting chitosan with 2-formyl-phenyl-boronic acid in a homogeneous medium, at 55 °C during 3 h, as described in reference [14]. Briefly, a 1% 2-formyl-phenyl-boronic acid in ethanol was slowly dropped into a 2% solution of chitosan in 0.7% acetic acid solution, under vigorous stirring at 55 °C for 3 h, when transparent yellowish hydrogels were obtained. Varying the ratio between the two reagents, two hydrogels containing 0.071% (coded C0.071) and 0.142% (coded C0.142) 2-formyl-phenyl-boronic were prepared.

2.3. Methods

The morphology of the hydrogels was investigated with a field emission Scanning Electron Microscope SEM EDAX Quanta 200 (FEI Company, U.S., Hillsboro, Oregon) at an accelerated electron energy of 20 KeV on small pieces of lyophilized hydrogel.

The supramolecular architecture of the hydrogels was assessed by polarized light microscopy on small hydrogel pieces placed between two lamellae, with an Olympus BH-2 polarized light microscope (Olympus BH-2 (Olympus company) Japan, Tokyo).

The swelling behavior of the two hydrogels was investigated by measuring the mass equilibrium swelling in an acidic medium of pH 4.2 [14].

The in vitro release of the antifungal aldehyde by shifting the imination equilibrium of the reagents was observed in an acidic medium of pH 4.2 similar to the vagina environment, by monitoring the absorbance of the aldehyde by UV-Vis and fitting it on a previously drawn calibration curve. Briefly, pieces of hydrogels were immersed in a buffer solution of pH 4.2. From time to time, 1 mL of supernatant was withdrawn and replaced with fresh buffer. The extracted supernatant was analyzed by UV-Vis spectroscopy and its concentration was found by fitting the absorbance on the calibration curve. Based on the obtained data, the release profile was drawn.

3. Empirical Data

Two imino-chitosan derivatives were synthetized as hydrogels by reaction of chitosan with 2-formyl-phenyl-boronic acid through imination reaction. The hydrogelation occurred by two concurrent processes: (i) formation of imine units, and (ii) their supramolecular self-ordering into ordered layers [14]. This hydrogelation pathway was supported by the polarized light microscopy measurements, which displayed fine banded textures characteristic to the layered architectures, in line with the formation of ordered clusters of imino-chitosan, similar to smectic mesophases (Figure 1a,b) [30,31]. The hydrogel nature of the imino-chitosan derivatives was further demonstrated by scanning electron microscopy images, which showed a microporous morphology with pores of diameter in the 10–30 μm range, suitable for reaching sink conditions, during the release of antifungal aldehyde (Figure 1c,d).

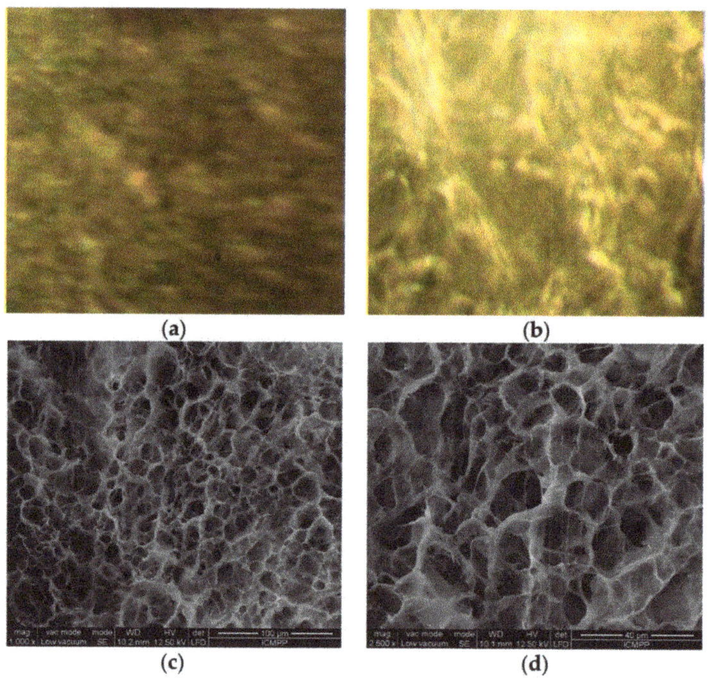

Figure 1. Representative (**a**,**b**) POM and (**c**,**d**) SEM images of the imino-chitosan hydrogels.

The hydrogels swelled in an acidic buffer of pH 4.2, reaching a mass equilibrium swelling of approximately 27, in line with a moisture medium inside the hydrogels that is favorable for the antifungal aldehyde release [32]. The in vitro investigation of the antifungal activity showed an excellent effect against *Candida albicans* and *Candida glabrata* strains [14], two virulent fungi accounting for systemic vulvovaginitis infections affecting women's health [33]. While the chitosan reference sample only reduced the fungi growth, the two hydrogels showed a progressive killing of the strains, causing their almost complete extinction in the case of the C0.142 sample (Figure 2a). The decrease of the fungi population along with the increase of the amount of antifungal aldehyde in the samples indicated the aldehyde as a promoter of the antifungal activity. To confirm this, the in vitro release of aldehyde from the hydrogel samples has been monitored in a synthetic vagina-simulative medium, using the UV-Vis method. The release curves (Figure 2b) displayed a burst release in the first four hours, followed by a prolonged release. This profile shows that the hydrogel samples can assure the initial bolus dose of antifungal agent and further a constant dose necessary to kill the strains [34]. Comparing the in vitro antifungal effect with the in vitro release rate of the aldehyde, a close correlation can be clearly seen, meaning that the higher release rate of aldehyde from the C0.142 sample impacted a higher killing rate of the fungi. This means that the key role for the antifungal activity was played by the slow release of the antifungal aldehyde, mainly triggered by its reversible bonding on the chitosan under the pH stimulus. For a deeper understanding of this phenomenon, a mathematical model has been developed.

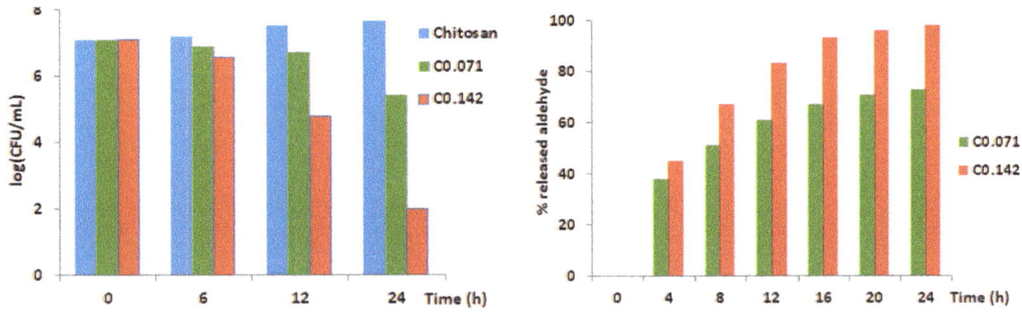

Figure 2. (**a**) Graphic representation of the in vitro fungicidal activity against planktonic yeast of *Candida albicans*; (**b**) The percent of antifungal aldehyde in vitro released during 24 h.

4. Theoretical Considerations

Taking into account the theoretical aspects presented in Section 1 by pursuing the transition approach from the scale space to the usual one, the imino-chitosan system release dynamics will imply the following:

(i) Aldehyde compound release dynamics in the scale space, dynamics which will be assimilated with the fungicidal activity;
(ii) Aldehyde compound release dynamics in the usual space, dynamics which will be assimilated with the antifungal aldehyde release;
(iii) Aldehyde compound release dynamics, associated with the transition from the scale space to the usual space, as a global release mechanism.

4.1. Aldehyde Compound Delivery Dynamics in Scale Space

Let a multifractal function $F(x)$ be considered with $x \in [a, b]$, which can be associated with any multifractal variable that describes drug delivery dynamics. Now, the sequence of the values of the variable x is considered:

$$x_a = x_0,\ x_1 = x_0 + \varepsilon, \ldots,\ x_k = x_0 + k\varepsilon,\ \ldots,\ x_n = x_0 + n\varepsilon = x_b \quad (1)$$

$F(x, \varepsilon)$ shall denote the fractured (broken) line connecting the points:

$$F(x_0),\ \ldots,\ F(x_k),\ \ldots,\ F(x_n) \quad (2)$$

The broken line will be considered as an approximation that is different from the one used before. Let it be noted that $F(x, \varepsilon)$ is an ε-approximation scale. Now, the $\bar{\varepsilon}$-approximation scale $F(x, \bar{\varepsilon})$ of the same function is considered. Since $F(x)$ is self-similar almost everywhere, if ε and $\bar{\varepsilon}$ are small enough, then the two approximations $F(x, \varepsilon)$ and $F(x, \bar{\varepsilon})$ must lead to the same results when a multifractal drug release dynamics by approximations is studied. If the two cases are compared, then to an infinitesimal increase or decrease $d\varepsilon$ of ε, an increase or decrease $d\overline{\varepsilon}$ for $\overline{\varepsilon}$ corresponds, if the scale is dilated or contracted. In this case:

$$\frac{d\varepsilon}{\varepsilon} = \frac{d\overline{\varepsilon}}{\overline{\varepsilon}} \tag{3a}$$

that is,

$$\frac{d\varepsilon}{\varepsilon} = d\mu \tag{3b}$$

is the ratio of the scale $\varepsilon + d\varepsilon$ and $d\varepsilon$ must be preserved. Then, it is possible to consider the infinitesimal transformation of the scale as:

$$\varepsilon' = \varepsilon + d\varepsilon = \varepsilon + \varepsilon d\mu \tag{4}$$

By such a transformation, in the case of the function $F(x, \varepsilon)$, the following results:

$$F(x, \varepsilon') = F(x, \varepsilon + \varepsilon d\mu) \tag{5}$$

Respectively, if a stop is made after the first approximation:

$$F(x, \varepsilon') = F(x, \varepsilon) + \frac{\partial F}{\partial \varepsilon}(\varepsilon' - \varepsilon) \tag{6}$$

that is:

$$F(x, \varepsilon') = F(x, \varepsilon) + \frac{\partial F}{\partial \varepsilon}\varepsilon d\mu \tag{7}$$

Let it be considered that for an arbitrary but fixed ε_0:

$$\frac{\partial \ln\left(\frac{\varepsilon}{\varepsilon_0}\right)}{\partial \varepsilon} = \frac{\partial(\ln \varepsilon - \ln \varepsilon_0)}{\partial \varepsilon} = \frac{1}{\varepsilon} \tag{8}$$

Thus, Equation (7) becomes:

$$F(x, \varepsilon') = F(x, \varepsilon) + \frac{\partial F(x, \varepsilon)}{\partial \ln\left(\frac{\varepsilon}{\varepsilon_0}\right)} d\mu \tag{9}$$

Finally:

$$F(x, \varepsilon') = \left(1 + \frac{\partial}{\partial \ln\left(\frac{\varepsilon}{\varepsilon_0}\right)} d\mu\right) F(x, \varepsilon) \tag{10}$$

The operator:

$$\hat{D} = \frac{\partial}{\partial \ln\left(\frac{\varepsilon}{\varepsilon_0}\right)} \tag{11}$$

is a dilation or contraction operator.

This is the well-known form of the infinitesimal dilation operator, obtained above through the Gell–Mann–Levy method, which allows the finding of the currents corresponding for a given symmetry [35].

This clearly shows that the natural variable for the resolution is $\ln\left(\frac{\varepsilon}{\varepsilon_0}\right)$ and that the expected new differential equations involve quantities like $\frac{\partial F(x,\varepsilon)}{\partial \ln\left(\frac{\varepsilon}{\varepsilon_0}\right)}$.

In the previous context,

$$\frac{\partial F(x, \varepsilon)}{\partial \ln\left(\frac{\varepsilon}{\varepsilon_0}\right)} = 0 \tag{12}$$

corresponds to the scale symmetry invariance, while the relation

$$\frac{\partial F(x,\varepsilon)}{\partial \ln\left(\frac{\varepsilon}{\varepsilon_0}\right)} \neq 0 \tag{13}$$

corresponds to the scale symmetry breaking. Since various problems related to complex system dynamics can be reduced to a Riccati-type gauge [15–17,28,29], in (13) such a gauge will be used, in the form of a Riccati-type equation:

$$\frac{dw}{ds} - \frac{w^2}{M} + \frac{2R}{M}w - K = 0 \tag{14}$$

where:

$$F(x,\varepsilon) \equiv w(s),\ s \equiv \ln\left(\frac{\varepsilon}{\varepsilon_0}\right) \tag{15}$$

and M, R, and K are constants of multifractal type introduced through external constrictions. Since the roots of the polynomial

$$P(w) = \frac{w^2}{M} + \frac{2R}{M}w - K \tag{16}$$

can be written in the form

$$w_0 = R + iM\Omega, \overline{w}_0 = R - iM\Omega,\ \Omega^2 = \frac{K}{M} - \left(\frac{R}{M}\right)^2 \tag{17}$$

the change of variable:

$$z = \frac{w - w_0}{w - \overline{w}_0} \tag{18}$$

transforms (14) into:

$$\dot{z} = 2i\Omega z \tag{19}$$

of the solution:

$$z(\tau) = z(0)e^{2i\Omega s} \tag{20}$$

As such, if the initial condition $z(0)$ is conveniently expressed, and then it is possible to construct the general solution of (14), by writing the transformation (18) in the form:

$$w_0 = \frac{w_0 + re^{2i\Omega s}\overline{w}_0}{1 + re^{2i\Omega s}} \tag{21}$$

where r is an integration constant of multifractal type. Using (17) it is possible to write this solution in real terms in the form:

$$z = R + M\Omega\left\{\frac{2r\sin[2\Omega s]}{1 + r^2 + 2r\cos[2\Omega s]} + i\frac{1 - r^2}{1 + r^2 + 2r\cos[2\Omega s]}\right\} \tag{22}$$

which can highlight various modes of scale symmetry breaking. These various modes are presented in Figure 3a–l. In the present context, it results that the natural transition in the aldehyde compound release dynamics is to evolve from a normal period doubling state towards damped oscillating and strong modulated dynamics. The imino-chitosan system never reaches a chaotic state, but it permanently evolves towards that state. There is a periodicity to the whole series of transitions and the imino-chitosan system evolves through period doubling, damped oscillations even reaching in some cases an intermittence state, but it never reaches a pure chaotic state. The evolution of the imino-chitosan system sees a "jump" into a period doubling oscillation state and the transition resumes towards a quasi-chaotic state. These dynamics are characteristic for short times and small distances, and are not usually seen for long time measurements. Such dynamics are true for a specific part of the release scenario being seen usually at the release interface and are strongly related with the polymeric structure and the multifractality of the matrix. The impossibility of reaching a chaotic state means that for a wide range of imino-chitosn systems, the release can be controlled to a high degree and there is no possibility for chaotic release and possible damage.

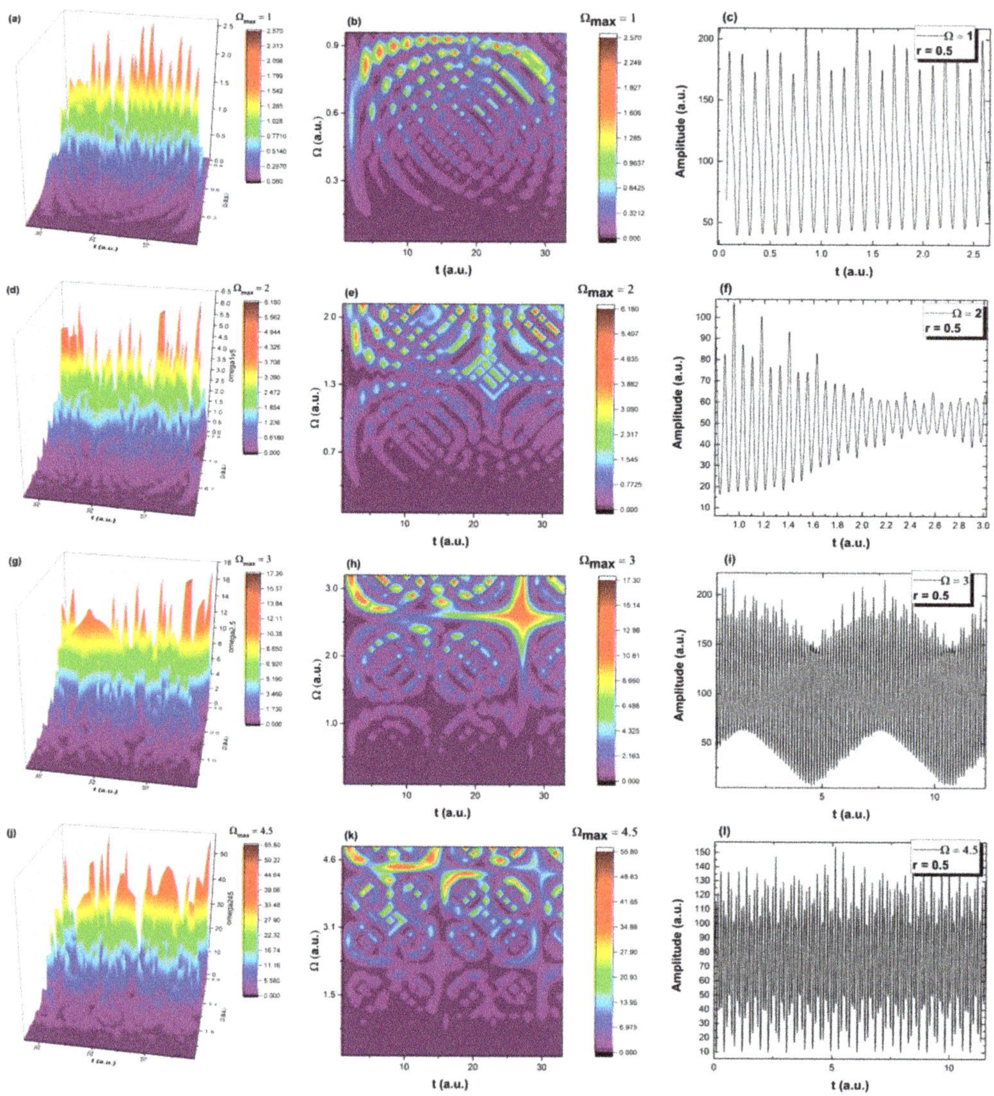

Figure 3. Transition scenarios in the aldehyde compound release dynamics and mimed as various modes of scale symmetry breaking (3D, contour plot and time series representation for $s \equiv t$: period doubling (**a–c**), damped oscillation region (**d–f**), signal modulation (**g–i**) and chaotic behavior (**j–l**)).

To better highlight this important stability property of the imino-chiotosan system, a study on the bifurcation map was attempted. It is observed that the imino-chitosan system starts from a steady state (double period state) and evolves towards pseudo-chaotic states ($\Omega_{max} = 2, 2.5, 3\ldots$) but it never reaches that state. Here, the pseudo-chaos is defined by a high density of oscillations frequencies on which the imino-chitosan system oscillates simultaneously. This could be a measure of aldehyde release processes stability at a small space-time scale. Each train of scenarios follows the same path presented in Figure 3: double period oscillations, damped, modulated and quasi-chaotic oscillation, without ever reaching chaos. This means that at a microscale, the release could be different from the traces seen experimentally. The continuous increase in the amplitude of the oscillation reflects well the increase in released aldehyde compound mass as time evolves.

Let the implementation of the multifractal model at large scales be explored further, by introducing the following notation:

$$r = \coth\mu \tag{23}$$

Then, (22) becomes:

$$z = R + M\Omega h \tag{24}$$

with

$$h = -i\frac{\cosh\mu + re^{2i\Omega s}\sinh\mu}{\cosh\mu + re^{2i\Omega s}\sinh\mu} \tag{25}$$

The significance of this parameter will be addressed later when aldehyde compound release dynamics will be analyzed. For the moment, it will be noted that the previous transition scenarios in the aldehyde compound release dynamics mimed as various modes of the scale symmetry breaking can be interpreted as phase self-modulation processes at various scale resolutions.

Let it be noted that the Ricatti-type differential Equation (14), rewritten in the notations:

$$M = 2\mu, \ \ R = \pm c, \ \ KM = -d$$

in the form:

$$2\mu\frac{dw}{ds} = w^2 \pm 2cu - d \tag{26}$$

admits, with the restriction $d + c^2 = -KM + R^2 > 0$, the bounded solution

$$s = -\frac{2\mu}{(d+c^2)^{\frac{1}{2}}}\tanh^{-1}\left[\frac{w \pm c}{(d+c^2)^{\frac{1}{2}}}\right] = -\frac{M}{(-KM+R^2)^{\frac{1}{2}}}\tanh^{-1}\left[\frac{w \pm R}{(-KM+R^2)^{\frac{1}{2}}}\right] \tag{27}$$

By setting:

$$A = \left(d + c^2\right)^{\frac{1}{2}} = \left(-KM + R^2\right)^{\frac{1}{2}} \tag{28}$$

Then

$$w(s) = c \pm A\tanh\left(\frac{As}{2\mu}\right) = R \pm \left(-KM + R^2\right)^{\frac{1}{2}}\tanh\left[\frac{(-KM+R^2)^{\frac{1}{2}}s}{M}\right] \tag{29}$$

Thus, there are specified behaviors of both multifractal kink type (+) as well as multifractal antikink type (−) in the drug release dynamics. For details on the standard kink and antikink solution see [28,29]. In Figure 4a,b, the multifractal kink and antikink aldehyde compound release modes of the imino-chitosan system are presented. The kink and antikink solutions found here describe two facets of the same problem. In the fractal paradigm, the aldehyde compound release processes can be best defined by the kink-type behavior. The simultaneous existence of both types of solution means that in the multifractal representation of the aldehyde release phenomena of the imino-chitosan system, ripples of the localized chemical interaction between the aldehyde compound and the medium can be present. For the case presented in this article, the decrease of the fungi population is reflected by the antikink type solution. Let it be noted that both solutions may exist at the same time, each contributing to create the complete image of the aldehyde release in the context of antifungal application.

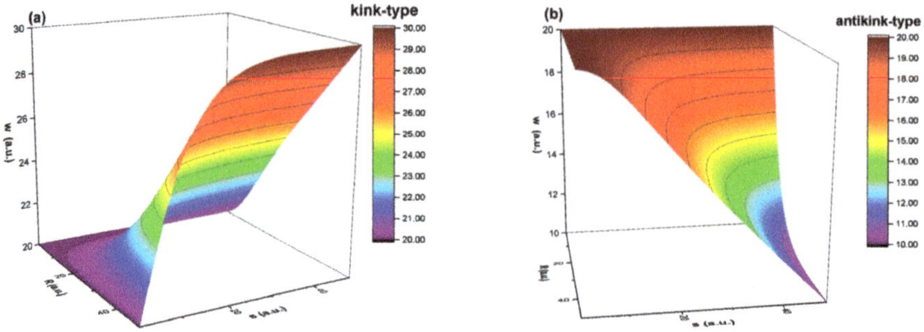

Figure 4. The multifractal kink (**a**) and antikink aldehyde (**b**) release modes.

Moreover, if the Ricatti-type differential Equation (14) is rewritten in the notations:

$$M = -\frac{1}{f}, \ R = \frac{1}{2}, \ K = 0 \tag{30}$$

It will form logistic-type equations:

$$\frac{dw}{ds} = fw(1 - w) \tag{31}$$

Its linearization by means of multiplication with $\frac{1}{w^2}$, i.e.,

$$\frac{dn}{ds} = fn(1 - n), \ n = \frac{1}{w} \tag{32}$$

implies, by means of integration, the solution:

$$w = \frac{1}{1 - \left(1 - \frac{1}{w_0}\right)exp[-fs]} \tag{33}$$

where w_0 is an integration constant of multifractal type. Thus, the increase (w) of aldehyde release dynamics of the imino-chitosan system is restricted by the self-interaction effects (w^2) (the "finite world" effect). In Figure 5 the logistic-type dependence of aldehyde compound release dynamics from the imino-chitosan system are presented. It can be seen that around a fixed value for w_0 it is possible to observe the complete representation of the process. As the system reaches this optimal value, the aldehyde compound release is enhanced and reaches a maximum at $w_0 - \Delta w$. It can be seen that there is a negative contribution at $w_0 + \Delta w$ following the increase in aldehyde compound release, which means the medium changes (i.e., the fungal strains are killed changing the fractality of the background media in which the aldehyde compound are released). The symmetry of the solution highlights the synchronicity of the two phenomena and the coupling between them.

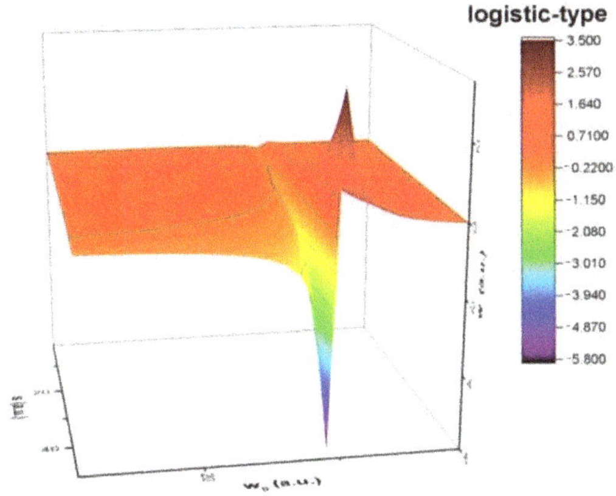

Figure 5. The logistic-type dependence of aldehyde release dynamics.

4.2. Aldehyde Compound Delivery Dynamics in the Usual Space

Considering the important role of the Riccati-type gauge in the dynamic analysis of drug release scenario in the space scale, in the following, such an analysis is extended for the aldehyde compound release dynamics in the usual space. Such an approach can be implemented through a Schrödinger equation of multifractal type:

$$\lambda^2 (dt)^{\left[\frac{4}{f(\alpha)}\right]-2} \partial^l \partial_l \Psi + i\lambda (dt)^{\left[\frac{2}{f(\alpha)}\right]-1} \partial_t \Psi = 0 \tag{34}$$

where:

$$\partial_t = \frac{\partial}{\partial_t},\ \partial_l = \frac{\partial}{\partial X^l},\ \partial_l \partial^l = \frac{\partial}{\partial X^l}\left(\frac{\partial}{\partial X^l}\right) \tag{35}$$

based on a special invariance of the above equation, through the transformations [36]:

$$X' = \frac{X}{\gamma t + \delta},\ t' = \frac{\alpha t + \beta}{\gamma t + \delta} \tag{36}$$

where α, β, γ, and δ are real elements.

In the above relation, ψ is the multifractal state function, X^l with $l = 1,2,3$ are the multifractal spatial coordinates, t is a non-multifractal temporal coordinate having the affine parameter role on the movement curves, dt is the scale resolution, λ is a coefficient associated to the multifractal-non-multifractal transition, $f(\alpha)$ is the singularity spectrum of order α, and α is the singularity index through which the fractal dimension D_F is specified (for D_F it is possible to use use any definitions—Kolmogorov fractal dimension, Hausdorff–Besikovich fractal dimension, etc. [37]; it is regularly found that $D_F < 2$ for correlative processes and $D_F > 2$ for non-correlative processes). From such a perspective, through $f(\alpha)$ it is possible to identify not only the aldehyde compound release volumes that are characterized by a certain fractal dimension (i.e., the case of monofractal drug release dynamics) but also the aldehyde compound release quantity for which the fractal dimension is situated in an interval of values (i.e., the case of multifractal drug release dynamics). Moreover, for the same $f(\alpha)$, it is possible to identify classes of universality in the aldehyde compound release dynamics laws, even when regular or strange attractors have various aspects [28,29].

Let it be observed that the transformation (36b) represents the homographic action of a matrix:

$$\hat{\alpha} = \begin{pmatrix} \alpha & \beta \\ \gamma & \delta \end{pmatrix} \tag{37}$$

In such a context, the aldehyde release dynamics analysis in usual space is reduced to the obtainment of a relation between the matrix ensemble $\hat{\alpha}$ and an ensemble of t values through which t' remains constant. Geometrically, this implies the searching of the ensemble of points (α, β, γ, δ), univocally corresponding to the values of the parameter t. Using (36b), the solution of the problem is reduced to a Riccati-type gauge in the form:

$$dt + \omega_1 t^2 + \omega_2 t + \omega_3 = 0 \tag{38}$$

where the following notations are used [36]:

$$\omega_1 = \frac{\gamma d\alpha - \alpha d\gamma}{\Delta}, \omega_2 = \frac{\delta d\alpha - \alpha d\delta + \gamma d\beta - \beta d\gamma}{\Delta}, \omega_3 = \frac{\delta d\beta - \beta d\delta}{\Delta} \tag{39}$$

with

$$\Delta = \alpha\,\delta - \gamma\beta \tag{40}$$

It is easy to verify the fact that the metric:

$$ds^2 = \frac{(\delta d\alpha + \alpha d\delta - \gamma d\beta - \beta d\gamma)^2}{4\Delta^2} - \frac{d\alpha d\delta - d\beta d\gamma}{\Delta} \tag{41}$$

is in a direct relation with the discriminant from of the quadratic polynom (40):

$$ds^2 = \frac{1}{4}\left(\omega_2^2 - 4\omega_1\omega_3\right) \tag{42}$$

The three differentiable 1-forms from (39) completely define the coframe in every point of the absolute space. This coframe allows the translation of all geometric properties of the absolute space in the algebraic properties linked to (38).

The simplest property refers to dynamics on the metric geodesics, which can be correlated directly into statistical properties and from here to the multi-fractalization through stochasticization. In this case, the 1-forms ω_1, ω_2, ω_3 are differentiable in the same parameters, i.e.,:

$$\omega_1 = a_1 d\tau,\ \omega_2 = 2a_2 d\tau, \omega_3 = a_3 d\tau \tag{43}$$

Along this geodesic, (38) becomes a Riccati-type equation:

$$\frac{dt}{d\tau} = a_1 t^2 + 2a_2 t + a_3 \tag{44}$$

with a_1, a_2, a_3 being constants that characterize certains geodesics from the family. From here, mathematical procedures can be implemented to obtain solutions similar to the ones presented in Section 4.2.

4.3. Aldehyde Compound Release Transitions from the Space Scale to the Usual One

Since any descriptions of aldehyde compound release dynamics are calibrated to Riccati-type equations, it implies algebraic structures of the SL(2R) type [29]. It results that the algebraic structure associated with the release dynamics in the scale space has to be isomorphic with the algebraic structure associated with the release dynamics in the usual space. Therefore, through the theorem of Stoka, the idea is to establish the joint invariant function that will also be the base of the aldehyde compound release transition from the space scale to the usual one. Thus, if the differential operators of the algebraic structure of SL(2R)-type associated with the scale space release dynamics are:

$$\hat{A}_1 = \frac{\partial}{\partial h} + \frac{\partial}{\partial \bar{h}},\ \hat{A}_2 = h\frac{\partial}{\partial h} + \bar{h}\frac{\partial}{\partial \bar{h}},\ \hat{A}_3 = h^2\frac{\partial}{\partial h} + \bar{h}^2\frac{\partial}{\partial \bar{h}} + \left(h - \bar{h}\right)k\frac{\partial}{\partial k} \tag{45}$$

with the structure:

$$[\hat{A}_1, \hat{A}_2] = \hat{A}_1,\ [\hat{A}_2, \hat{A}_3] = \hat{A}_3,\ [\hat{A}_3, \hat{A}_1] = -2\hat{A}_2 \tag{46}$$

and the differentiable operators of the algebraic structure of SL(2R)-type associated to usual space release dynamics:

$$\hat{B}_1 = \frac{\partial}{\partial z} + \frac{\partial}{\partial \overline{z}},\ \hat{B}_2 = z\frac{\partial}{\partial z} + \overline{z}\frac{\partial}{\partial \overline{z}},\ \hat{B}_3 = z^2\frac{\partial}{\partial z} + \overline{z}^2\frac{\partial}{\partial \overline{z}} \tag{47}$$

with the structure:

$$[\hat{B}_1, \hat{A}_2] = \hat{B}_1,\ [\hat{B}_2, \hat{B}_3] = \hat{B}_3,\ [\hat{B}_3, \hat{B}_1] = -2\hat{B}_2 \tag{48}$$

Then Stoka's system becomes [34]:

$$\hat{A}_i f + \hat{B}_i f = 0,\ i = 1,2,3,\ f = f\left(z, \overline{z}; h, \overline{h}, k\right) \tag{49}$$

The general solution is a function based on the following algebraic functions:

$$\frac{h-z}{h-\overline{\overline{z}}} : \frac{\overline{h}-z}{\overline{h}-\overline{\overline{z}}} \equiv \rho^2,\ \frac{h-z}{\overline{h}-z}k \tag{50a,b}$$

Any joint invariant function is here a regular function of (50a,b). In (45) and (47), $\overline{h}$ is the complex conjugate of h, $\overline{z}$ is the complex conjugate of z, and k is a unidimensional factor.

For the case in which the joint invariant function is dependent online on (51a) with $\rho^2 > 0$, then for $\rho = \tanh\mu$, z is by (50a) related to h through the linear relation:

$$z = u + vh_0 \tag{51}$$

where $h = u + iv$ is taken, provided h_0 is given by (25).

This isomorphism is responsible for the aldehyde compound release transitions between the scale space and the usual one. As such, aldehyde release dynamics can be interpreted through a joint invariant function of two isomorphic groups of SL(2R)-type, as drug release dynamics in the form of kink–antikink pairs' dynamics of a multifractal type. The drug release dynamics through kink–antikink pairs of a multifractal type are presented in Figure 6.

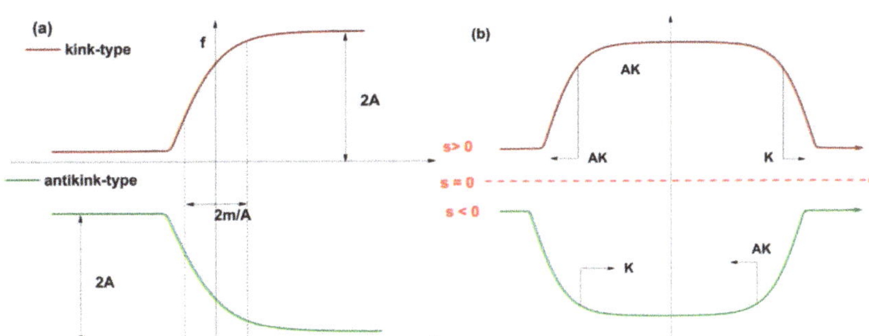

Figure 6. The dynamics of the kink–antikink pairs of multifractal type (**a**,**b**).

In Figure 7, the calibration of the experimental data relative to the proposed theoretical model is presented. A good consistency of data between the theoretical model and the experimental data can be observed. From Figure 2 it can be observed that the imino-chitosan derivatives have a porous quasi-fractal structure. This morphology is expected to affect the drug release dynamics as well. A higher fractality on the surface will be reflected in the fractality of the system. As with any fractal system, the scale transition from polymer to the in vitro system is dependent on the fractality of the release process. Therefore, the C0.071 system will be described by a higher fractality degree and a faster release dynamic coupled with enhanced fungal activity. The C0.142 case is described according to the authors' proposed model, by a fractality degree of 1.7, which is a factor of 3.2 lower than the C0.071 system. The lower fractality will induce a lower release rate and lower fungal activity. The good correlation between the fractality degree of the system and the drug-release behavior can have a potentially large impact on tailoring new polymer–drug configurations and improving their properties.

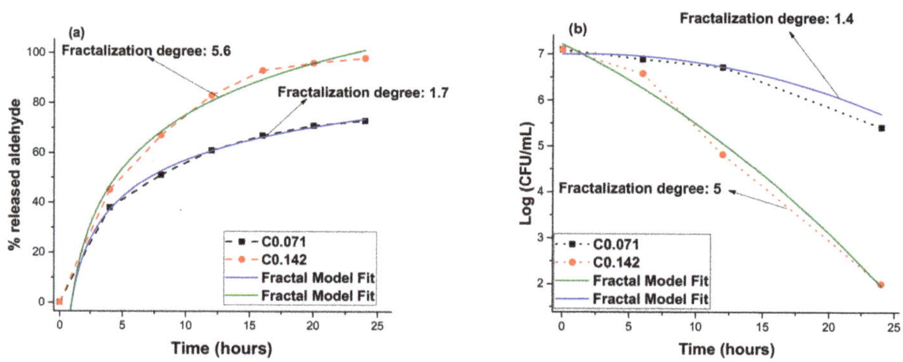

Figure 7. Fractal model fit on the empirical data for aldehyde release (**a**) and fungal activity (**b**).

5. Conclusions

The paper reports for the first time a theoretical model for the release of a bioactive compounds covalently bonded to a matrix, i.e., on an imine-chitosan system obtained by covalent bonding of an antifungal aldehyde to the chitosan via reversible imine linkages. The premises for creating this model were the fungicidal activity on Candida strains of the system and the in vitro release curves of the antifungal aldehyde. Their good agreement proved that the reversible bonding of aldehyde triggers its controlled release. Starting from these premises, the model has been created considering the imino-chitosan system as a multifractal object—a mathematical model for the release of bioactive aldehyde due to the reversible imine linkages. The release dynamics were analyzed in the framework of the Scale Relativity Theory, and more precisely in the form of the Multifractal Theory of Motion. Through Riccati-type gauges, two synchronous dynamics, one in the scale space, associated with the fungicidal activity, and the other in the usual space, associated with the antifungal aldehyde release, become operational. Their synchronicity, reducible to the isomorphism of two SL(2R)-type groups, implies, by means of its joint invariant functions, aldehyde compound release dynamics in the form of kink–antikink pairs' dynamics of a multifractal type. The synchronicity of the two phenomena and the coupling between them was highlighted by the symmetry of the solutions of the two operational procedures. The theoretical model has been validated through its good consistency with the experimental data.

Author Contributions: Conceptualization, L.M., M.A. and M.P.; methodology, A.A. and M.P., software, S.-A.I.; validation, L.H. and S.-A.I.; formal analysis, L.M., A.A., D.V. and S.-A.I.; investigation, L.M., A.A., D.V., M.A. and S.-A.I.; writing—original draft preparation, T.-C.P., L.M., S.-A.I. and M.A.; writing—review and editing, M.P. and L.H.; visualization, M.P., L.M. and S.-A.I. supervision, M.A. and M.P. All authors have read and agreed to the published version of the manuscript.

Funding: This research received no external funding.

Institutional Review Board Statement: Not applicable.

Informed Consent Statement: Not applicable.

Data Availability Statement: The data will be available upon request by the corresponding authors.

Conflicts of Interest: The authors declare no conflict of interest.

Sample Availability: Samples of the compounds are not available from the authors.

References

1. Crini, G. Historical review on chitin and chitosan biopolymers. *Environ. Chem. Lett.* **2019**, *17*, 1623–1643. [CrossRef]
2. Pokhrel, S.; Yadav, P.N. Functionalization of chitosan polymer and their applications, Journal of Macromolecular Science Part A. *Pure Appl. Chem.* **2019**, *56*, 450–475.
3. Marin, L.; Ailincai, D.; Mares, M.; Paslaru, E.; Cristea, M.; Nica, V.; Simionescu, B.C. Imino-chitosan biopolymeric films. Obtaining self-assembling, surface and antimicrobial properties. *Carbohyd. Polym.* **2015**, *117*, 762–770. [CrossRef] [PubMed]
4. Jommanee, N.; Chanthad, C.; Manokruang, K. Preparation of injectable hydrogels from temperature and pH responsive grafted chitosan with tuned gelation temperature suitable for tumor acidic environment. *Carbohyd. Polym.* **2018**, *198*, 486–494. [CrossRef] [PubMed]
5. Wu, Y.; Rashidpour, A.; Almajano, M.P.; Metón, I. Chitosan-Based Drug Delivery System: Applications in Fish Biotechnology. *Polymers* **2020**, *12*, 1177. [CrossRef] [PubMed]
6. Iftime, M.M.; Ailiesei, G.L.; Ungureanu, E.; Marin, L. Designing chitosan based eco-friendly multifunctional soil conditioner systems with urea controlled release and water retention. *Carbohyd. Polym.* **2019**, *223*, 115040. [CrossRef] [PubMed]
7. Chaiwong, N.; Leelapornpisid, P.; Jantanasakulwong, K.; Rachtanapun, P.; Seesuriyachan, P.; Sakdatorn, V.; Leksawasdi, N.; Phimolsiripol, Y. Antioxidant and Moisturizing Properties of Carboxymethyl Chitosan with Different Molecular Weights. *Polymers* **2020**, *12*, 1445. [CrossRef]
8. Marin, L.; Dragoi, B.; Olaru, N.; Perju, E.; Coroaba, A.; Doroftei, F.; Scavia, G.; Destri, S.; Zappia, S.; Porzio, W. Nanoporous furfuryl-imine-chitosan fibers as a new pathway towards eco-materials for CO_2 adsorption. *Eur. Polym. J.* **2019**, *120*, 109214. [CrossRef]
9. Wan Ngah, W.S.; Teong, L.C.; Hanafiah, M.A.K.M. Adsorption of dyes and heavy metal ions by chitosan composites: A review. *Carbohyd. Polym.* **2011**, *83*, 1446–1456. [CrossRef]
10. Bejan, A.; Doroftei, F.; Cheng, X.; Marin, L. Phenothiazine-chitosan based eco-adsorbents: A special design for mercury removal and fast naked eye detection. *Int. J. Biol. Macromol.* **2020**, *162*, 1839–1848. [CrossRef]
11. Lehn, J.M. Perspectives in Chemistry—Aspects of Adaptive Chemistry and Materials. *Angew. Chem.* **2015**, *54*, 3276–3289. [CrossRef]
12. Belowicha, M.E.; Stoddart, J.F. Dynamic imine chemistry. *Chem. Soc. Rev.* **2012**, *41*, 2003–2024. [CrossRef]
13. Zhang, Y.; Pham, C.Y.; Yu, R.; Petit, E.; Li, S.; Barboiu, M. Dynamic Hydrogels Based on Double Imine Connections and Application for Delivery of Fluorouracil. *Front. Chem.* **2020**, *8*, 739. [CrossRef] [PubMed]
14. Ailincai, D.; Marin, L.; Morariu, S.; Mares, M.; Bostanaru, A.C.; Pinteala, M.; Simionescu, B.C.; Barboiu, M. Dual crosslinked iminoboronate-chitosan hydrogels with strong antifungal activity against Candida planktonic yeasts and biofilms. *Carbohyd. Polym.* **2016**, *152*, 306–316. [CrossRef]
15. Bar-Yam, Y. *Dynamics of Complex Systems. The Advanced Book Program*; Addison-Wesley: Reading, MA, USA, 1997.
16. Mitchell, M. *Complexity: A Guided Tour*; Oxford University Press: Oxford, UK, 2009.
17. Badii, R. *Complexity: Hierarchical Structures and Scaling in Physics*; Cambridge University Press: Cambridge, UK, 1997.
18. Flake, G.W. *The Computational Beauty of Nature*; MIT Press: Cambridge, MA, USA, 1998.
19. Singhvi, G.; Singh, M. Review: In-Vitro drug release characterization models. *Int. J. Pharm. Stud. Res.* **2011**, *2*, 77–84.
20. Băceanu, D.; Diethelm, K.; Scalas, E.; Trujillo, H. *Fractional Calculus, Models and Numerical Methods*; World Scientific: Singapore, 2016.
21. Ortigueria, M.D. *Fractional Calculus for Scientists and Engineers*; Springer: New York, NY, USA, 2011.
22. Craciun, A.-M.; Barhalescu, M.L.; Agop, M.; Ochiuz, L. Theoretical Modeling of Long-Time Drug Release from Nitrosalicyl-Imine-Chitosan Hydrogels through Multifractal Logistic Type Laws. *Comp. Math. Method. Med.* **2019**, *2019*, 4091464. [CrossRef] [PubMed]
23. Nottale, L. *Scale Relativity and Fractal Space-Time: A New Approach to Unifying Relativity and Quantum Mechanics*; Imperial College Press: London, UK, 2011.
24. Merches, I.; Agop, M. *Differentiability and Fractality in Dynamics of Physical Systems*; World Scientific: Hackensack, NJ, USA, 2016.
25. Agop, M.; Paun, V.P. *On the New Perspectives of the Fractal Theory. Fundaments and Applications*; Romanian Academy Publishing House: Bucharest, Romania, 2017.
26. Ailincai, D.; Agop, M.; Marinas, I.C.; Zala, A.; Irimiciuc, S.A.; Dobreci, L.; Petrescu, T.-C.; Volovat, C. Theoretical model for the diclofenac release from PEGylated chitosan hydrogels. *Drug Deliv.* **2021**, *28*, 261–271. [CrossRef]
27. Iftime, M.M.; Dobreci, D.L.; Irimiciuc, S.A.; Agop, M.; Petrescu, T.; Doroftei, B. A theoretical mathematical model for assessing diclofenac release from chitosan-based formulations. *Drug Deliv.* **2020**, *27*, 1125–1133. [CrossRef] [PubMed]

28. Jackson, E.A. *Perspective of Nonlinear Dynamics, Volume 2*; Cambridge University Press: Cambridge, UK, 1991.
29. Cristescu, C.P. *Nonlinear Dynamics and Chaos. Theoretical Fundaments and Applications*; Academy Publishing House: Bucharest, Romanian, 2008.
30. Marin, L.; Zabulica, A.; Sava, M. Symmetric Liquid Crystal Dimers Containing a Luminescent Mesogen: Synthesis, Mesomorphic Behavior and Optical Properties. *Soft. Mater.* **2013**, *11*, 32–39. [CrossRef]
31. Liu, X.; Guo, N.; Guo, Z.; Hu, Z.; Chen, Z.; Hu, J.; Yang, L. Synthesis and liquid crystal properties of new cyclic carbonate monomers functionalised with cholesteryl moiety. *Liq. Cryst.* **2018**, *45*, 834–1843. [CrossRef]
32. Craciun, A.M.; Mititelu Tartau, L.; Pinteala, M.; Marin, L. Nitrosalicyl-imine-chitosan hydrogels based drug delivery systems for long-term sustained release in local therapy. *J. Colloid Interface Sci.* **2019**, *536*, 196–207. [CrossRef] [PubMed]
33. Sobel, J.D. Recurrent vulvovaginal candidiasis. *Am. J. Obst. Gynecol.* **2016**, *214*, 115–121. [CrossRef] [PubMed]
34. Mansour, H.M.; Sohn, M.J.; Al-Ghananeem, A.; DeLuca, P.P. Materials for Pharmaceutical Dosage Forms: Molecular Pharmaceutics and Controlled Release Drug Delivery Aspects. *Int. J. Mol. Sci.* **2010**, *11*, 3298–3322. [CrossRef] [PubMed]
35. Aitchison, J.R. *An Informal Introduction to Gauge Field Theories*; Cambridge University Press: Cambridge, UK, 1982.
36. Mazilu, N.; Agop, M. *Skyrmions: A Great Finishing Touch to Classical Newtonian Philosophy*; Nova Science Publisher: New York, NY, USA, 2014.
37. Mandelbrot, B.B. *The Fractal Geometry of Nature*; W.H. Freeman and Co.: San Fracisco, CA, USA, 1982.

molecules

MDPI

Article

New Grafted Copolymers Carrying Betaine Units Based on Gellan and N-Vinylimidazole as Precursors for Design of Drug Delivery Systems

Stefania Racovita [1], Nicolae Baranov [2,3], Ana Maria Macsim [1], Catalina Lionte [4], Corina Cheptea [5], Valeriu Sunel [3], Marcel Popa [2,6], Silvia Vasiliu [1] and Jacques Desbrieres [7,*]

1 "Petru Poni" Institute of Macromolecular Chemistry, Grigore Ghica Voda Alley, No. 41A, 700487 Iasi, Romania; stefania.racovita@icmpp.ro (S.R.); macsim.ana@icmpp.ro (A.M.M.); silvia.vasiliu@icmpp.ro (S.V.)
2 Department of Natural and Synthetic Polymers, Faculty of Chemical Engienering and Environmental Protection, "Gheorghe Asachi" Technical University of Iasi, Prof. Dr. Docent Dimitrie Mangeron Street, No. 73, 700050 Iasi, Romania; Baranov_nicolae@yahoo.com (N.B.); marpopa2001@yahoo.fr (M.P.)
3 Faculty of Chemistry, "Al. I. Cuza" University, Carol 1 Bvd., No. 11, 700506 Iasi, Romania; vsunel@uaic.ro
4 Faculty of Medicine, "Gr. T. Popa" University of Medicine and Pharmacy, Universitatii Street, No.16, 700115 Iasi, Romania; clionte@yahoo.com
5 Department of Biomedical Sciences, Faculty of Biomedical Bioengineering, "Gr. T. Popa" University of Medicine and Pharmacy, Kogalniceanu Street No. 9-13, 700454 Iasi, Romania; coricheptea@yahoo.com
6 Academy of Romanian Scientists, Splaiul Independentei Street No. 54, 050085 Bucuresti, Romania
7 Institut des Sciences Analytiques et de Physico-Chimie Pour l'Environnement et les Materiaux (IPREM), Pau and Pays de l'Adour University (UPPA), UMR CNRS 5254, Helioparc Pau Pyrenees, 2, av. President Angot, 64053 Pau CEDEX 09, France
* Correspondence: jacques.desbrieres@univ-pau.fr; Tel.: +33-4-76-07-15-86

Academic Editor: Sonia Trombino
Received: 24 October 2020; Accepted: 20 November 2020; Published: 20 November 2020

Abstract: New grafted copolymers possessing structural units of 1-vinyl-3-(1-carboxymethyl) imidazolium betaine were obtained by graft copolymerization of *N*-vinylimidazole onto gellan gum followed by the polymer-analogous reactions on grafted polymer with the highest grafting percentage using sodium chloroacetate as the betainization agent. The grafted copolymers were prepared using ammonium persulfate/*N*,*N*,*N'*,*N'* tetramethylethylenediamine in a nitrogen atmosphere. The grafting reaction conditions were optimized by changing one of the following reaction parameters: initiator concentration, monomer concentration, polymer concentration, reaction time or temperature, while the other parameters remained constant. The highest grafting yield was obtained under the following reaction conditions: $c_i = 0.08$ mol/L, $c_m = 0.8$ mol/L, $c_p = 8$ g/L, $t_r = 4$ h and T = 50 °C. The kinetics of the graft copolymerization of *N*-vinylimidazole onto gellan was discussed and a suitable reaction mechanism was proposed. The evidence of the grafting reaction was confirmed through FTIR spectroscopy, X-ray diffraction, ^{1}H-NMR spectroscopy and scanning electron microscopy. The grafted copolymer with betaine structure was obtained by a nucleophilic substitution reaction where the betainization agent was sodium chloroacetate. Preliminary results prove the ability of the grafted copolymers to bind amphoteric drugs (cefotaxime) and, therefore, the possibility of developing the new sustained drug release systems.

Keywords: graft polymerization; *N*-vinylimidazole; gellan gum; betaine structure

1. Introduction

In recent years, the chemical modification of natural polymers, polysaccharides in particular, by graft polymerization or by introduction of some functional groups, represents one of the most accesible and attractive method to obtain the polymeric materials with desired properties [1].

Since 1978, when gellan gum (GLL) was isolated for the first time from *Pseudomonas elodea*, this microbial polysaccharide has been used in various applications, such as in food industry as a thickening agent [2] or in biomedical and pharmaceutical fields [3,4].

This polysaccharide can be chemically modified due to the presence of many hydroxyl groups that can act as possible sites for grafting reaction. In the literature, some studies were found regarding the grafting of different vinyl monomers onto gellan gum chains using various sources of initiator radicals such as persulfate, redox systems or microwave irradiations [5,6]. Acrylic and methacrylic monomers such as, acrylamide [7], methacrylamide [8], *N,N*-dimethylacrylamide [5] or 2-(dimethylamino)ethyl methacrylate [9] were grafted onto gellan backbone in order to obtain polymeric materials with applications in industry as adsorbents/flocculants or in pharmaceutical field as sustained/controlled drug delivery systems. Verma et al. [10] combined the properties of *N*-vinyl-2-pyrrolidone and gellan gum by a grafting reaction, obtaining a new product that can be used as a coating material, superadsorbent or flocculating agent in the mining industry.

Among the vinyl monomers, *N*-vinylimidazole (NVI) has gained much attention in past years because the grafted polymers containing the imidazole ring are biocompatible, biodegradable and show antibacterial activity [11–15]. They were successfully used in medical and pharmaceutical fields. Moreover, chemical modification of imidazole-based polymers opens a huge opportunity for the development of functional polymers with complex structures [16].

The term polymers carrying betaine units (PB), or polyzwitterions, refers to polymers containing both negative (carboxylate group, sulfonate group or phosphonate/phosphate/phosphinate group) and positive (*onium* group without hydrogen atoms) charges located in the same repeat unit and separated by an alkylene group [17–20].

The grafted polymer carrying betaine units (PGB) can be obtained by one step and multistep processes [21,22].

One step process can be achieved by grafting monomers with a betaine structure onto the polymer backbone, while the multistep process consists of grafting monomers containing both nitrogen and vinyl group followed by polymer-analogous reactions onto grafted polymers in the presence of betainization agents.

The commercial interest for these polymers is determined by their possibility to be used as polymeric sorbents, oil recovery agents, fungicides, flame retardant polymers, emulsifying agents, wetting agents, cleaning agents and cryoprotectants [23–25]. Polymers having various betaine structure were synthesized in order to obtain materials sensitive to different stimuli (pH, temperature, etc.) [26].

The most interesting properties of PB are bio and hemocompatibility. Various studies have shown that polymer coatings based on poly(phosphobetaine) can improve the biocompatibility of some ocular devices by decreasing the adhesion of some microorganisms and eukaryotic cells [27,28]. Poly (methyl methacrylate) discs coated with a layer of sulfobetainic copolymer were exposed to *Staphylococcus epidermidis, Staphylococcus aureus* and *Pseudomonas aeruginosa* and it was found that the number of bacteria adhering to the surface of the discs coated with a layer of polybetaine decreased compared to the uncovered discs [29].

Due to the presence of vinyl groups capable of grafting polymerization and tertiary nitrogen suitable for the polymer-analogous reactions, the *N*-vinylimidazole is a very good candidate to obtain polymers carrying betaine units [30,31].

In this context, the work described in this paper follows several aspects: (1) synthesis of new grafted copolymers starting from gellan gum and *N*-vinylimidazole (PG); (2) finding the optimal conditions of the grafting reaction; (3) evidence of the grafting reaction by different methods; (4) chemical modification of new grafted copolymers containing an imidazole ring with the highest grafting yield by polymer analogous reactions in order to obtain the grafted copolymer with a betaine structure (PGB1); (5) immobilization and drug release studies of an antibiotic drug.

2. Results and Discussion

The preparation of grafted copolymers with betaine structure took place in two steps.

1. Preparation of PG and finding the optimal conditions for obtaining the grafted copolymer with maximum grafting yield.
2. Betainization reaction of PG with the highest grafting yield in the presence of sodium chloroacetate (Figure 1).

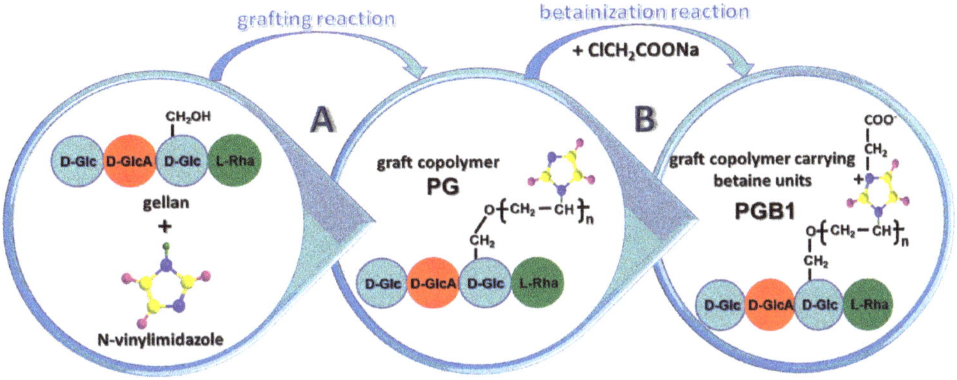

Figure 1. Graphical representation of the synthesis pathway of PG and grafted polymer carrying betaine units (PGB1) copolymers.

2.1. Optimal Conditions for the Preparation of PG

Various parameters such as monomer, initiator and polymer concentration, temperature and reaction time were investigated in order to optimize the reaction conditions in graft copolymerization of NVI onto gellan gum, as well as to improve the copolymer yield. Optimal reaction conditions were found by modifying one of the above-mentioned reaction parameters keeping the others constant, the reaction taking place under vigorous stirring and under a N_2 atmosphere.

2.1.1. Influence of Initiator Concentration

In the initiation stage, the rate of grafting can be influenced by a number of factors such as the nature and the concentration of the initiator, as well as the solubility of the initiator in the reaction medium. Among the thermal initiators, ammonium persulfate (APS) is preferred due to its high solubilization in water. In this study a redox initiator APS and *N,N,N′,N′* tetramethylethylenediamine (TEMED) (molar ratio 1:1) was used, and the influence of the initiator concentration on the grafting parameters was performed by modifying the APS/TEMED concentration in the range of 0.02–0.1 mol/L, keeping the monomer and gellan concentrations (c_m = 0.5 mol/L; c_p = 10 g/L), temperature (T = 60 °C) and reaction time (t = 4 h) constant. The graphical representation of grafting parameters versus initiator concentration is presented in Figure 2a.

As can be seen from Figure 2a grafting parameters like grafting yield (GY, %), grafting efficiency (GE, %) and conversion (C, %) increase with initiator concentration, reaching the maximal value at 0.08 mol/L APS/TEMED. After that, a slight decrease of grafting parameters is observed. This behavior could be explained in the following manner: at the beginning of the initiation reaction, the number of active sites on the gellan chain increases with the increase of initiator concentration, leading both to the formation of gellan macroradicals as well as the initiation of the grafting reaction. Then, at an initiator concentration greater than 0.08 mol/L, a decrease of grafting parameter values is observed probably because of an increase in the rate of homopolymerization compared to the grafting rate, this observation being proven by the increase in the value of homopolymer yield (H%). Another explanation of the decrease of grafting parameters can be attributed to a competition between initiation and termination reactions, the latter being achieved through chain-transfer to initiator or by a coupling reaction between initiator radicals [32]. The same behaviors were observed in the case of graft copolymerization of *N*-vinylimidazole onto various polysaccharides: xanthan [14], hyaluronic acid [11] and carboxymethyl starch [32].

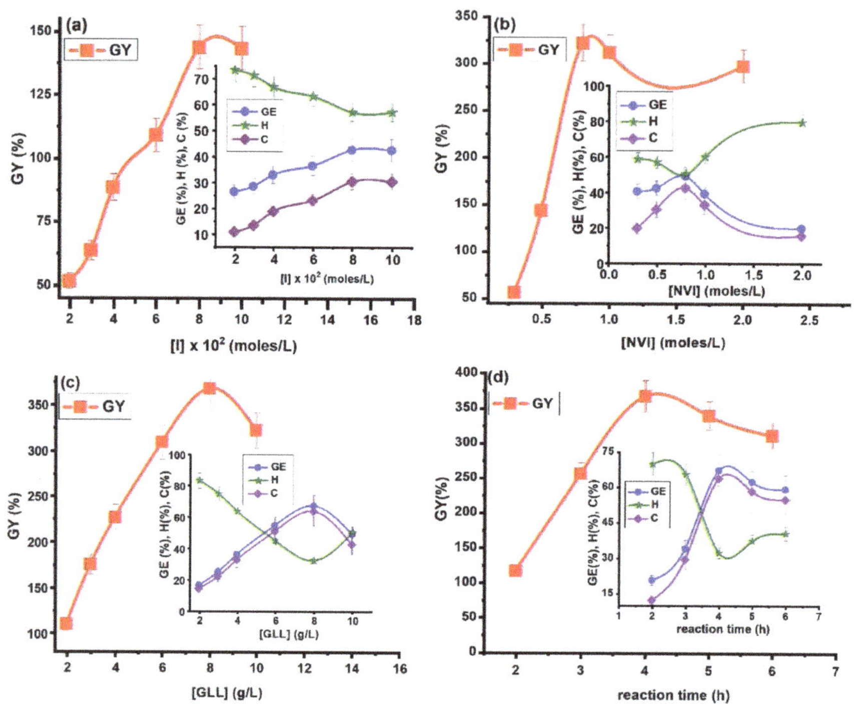

Figure 2. Influence of various factors on grafting parameters: (**a**) initiator concentration (c_m = 0.5 mol/L; c_p = 10 g/L, T = 60 °C and t = 4 h); (**b**) monomer concentration (c_i = 0.08 mol/L, c_p = 10 g/L, T = 60 °C and t_r = 4 h); (**c**) gellan concentration (c_i = 0.08 mol/L; c_m = 0.8 mol/L; T = 60 °C and t_r = 4 h); (**d**) reaction time (c_i = 0.08 mol/L; c_m = 0.8 mol/L; c_p = 8 g/L; T = 60 °C).

2.1.2. Influence of Monomer Concentration

Regarding the influence of the monomer concentration on the grafting yield, there are several factors that must be taken into account: monomer reactivity, polarity, steric hindrance, stability of the monomer radicals and concentration of monomer. In this study, the monomer concentration was varied within the range 0.3–2 mol/L, keeping the other parameters constant: c_i = 0.08 mol/L, c_p = 10 g/L, T = 60 °C and t_r = 4 h. The influence of NVI concentration on the grafting reaction is illustrated in Figure 2b. The results revealed an increase of GY%, GE% and C% with increasing NVI concentration up to 0.8 mol/L due to the greater availability of NVI molecules in the immediate vicinity of active sites located on the gellan chain, leading both to chain initiation and the formation of free radical donor that participate in the propagation reaction.

Thereafter, an increase in the monomer concentration above 0.8 mol/L caused a decrease of grafting parameter values, an exception being observed for the H% values that were found to increase. This behavior can be explained on the one hand by degradative chain-transfer to NVI as proposed in literature by Bamford and Schofield [33], and on the other hand by the poor diffusion of the monomer to the active sites situated on the polysaccharide chains leading to the homopolymer formation. A similar trend was reported in the literature for grafting of various vinyl monomers onto polysaccharides [11,32].

2.1.3. Influence of Gellan Concentration

The influence of polysaccharide concentration on grafting parameters is shown in Figure 2c. The gellan concentration was modified in the range 2–10 g/L, keeping constant the other parameters: c_i = 0.08 mol/L; c_m = 0.8 mol/L; T = 60 °C and t_r = 4 h. As can be seen from Figure 2c the grafting parameters (GY%, GE% and C%) increased with increasing of gellan concentration until 8 g/L and then a decrease of these parameters was observed. The increase of the grafting parameters until c_p = 8 g/L could be attributed to the increased number of grafting active centers situated on the polysaccharide chains leading to the formation of graft copolymers. An increase of polymer concentration above 8 g/L determines the increase of the viscosity of the reaction medium which hinders the possibility of e NVI movement toward the active sites situated on the gellan chains. The same behavior was observed in case of grafting on *N,N*-dimethylacrylamide onto gellan gum [5].

2.1.4. Influence of Reaction Time

The influence of reaction time on the grafting parameters is presented in Figure 2d and was determined by changing the time period of grafting reaction from 2 h to 6 h, keeping all the other reaction parameters constant (c_i = 0.08 mol/L; c_m = 0.8 mol/L; c_p = 8 g/L; T = 60 °C).

The obtained results revealed that the values of grafting parameters (GY%, GE% and C%) increased with increasing reaction time up to 4 h, due to the increase of active sites number situated on the gellan chains. After that, an increase in reaction time led to a slow decrease of the values of the grafting parameters. However, there was a slight increase in the homopolymer yield, suggesting the preference of the monomer to its own macroradical than for the one on the grafted chain, probably due to some steric hindrances. The same behavior was observed by other investigators whose results have been published in the literature [34,35].

2.1.5. Influence of Temperature

Temperature is one of the most important factors that influence the kinetics of graft copolymerization. The effect of temperature on the grafting parameters is shown in Figure 3a.

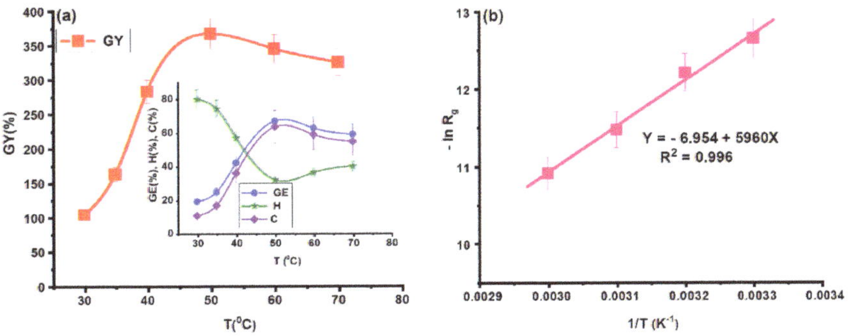

Figure 3. Influence of temperature on grafting of *N*-vinylimidazole (NVI) onto gellan gum (**a**) (c_i = 0.08 mol/L; c_m = 0.8 mol/L; c_p = 8 g/L and t_r = 4 h) and (**b**) the activation energy of the grafting reaction.

Temperature was varied from 30 to 80 °C, keeping constant c_i = 0.08 mol/L; c_m = 0.8 mol/L; c_p = 8 g/L and t_r = 4 h. As can be observed from Figure 3a the optimal temperature for maximum grafting yield was 50 °C and increasing temperature led to enhancement of the grafting parameters. The increase of grafting parameters until 50 °C can be attributed to:

(1) an increase in the production of free radicals resulting from thermal decomposition of initiator leading to increase in the number of the active sites on the polysaccharide chains;
(2) a decrease of reaction medium viscosity;

(3) fast diffusion of the monomer toward the active sites from the polysaccharide backbone;
(4) increase in rate of initiation and propagation steps [13,36].

Above 50 °C, the values of the grafting parameters decreased, probably due to the increase of the rate of the termination step by a combination of the monomer radicals and formation of the homopolymer. The effect may also have been due to the slower diffusion of the monomer to growing grafts due to steric hindrances, and its preference for growing homopolymer macroradicals.

The activation energy is another important parameter that can give information about the grafting process and can be determined from Arrhenius equation, as follows:

$$k = A \cdot e^{-E_a/RT} \tag{1}$$

where k is the rate constant, A is the pre-exponential factor, R is the gas constant (8.314 $J \cdot K^{-1} \cdot mol^{-1}$) and T is the absolute temperature (K).

The general rate equation of the grafting reaction can be written:

$$R_g = A \cdot [I]^a \cdot [M]^b \cdot [P]^c \cdot e^{-E_a/RT} \tag{2}$$

$$\ln R_g = \ln k' - \frac{E_a}{RT} \tag{3}$$

$$k' = A \cdot [I]^a \cdot [M]^b \cdot [P]^c \tag{4}$$

where $[I]$ = initiator concentration; $[M]$ = monomer concentration; $[P]$ = polymer concentration; a, b, and c—reaction order with respect to initiator, monomer and polymer.

The slope ($-E_a/RT$) of the ln R_g plots versus $1/T$ is used to find the E_a value (Figure 3b). The activation energy for PG was 49.5 KJ/mol. Similar results were observed in the case of grafting of various vinyl monomers (acrylamide, acrylic acid and methyl methacrylate) onto starch [37].

From the above studies it can be said that the optimized reaction conditions for grafting reaction of *N*-vinylimidazole onto gellan gum are as follows: initiator concentration = 0.08 mol/L; monomer concentration = 0.8 mol/L; polymer concentration = 8 g/L, reaction temperature = 50 °C and reaction time = 4 h.

2.2. Kinetics and Mechanism of graft Copolymerization

Generally, the rate equation of radical polymerization reaction is:

$$R_g = k \cdot [I]^{1/2} \cdot [M] \cdot [P]^{1/2} \tag{5}$$

The rate of graft copolymerization depends on several parameters such as initiator, monomer and polymer concentrations. In this study the graft copolymerization of *NVI* onto gellan was studied by modifying one of the parameters while the other parameters remained constant. In this context, the rate equations are:

$$\ln R_g = \ln k_1 + a \cdot \ln[I],\ k_1 = K \cdot [M]^b \cdot [P]^c \tag{6}$$

$$\ln R_g = \ln k_2 + b \cdot \ln[M],\ k_2 = K \cdot [I]^a \cdot [P]^c \tag{7}$$

$$\ln R_g = \ln k_3 + c \cdot \ln[P],\ k_3 = K \cdot [I]^a \cdot [M]^b \tag{8}$$

The graphical representations of ln R_g versus ln $[I]$, ln $[M]$ and ln $[P]$, respectively represent a straight line and are shown in Figure 4.

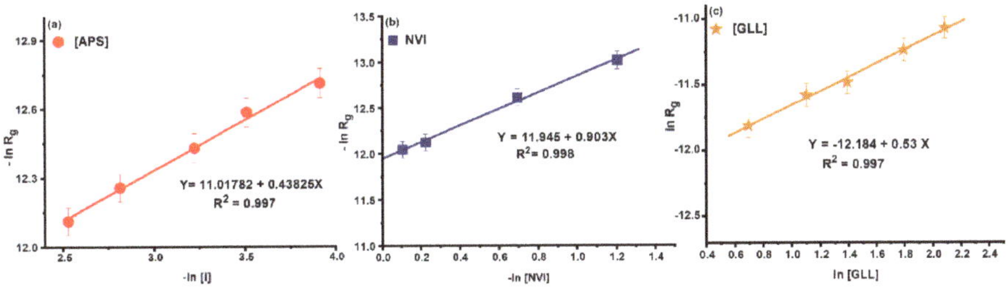

Figure 4. Plotting ln R_g versus (**a**) ln $[I]$; (**b**) ln $[NVI]$; and (**c**) ln $[GLL]$.

From Figure 4 the values of the slopes of the plots suggest that the reaction order with respect to initiator, monomer and polymer are 0.44, 0.90 and 0.53, respectively. Thus, the rate equation of grafting reaction of *N*-vinylimidazole onto gellan gum is:

$$R_g = k \cdot [APS/TEMED]^{0.44} \cdot [NVI]^{0.903} \cdot [GLL]^{0.53} \tag{9}$$

Equation (13) is very similar to Equation (9) and, for this reason, the mechanism of the grafting reaction of NVI onto gellan involves the same elementary reactions encountered in the free radical polymerization, namely, initiation, chain growth or chain propagation, chain transfer and termination of polymer chains [38].

The first step of the radical polymerization reaction corresponds to the formation of active species (radicals) followed by their addition to the monomer molecule (Figure 5).

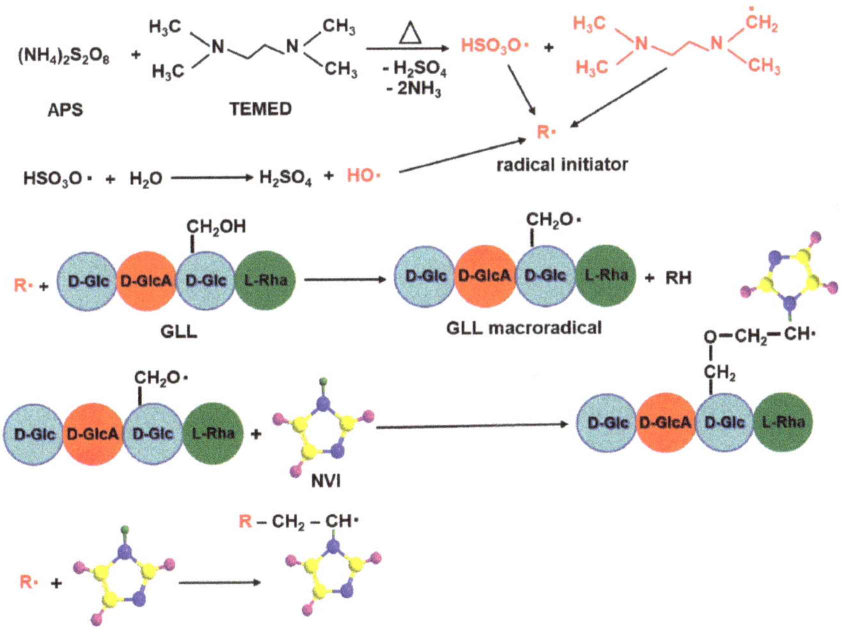

Figure 5. Initiation step of grafting copolymerization.

When the initiation system is heated, it decomposes in radical species that are able to initiate the polymerization reaction. In the case of the APS and TEMED system, two types of radicals (alkylaminoethyl radical derived from TEMED and sulfate radical from APS) [39] are formed and are able to initiate the graft copolymerization of *N*-vinylimidazole onto the gellan backbone. The propagation step consists of successive addition of a large number of monomer molecule to the promoter generated in the initial step (Figure 6).

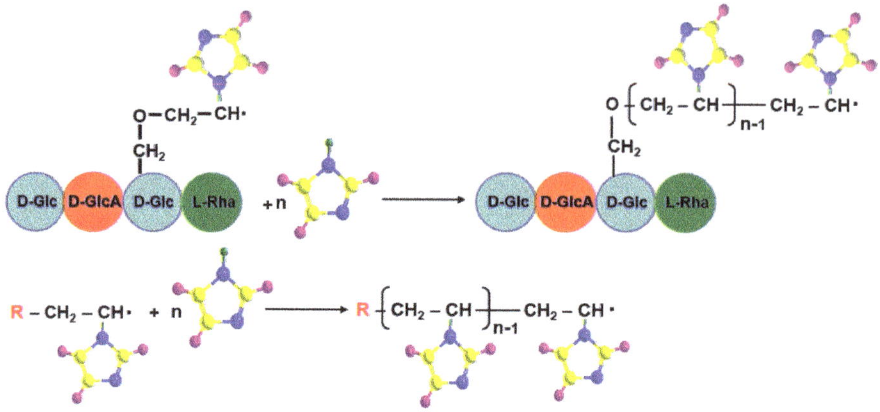

Figure 6. Propagation step of grafting copolymerization.

The termination reactions depend on the size, activity and structure of the macroradical, the viscosity of the medium, the temperature and the composition of the reaction mixture (Figure 7). Depending on these factors, the mechanism of the termination reaction is different and occurs as follows: (a) reaction with the initiator; (b) coupling/combination and disproportionation reaction; (c) inactivation of the growth radicals by inhibitors.

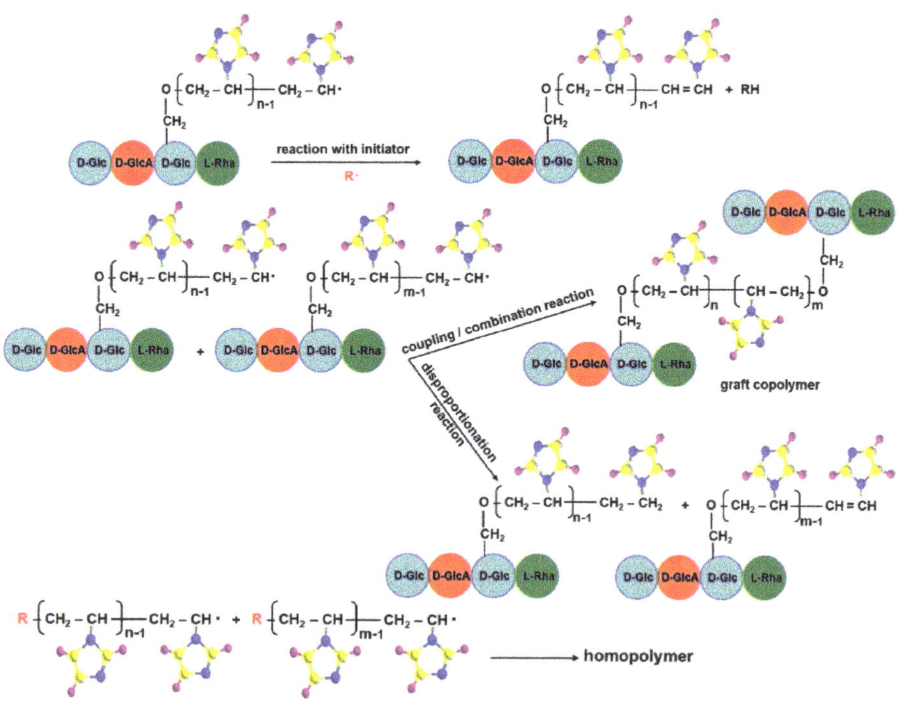

Figure 7. Termination steps of graft copolymerization.

2.3. *Synthesis of Copolymer Carrying Betaine Structure*

The second stage for preparation of grafted copolymer carrying structural units of 1-vinyl-3(1-carboxymethyl) imidazolium betaine consists in polymer-analogous reactions on grafted copolymer with the highest grafting yield in the presence of sodium chloroacetate as a betainization agent. From the reaction mechanism point of view, the grafted betaine copolymer with one methylene group between the opposite charges is achieved by the nucleophilic substitution reaction. The chemical structure of PGB1 is presented in Figure 8.

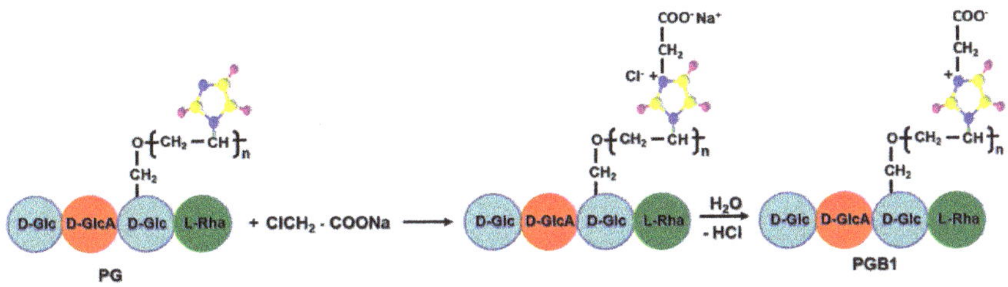

Figure 8. Betainization reaction of PG copolymer with sodium chloroacetate.

Different reaction parameters (temperature, reaction time, concentration of betainization agent) were modified in order to find the optimal value of betainization degree. It was found that the increase of all parameters mentioned above, up to a certain value, led to an increase of betainization degree. The optimal conditions for synthesis of graft copolymers carrying betaine structure was found to be: concentration of betainization degree = 20%, T = 60 °C and the reaction time = 72 h.

The degree of betainization was obtained by FTIR spectroscopy using the relative ratio between integrated area of betainized (1637 cm^{-1}) and chemically unmodified imidazole ring (1500 cm^{-1}). The curve fitting of the FTIR spectrum of PGB1 in the region 1700–1420 cm^{-1} was realized using OPUS Software on the basis of a linear regression Levenberg-Marquardt model, the method being similar to that mentioned in our previous paper [40]. In this case, the betainization degree was found to be 89.97%.

2.4. *Characterization of PG and PGB1*

2.4.1. ^{1}H-NMR Spectroscopy

^{1}H-NMR spectroscopy was used to elucidate only the structure of PG because the PGB1 is insoluble in the solvent used for PG, poly(*N*-vinylimidazole) (PNVI) and GLL samples. The structure of PG was elucidated by comparing the PNVI and GLL spectra with the spectrum of the grafting copolymer (Figure 9).

It is well known that for PNVI (Figure 9b) the following characteristic signals can be observed:

(1) multiplet signals at δ = 6.61–7.08 ppm assigned to the protons (H_2, H_4 and H_5) belonging to the imidazole ring; (2) multiplet signals at δ = 3.7–3.9 ppm due to the methine protons; (3) doublet signal at δ = 2.07–2.13 ppm assigned to the backbone methylene protons; (4) triplet signal at δ = 2.58–2.87 ppm related to the splitting chain -CH group (isotactic, heterotactic and syndiotactic triads) [41].

The spectrum of GLL (Figure 9a) contains characteristic peaks of the tetrasaccharide repeating units as follows: -CH of glycosidic bonds in sugar at δ = 5.83 ppm; -CH of rhamnose at δ = 5.15 ppm; -CH of glucose and glucuronic acid at δ = 3.36–4.61 ppm; -CH_3 of rhamnose at δ = 1.29–1.31 ppm [42,43].

The ^{1}H-NMR spectrum presented in Figure 9c proved the synthesis of the PG copolymer because in this spectrum the characteristic signals of both PNVI (δ = 6.72–7.13 ppm—protons belonging to the imidazole ring; δ = 2.14–2.35 ppm—backbone methylene protons) and GLL (δ = 3.37–4.61 ppm—protons of the tetrasaccharide repeating units; δ = 1.29–1.31—CH_3 of rhamnose) polymers are found.

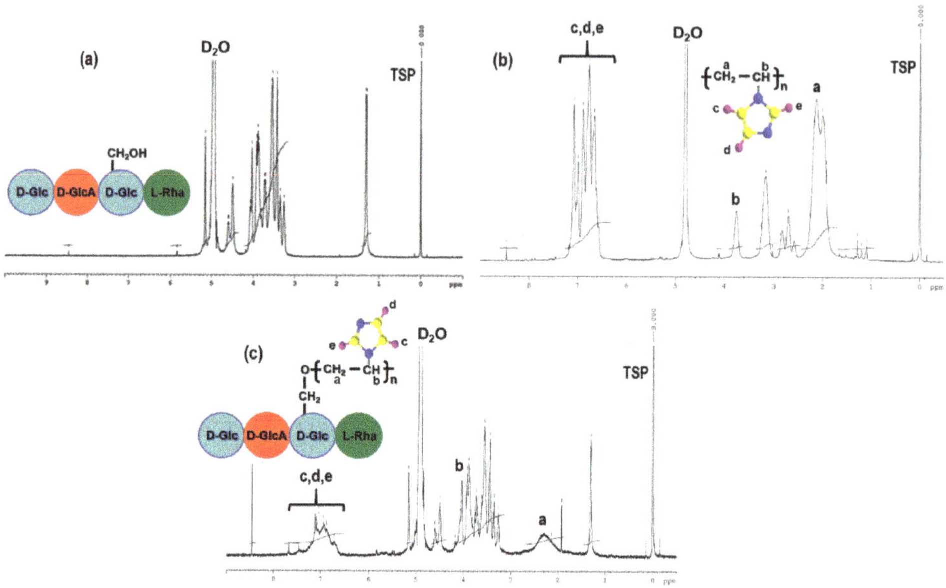

Figure 9. ^{1}H-NMR spectra of GLL (**a**); PNVI (**b**) and PG samples (**c**).

2.4.2. FTIR Spectroscopy

The infrared spectra of PNVI, GLL PG and PGB1 samples are presented in Figure 10.

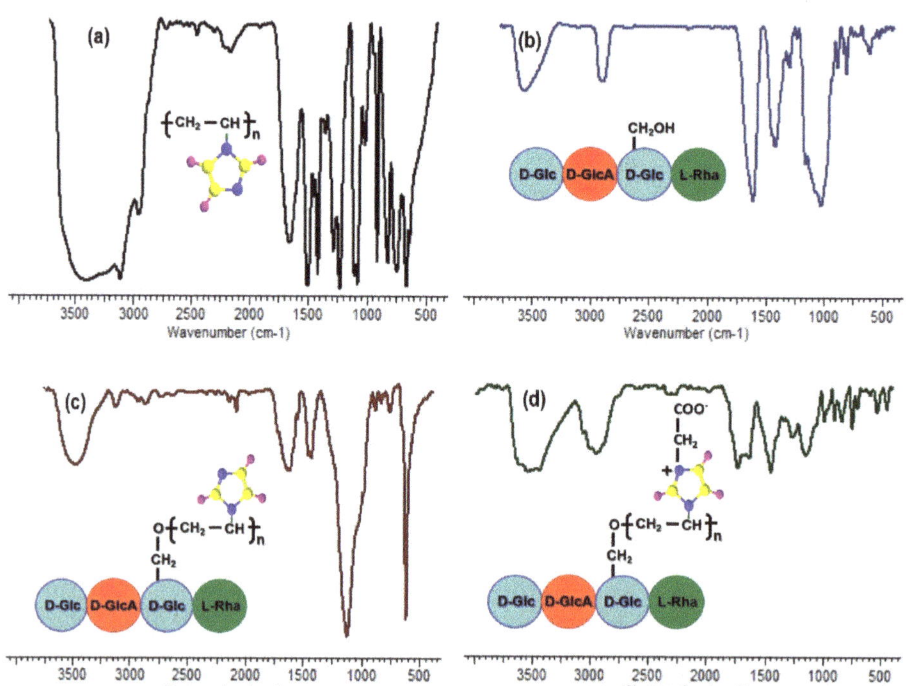

Figure 10. The infrared spectra of PNVI (**a**); GLL (**b**); PG (**c**) and PGB1 samples (**d**).

The infrared spectrum analysis of gellan gum (Figure 10b) showed the following adsorption bands: 3566 cm^{-1} attributed to the O-H stretching of hydroxyl groups of glucopyranose ring; 2892 cm^{-1} characteristic of the aliphatic-CH; 1611 and 1420 cm^{-1} assigned to the asymmetric and symmetric vibrations of the carboxylate group and 1027 cm^{-1} attributed to the C-O-C bonds.

In the PNVI spectrum (Figure 10a) the following characteristic absorption bands are observed: 3100 cm^{-1} assigned to C-H stretching vibration; 2957 cm^{-1} attributed to the C-H and CH_2 stretching vibrations of the backbone chain. A strong and broad band is observed at 1655 cm^{-1} that is characteristic of the C=C ring stretching vibrations. The bands observed at 1500, 1285, 1230, 1083 and 914 cm^{-1} are assigned to C-C, C=N ring stretching vibrations, C-H bending vibration, the ring C-H bending and C-C-C bending vibration of the backbone of the aliphatic chain. Two characteristic bands for the imidazole ring were observed at 1500 and 665 cm^{-1}, the last being attributed to the puckering vibration of imidazole ring [41].

In the FT-IR spectrum of the PG copolymer (Figure 10c) a shifting of -OH stretching vibration from about 3400 cm^{-1} (3392 cm^{-1} for PNVI and 3415 cm^{-1} for GLL) to 3479 cm^{-1} is observed, as well as a decrease of intensity of the absorption bands corresponding to the -OH group vibration, indicating the participation of hydroxyl groups in the grafting reaction. The bands at 1630 cm^{-1} (>C=C< stretching vibrations in imidazole ring), 1545 cm^{-1} (>$C=N^-$ stretching vibration) and 1124 cm^{-1} (in-plane bending vibration of the C-H bond inside imidazole ring) indicate the presence of PNVI in the structure of the grafted copolymer, and also the confirmation of grafting of the vinyl monomer onto the gellan backbone.

If the FT-IR spectra of PG and PGB1 copolymers are compared (Figure 10d) the appearance of new absorption bands specific to the betaine structure can be observed as follows: the band at 3439 cm^{-1} appears due to the >$C=N^+$ group, and the shift of the adsorption band from 1630 cm^{-1} to 1637 cm^{-1} is due to the overlap the vibration band of >$C=N^-$ group from PNVI with the absorption band corresponding to the asymmetric COO^- group. The absorption bands at 1340 and 1266 cm^{-1} may be attributed to the stretching vibration of carboxylate group.

2.4.3. Surface Morphology Analysis

Surface morphology images at high magnification (2000×) of PNVI, GLL, PG and PGB1 samples were analyzed using a scanning electron microscope coupled with an energy dispersive X-ray system and are presented in Figure 11.

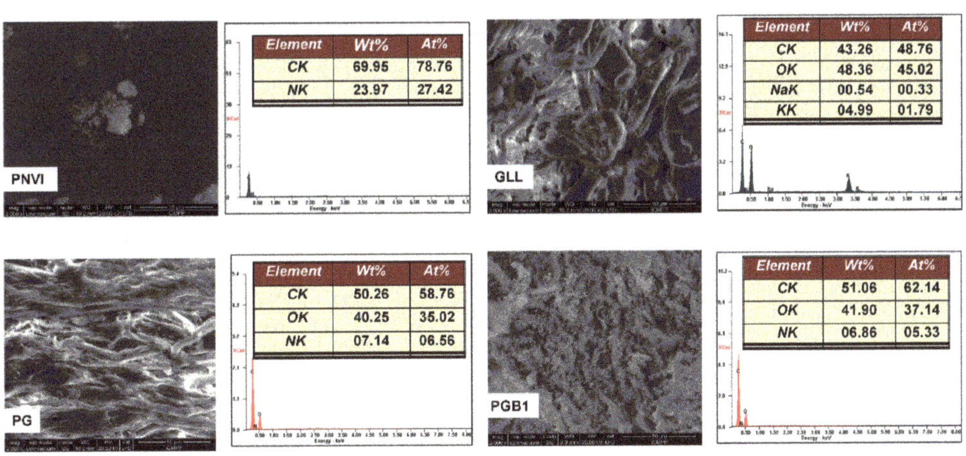

Figure 11. Scanning electron microscope (SEM) images and energy dispersive X-ray (EDAX) analysis of PNVI, GLL, PG and PGB1 samples.

The PNVI sample showed a compact porous structure while GLL and PG samples seemed to have a fibrous structure. When NVI was grafted onto the GLL backbone, microstructural changes were observed, the PG copolymer having a more ordered structure compared to that of the gellan. The reaction between PG copolymer and sodium chloroacetate led to the preparation of the grafting copolymer with betaine structure having a porous structure.

The EDAX analysis confirmed both grafting of *N*-vinylimidazole onto gellan gum and the formation of graft copolymer carrying betaine units as follows:

(1) the presence of nitrogen belonging only to NVI on the surface of PG and PGB1 copolymers;
(2) increase of C% value of the PG molecule compared to that from GLL indicating the presence of PNVI in the PG structure;
(3) increase of C% and O% values, as well as the decrease of N% values, of the PGB1 copolymer compared those from the PG copolymer due to the formation of betaine units.

2.4.4. X-Ray Diffractions Analysis (XRD)

XRD patterns of GLL, PNVI, PG and PGB1 samples are illustrated in Figure 12.

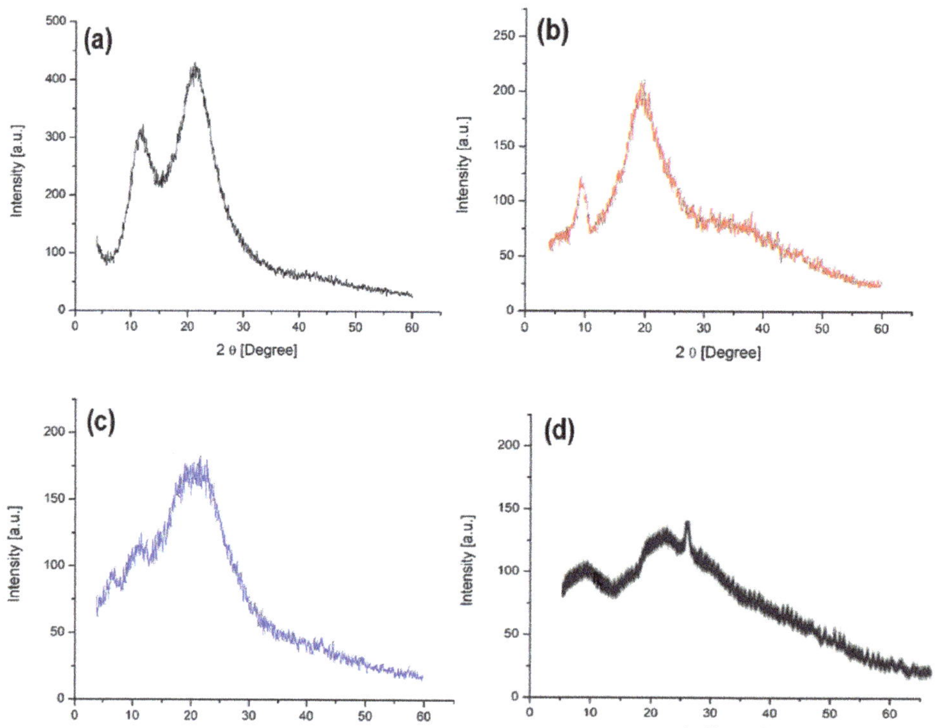

Figure 12. XRD patterns of PNVI (**a**); GLL (**b**); PG (**c**) and PGB1 samples (**d**).

X-ray diffraction spectra of PNVI and GLL show two broad diffraction peaks at lower diffraction angle values (2θ = 20° and 10°) indicating the amorphous structure of both polymers. The XRD pattern of PG copolymer presents the combined signals of started polymers leading to the conclusion that the grafting of PNVI onto gellan gum was successfully completed. The PGB1 copolymers presented an amorphous structure.

2.4.5. Thermogravimetric Studies

Thermogravimetric studies were realized to prove the preparation of both grafted copolymers and grafted copolymers carrying betaine units. The thermal stability of PNVI, GLL, PG and PGB1 samples was discussed according to the temperature at which intensive degradation took place and the results are presented in Table 1.

Table 1. Thermogravimetric analysis data for main stage of decomposition.

Sample Codes	Decomposition Temperature			Weight Loss (%)	Residual Mass (%)
	T_i (°C)	T_m (°C)	T_f (°C)		
PNVI	406	433	519	47.3	3.9
GLL	241	250	322	45.0	16.1
PG	429	454	548	15.9	53.3
PGB1	256	278	410	40.9	31.2

Thermogravimetric (TG) and derivative thermogravimetric (DTG) curves of PNVI, GLL, PG and PGB1 samples (Figure 13) showed that thermal degradation was characterized by four (PNVI), three (GLL), five (PG) and three (PGB1) stages of degradation.

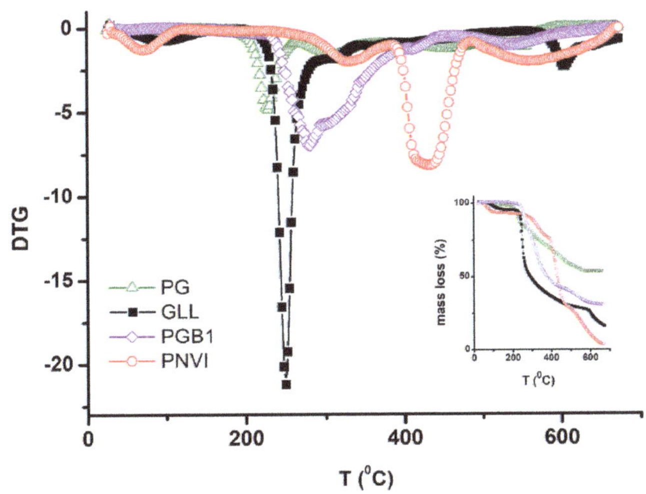

Figure 13. TG and DTG curves of PNVI, GLL, PG and PGB1 samples.

The main stages of degradation for PNVI and gellan occurred in the temperature ranges 406–519 °C and 241–322 °C, respectively being characterized by an important weight loss of 47.30% and 44.96%, probably due to the degradation of the polymer backbone. The degradation results of PG showed that the grafting of NVI onto gellan gum increased the thermal stability of the copolymer, the main stage of decomposition occurring in the range 429–548 °C with a weight loss of 15.85%. This behavior can be explained by the presence of an imidazole ring belonging to PNVI which is thermally stable. The same observations were reported by other authors for grafting of poly(*N*-vinylimidazole) onto carboxymethyl chitosan [32]. The introduction of betaine units led to decrease of thermal stability of PGB1, the main stage of decomposition taking place in the range 256–410 °C with a weight loss of 40.92%. After thermal treatment up to 700 °C the remaining residual masses were 3.92% (PNVI), 16.14% (GLL), 53.29% (PG) and 31.18% (PGB1), respectively. Taking into account the results previously presented, it can be said that the grafting reaction was realized and the grafting of NVI onto gellan gum led to an increase in thermal stability of the PG copolymers, whereas the introduction of betaine units induced a slight decrease in the thermal stability of PGB1 copolymers.

2.5. Immobilization and Drug Release

The synthesis of grafted copolymers with *N*-vinyl imidazole (PG), and subsequently functionalized with betaine structure (PGB1), was made in order to obtain precursors capable of binding amphoteric drugs through ionic interactions. Consequently, a preliminary study was performed on the immobilization; respectively the release of an amphoteric model drug (cefotaxime sodium salt) to verify our hypothesis. The ionic interaction of the drug is possible both with the copolymer PG, which contains carboxylic groups belonging to gellan, and especially with the copolymer PGB1, which contains carboxylic groups belonging to both the gellan and the betaine structure.

Cefotaxime sodium salt (CF) is a semisynthetic third generation cephalosporin with bactericidal activity, being more active against gram-negative bacteria than gram-positive bacteria [44].

Cefotaxime sodium salt was immobilized onto PG and PGB1 samples in a bach system. The maximum immobilization capacities of CF onto PG and PGB1 samples were found to be 420 mg/g and 473.2 mg/g, repectively. Obviously, the higher CF binding capacity was higher for the PGB1 copolymer which contains a higher number of carboxylic groups compared to the PG copolymer.

The release profiles of CF as a function of time are presented in Figure 14.

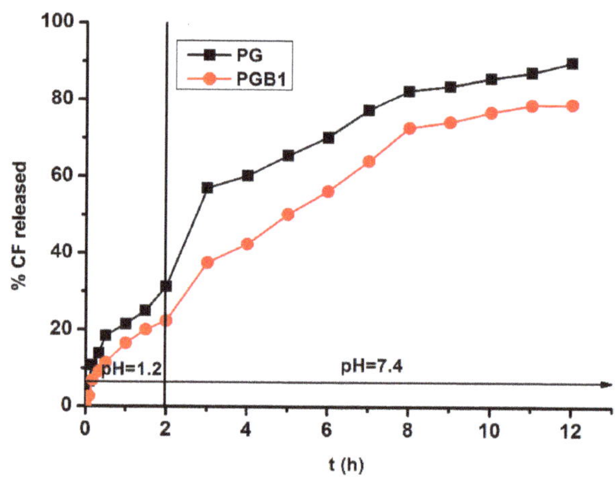

Figure 14. Release profile of cefotaxime sodium salt from PG and PGB1 samples.

From Figure 14, it can be observed that the amount of CF released from PG and PGB1 was higher at pH = 7.4 than pH = 1.2. In acidic pH values, the carboxylate groups of gellan became protonated, the repulsions between polymer chains being much diminished resulting in a decrease of swelling degree and finally in a decrease of the amount of drug released. At higher pH values, the carboxylate groups were ionized and the electrostatic repulsive forces between COO^- groups located on the copolymer backbone caused an increase of the swelling capacity leading to an increase of amount of drug released.

To analyze the type of release mechanism, the drug release data were fitted using two release kinetic models: the Higuchi and Korsmeyer-Peppas models.

The Higuchi model is based on the Fick's law and is used to describe the release of a water soluble drug from solid matrices [45]. The mathematical equation of Higuchi model is:

$$Q_t = k_H \cdot t^{\frac{1}{2}} \tag{10}$$

where k_H is Higuchi dissolution constant.

The Korsmeyer-Peppas model describes the mechanism of drug release from polymeric systems [46]:

$$\frac{M_t}{M_\infty} = k_r \cdot t^n \quad (11)$$

where Mt/M_∞ = fraction of drug released at time t; k_r = release rate constant that is characteristic to polymer-drug interactions; n = the diffusion exponent is characteristic to the different release mechanisms.

Depending on the values of n, the release can take place through several mechanisms such as: $n = 0.5$, Case I diffusion or Higuchi kinetic (diffusion/controlled drug release); $0.5 < n < 1$, anomalous diffusion; $n = 1$, Case II transport (swelling-controlled drug release); $n > 1$, super Case II transport.

The values of the release parameters of CF from PG and PGB1 are presented in Table 2.

Table 2. Kinetic parameters of CF release from PG and PGB1.

Sample Codes	Higuchi Model		Korsmeyer-Peppas Model		
	k_H (h$^{-1/2}$)	R^2	k_r (min^{-n})	n	R^2
PG	0.249	0.993	0.053	0.377	0.992
PGB1	0.236	0.990	0.023	0.550	0.991

Based on the release exponent n from the Korsmeyer-Peppas equation, the release mechanism of CF from the PG sample is controlled by a diffusion process while the release mechanism of CF from the PGB1 sample is more complex, being controlled by both swelling and diffusion processes.

The results presented in this study attest the ability of the two types of copolymers to bind amphoteric drugs through electrostatic interactions. An in-depth study on the immobilization of biologically active principles with cationic or anionic characteristics on grafted copolymers carrying betaine units is the subject of further work.

3. Materials and Methods

3.1. Materials

Gellan gum ($M_w = 1 \times 10^6$ g/mol), ammonium persulfate, *N,N,N′,N′*-tetramethylethylenediamine, acetone, ethanol, sodium chloroacetate, hydrochloric acid, potassium chloride, monobasic and dibasic sodium phosphate and cefotaxime sodium salt were supplied from Sigma-Aldrich, Germany and were used as received. *N*-vinylimidazole was purchased from Sigma Aldrich, Germany and was distilled under vacuum. Ultrapure grade water ($\Omega < 10^{-6}$ s/cm) was prepared by purifying deionized water with Millipore Simplicity-UV apparatus.

3.2. Synthesis of Grafted Polymers (PG)

The PG copolymers were prepared by using the free radical polymerization technique in presence of redox initiator (APS + TEMED) in nitrogen atmosphere.

60 mL of (2–10 g/L) gellan solution in ultrapure distilled water was added into a 250 mL three-necked round-bottom flask equipped with a magnetic stirrer and a thermostater water bath. To gellan solution various amounts of NVI (c_m = 0.3–2 mol/L) were added followed by the addition of different amounts of initiator solutions (c_i = 0.02–0.1 mol/L). The molar ratio between APS and TEMED was 1:1, which has been reported in literature to be the optimal value for obtaining free radicals as well as the polymer with highest grafting yield [47]. The grafting reactions were carried out at different reaction temperatures (30–80 °C) and at various periods of time (2–6 h). Finally, the grafting mixture was precipitated under vigourous stirring in cold acetone and then the grafted copolymers were separated by filtration under vacuum using borosilicate glass filter crucibles, porosity 4. Thereafter, the grafted copolymers were purified by extraction with ethanol in a Soxhlet apparatus in order to remove the homopolymer, and were dried in vacuum oven at 50 °C for 48 h.

3.3. Estimation of Grafting Parameters

The grafting parameters were calculated as follows [48,49].

$$GY\ (\%) = \frac{weight\ of\ grafted\ PNVI}{weight\ of\ gellan} \cdot 100 \quad (12)$$

$$GE\ (\%) = \frac{weight\ of\ grafted\ PNVI}{weight\ of\ polymer\ formed} \cdot 100 \quad (13)$$

$$C\ (\%) = \frac{weight\ of\ polymer\ formed}{weight\ of\ monomer\ fed} \cdot 100 \quad (14)$$

$$H\ (\%) = 100 - GE\ (\%) \quad (15)$$

3.4. Synthesis of Grafted Polymer Carrying Betaine Units

Grafted polymer carrying betaine units was prepared by the betainization reaction of PG with maximum grafting yield in the presence of sodium chloroacetate. Thus, 5 g of PG were swollen in water for 24 h at room temperature then centrifuged at 78 RCF for 10 min after which an aqueous solution of sodium chloroacetate (100 mL, c = 20%, *w*/*v*) was added. To calculate the amount of betainization agent required for the polymer-analogous reaction, the molar ratio between nitrogen and sodium chloroacetate was considered to be 1:1.5. The reaction mixture was gently stirred at T = 60 °C for 72 h. The final product was removed from the reaction medium by precipitation in acetone and then the precipitate was extracted with water in a Soxhlet apparatus. The final product was insoluble in water.

3.5. Infrared Spectroscopy

FTIR spectra were recorded on a Bruker Vertex 70 FTIR spectrometer (Wien, Austria) at a resolution of 2 cm^{-1} in the frequency range of 400–4000 cm^{-1}, by the KBr pellet technique (pellets obtained at a pressure of 2 tons for 1 min).

3.6. ^{1}H-NMR Analysis

^{1}H-NMR measurements of PNVI, gellan and PG were performed on a high-resolution liquid NMR 400 MHz spectrometer Bruker Neo-1 (Rheinsteitten, Germany) for direct detection probe, with 5 mm Quadra nuclei probes, QNP (four nuclei, $^{1}H/^{13}C/^{19}F/^{29}Si$). Sample solutions were prepared with deuterated water and 1N NaOH as solvent. TSP (trimethylsilylpropanoic acid) (δ = 0.0 ppm) was used as an internal standard. Detection temperature was set at 25 °C and the sample was scanned 64 times.

3.7. Scanning Electron Microscopy (SEM)

The surface morphologies of gellan, PNVI, PG and PGB1 in powder form were analyzed with an environmental scanning electron microscope type Quanta 200 at 25 kV with secondary electrons in low vacuum. The microscope was coupled with an energy dispersive X-ray system for qualitative and quantitative analysis.

3.8. X-Ray Diffraction Analysis (XRD)

The XRD patterns of PNVI, GLL, PG and PGB1 samples were recorded employing a D8 Advance Bruker AXS device using a CuKα radiation at a current/voltage of 36 mA/30 kV.

3.9. Thermogravimetric Analysis (TG/DTG)

The thermal stability of the crosslinked epoxy resins were thermogravimetrically analyzed using a STA 449 F1 Jupiter apparatus (Netzsch, Selb, Germany) coupled to a Vertex 70 spectrophotometer for FT-IR analysis and Aeölos QMS 403C mass spectrometer (Netzsch-Germany) for the mass spectroscopic analysis of the evolved gases. Samples of about 10 mg placed in Al_2O_3 crucibles were thermally degraded at a heating rate of 10 °C min^{-1}, under air atmosphere in the temperature range between 25 °C and 700 °C.

3.10. Immobilization and Drug Release

CF immobilization was realized as follows: 0.2 g of PG and PGB1 copolymers were weighed into 50 mL conical flasks and then 20 mL of CF ($c_{CF} = 3 \cdot 10^{-3}$ g/mL) were added. The samples were placed in a thermostatic shaker bath (Memmert MOO/M01, Schwabach, Germany) and shaken at 180 strokes/minute until equilibrium was reached. The flasks were removed from the shaker and the copolymers were centrifuged at 78 RCF for 10 min. The amount of CFR immobilized onto PG and PGB1 copolymers was determined by UV-VIS spectrophotometry (SPEKOL 1300 Spectrophotometer, Analytik Jena, Jena, Germany) at a wavelength of 236 nm, based on a calibration curve.

The amount of drug immobilized was obtained using the following equation:

$$q_e = \frac{(C_0 - C_e) \cdot V}{W} \quad (16)$$

where q_e is the amount of CF immobilized onto PG and PGB1 copolymers (mg/g), C_0 is the initial concentration of drug (mg/mL), C_e is the drug concentration at equilibrium (mg/mL), V is the volume of drug solution (mL) and W is the weight of the copolymers.

In vitro release studies of cefotaxime sodium salt were performed by immersing the PG and PGB1 samples (0.1 g) in 10 mL of simulated gastric fluid (pH = 1.2) for 2 h and phosphate buffer solution (pH = 7.4) for 10 h at 37 °C. The buffer solutions were prepared according to protocols well known in the literature [50,51]. The samples were placed in a thermostated shaker bath (Memmert M00/M01, Germany) under gentle shaking (50 strokes/minute). Withdrawal of a small volume (1 µL) of release solution was done and the CF solutions collected at different intervals of time were measured spectrophotometrically at a wavelength of 236 nm using a UN-VIS spectrophotometer (Nanodrop ND 100, Wilmington, DE, USA). The amount of CF released was calculated using a calibration curve.

4. Conclusions

Grafted polymers carrying betaine units were obtained by grafting *N*-vinylimidazole onto a gellan gum backbone followed by a betainization reaction of the grafted copolymer with maximum grafting yield in the presence of sodium chloroacetate. The grafted copolymers were successfully obtained by a free radical polymerization technique in the presence of redox initiator (APS + TEMED) and a nitrogen atmosphere. From the point of view of the reaction mechanism, the betainization reaction is a nucleophilic reaction and the betainization degree estimated by FTIR spectroscopy was found to be about 90%. The grafting parameter (GY%, GE%, H% and C%) could be adjusted by changing one of the reaction parameters. The optimized reaction conditions for the grafting reaction of NVI onto GLL were as follows: initiator concentration = 0.08 mol/L; monomer concentration = 0.8 mol/L; polymer concentration = 8 g/L; reaction temperature = 50 °C and the reaction time = 4 h. The mechanism of grafting reaction is similar with that of free radical polymerization.

FTIR spectroscopy, ^{1}HNMR spectroscopy, X-ray diffraction, thermogravimetric analysis and scanning electron microscopy confirmed the grafting reaction of NVI onto gellan, as well as the synthesis of grafted polymers having betaine structure.

In vitro release studies of CF proved the capacity of the new grafted copolymers to immobilize amphoteric drugs and highlighted the fact that the release mechanism of CF from PG and PGB1 samples is controlled by diffusion process or by a combination between diffusion and swelling processes. These results demonstrated that the grafted copolymer with betaine structure can be a potential candidate for developing sustained/controlled drug delivery systems.

Author Contributions: Conceptualization, S.R. and S.V.; methodology, S.R., N.B. and S.V.; validation, V.S., M.P. and J.D.; formal analysis, N.B., A.M.M. and C.C.; investigation, C.L. and V.S.; data curation, S.R., N.B. and S.V.; writing—original draft preparation, S.R. and S.V.; writing—review & editing, M.P. and J.D., visualization and supervision, S.V., M.P. and J.D. All authors have read and agreed to the published version of the manuscript.

Funding: This research received no external funding.

Conflicts of Interest: The authors declare no conflict of interest.

References

1. Garcia-Valdez, O.; Champagne, P.; Cunningham, M.F. Graft modification of natural polysaccharides via reversible deactivation radical polymerization. *Prog. Polym. Sci.* **2018**, *76*, 151–173. [CrossRef]
2. Cassanelli, M.; Prosapio, V.; Norton, I.; Mills, T. Acidified/basified gellan gum gels: The role of the structure in drying/rehydration mechanisms. *Food Hydrocoll.* **2018**, *82*, 346–354. [CrossRef]
3. Osmałek, T.; Froelich, A.; Tasarek, S. Application of gellan gum in pharmacy and medicine. *Int. J. Pharm.* **2014**, *466*, 328–340. [CrossRef]
4. Chang, S.J.; Huang, Y.-T.; Yang, S.-C.; Kuo, S.M.; Lee, M.-W. In vitro properties of gellan gum sponge as the dental filling to maintain alveolar space. *Carbohydr. Polym.* **2012**, *88*, 684–689. [CrossRef]
5. Pandey, V.S.; Verma, S.K.; Yadav, M.; Behari, K. Studies on graft copolymerization of gellan gum with *N*,*N*-dimethylacrylamide by the redox system. *Int. J. Biol. Macromol.* **2014**, *70*, 108–115. [CrossRef]
6. Giri, T.K.; Verma, P.; Tripathi, D.K. Grafting of vinyl monomer onto gellan gum using microwave: Synthesis and characterization of grafted copolymer. *Adv. Compos. Mater.* **2014**, *24*, 531–543. [CrossRef]
7. Vijan, V.; Kaity, S.; Biswas, S.; Isaac, J.; Ghosh, A. Microwave assisted synthesis and characterization of acrylamide grafted gellan, application in drug delivery. *Carbohydr. Polym.* **2012**, *90*, 496–506. [CrossRef]
8. Nandi, G.; Patra, P.; Priyadarshini, R.; Kaity, S.; Ghosh, L.K. Synthesis, characterization and evaluation of methacrylamide grafted gellan as sustained release tablet matrix. *Int. J. Biol. Macromol.* **2015**, *72*, 965–974. [CrossRef]
9. Karthika, J.S.; Vishalakshi, B. Microwave-Assisted Synthesis and Characterization of Poly(2-(dimethylamino)ethyl methacrylate) Grafted Gellan Gum. *Int. J. Polym. Anal. Charact.* **2014**, *19*, 709–720. [CrossRef]
10. Verma, S.K.; Pandey, V.S.; Behari, K. Gellan gum-g-*N*-vinyl-2-pyrrolidone: Synthesis, swelling, metal ion uptake and flocculation behavior. *Int. J. Biol. Macromol.* **2015**, *72*, 1292–1300. [CrossRef]
11. Abu Elella, M.H.; Mohamed, R.R.; Sabaa, M.W. Synthesis of novel grafted hyaluronic acid with antitumor activity. *Carbohydr. Polym.* **2018**, *189*, 107–114. [CrossRef]
12. Caner, H.; Yilmaz, E.; Yilmaz, O. Synthesis, characterization and antibacterial activity of poly(N-vinylimidazole) grafted chitosan. *Carbohydr. Polym.* **2007**, *69*, 318–325. [CrossRef]
13. Unal, H.I.; Inegollu, C.; Sanli, O. Graft copolymerization of N-vinylimidazole on poly(ethylene tetephthalate) fibers in a swelling solvent using azobisisobutyronitrile as initiator. *Turk. J. Chem.* **2003**, *27*, 403–415.
14. Abu Elella, M.H.; Mohamed, R.R.; Elhafeez, E.A.; Sabaa, M.W. Synthesis of novel biodegradable antibacterial grafted xanthan gum. *Carbohydr. Polym.* **2017**, *173*, 305–311. [CrossRef]
15. El-Hamshary, H.; Fouda, M.M.; Moydeen, M.; El-Newehy, M.H.; Al-Deyab, S.S.; Abdel-Megeed, A. Synthesis and antibacterial of carboxymethyl starch-grafted poly(vinyl imidazole) against some plant pathogens. *Int. J. Biol. Macromol.* **2015**, *72*, 1466–1472. [CrossRef]
16. Anderson, E.B.; Long, T.E. Imidazole- and imidazolium-containing polymers for biology and material science applications. *Polymer* **2010**, *51*, 2447–2454. [CrossRef]
17. Pineda-Contreras, A.; Hernández-Madrigal, J.V.; Vázquez-Vuelvas, O.F.; Fomine, S. Synthesis and ROMP of new sulfobetaine and carboxybetaine norbornene. *e-Polymers* **2016**, *16*, 181–188. [CrossRef]
18. Vasiliu, S.; Neagu, V.; Popa, M.; Luca, C. Materials with special properties based on polybetaines. *Mater. Plast.* **2008**, *45*, 177–183.
19. Barboiu, V.; Holerca, M.N.; Ştreba, E.; Luca, C. Reactions on polymers with amine groups. 3. Description of the addition reaction of pyridine and imidazole compounds with alpha, beta-unsaturated monocarboxylic acids. *J. Polym. Sci. Part A Polym. Chem.* **1996**, *34*, 261–270. [CrossRef]
20. Gonsior, N.; Mohr, F.; Ritter, H. Synthesis of mesomeric betaine compounds with imidazolium-enolate structure. *Beilstein J. Org. Chem.* **2012**, *8*, 390–397. [CrossRef]
21. Laschewsky, A. Structures and Synthesis of Zwitterionic Polymers. *Polymer* **2014**, *6*, 1544–1601. [CrossRef]
22. Liu, D.; Zhu, J.; Qiu, M.; He, C. Antifouling PVDF membrane grafted with zwitterionic poly(lysine methacrylamide) brushes. *RSC Adv.* **2016**, *6*, 61434–61442. [CrossRef]
23. Li, M.; Zhuang, B.; Yu, J. Functional Zwitterionic Polymers on Surface: Structures and Applications. *Chem. Asian J.* **2020**, *15*, 2060–2075. [CrossRef]

24. Blackman, L.D.; Gunatillake, P.A.; Cass, P.; Locock, K.E.S. An introduction to zwitterionic polymer behavior and applications in solution and at surfaces. *Chem. Soc. Rev.* **2019**, *48*, 757–770. [CrossRef]
25. Tarannum, N.; Singh, M. Advances in Synthesis and Applications of Sulfo and Carbo Analogues of Polybetaines: A Review. *Rev. Adv. Sci. Eng.* **2013**, *2*, 90–111. [CrossRef]
26. Didukh, A.G.; Koizhaiganova, R.B.; Khamitzhanova, G.; Bimendina, L.A.; Kudaibergenov, S.E. Stimuli-sensitive behaviour of novel betaine-type polyampholytes. *Polym. Int.* **2003**, *52*, 883–891. [CrossRef]
27. Tada, S.; Inaba, C.; Mizukami, K.; Fujishita, S.; Gemmei-Ide, M.; Kitano, H.; Mochizuki, A.; Tanaka, M.; Matsunaga, T. Anti-Biofouling Properties of Polymers with a Carboxybetaine Moiety. *Macromol. Biosci.* **2008**, *9*, 63–70. [CrossRef]
28. Lewis, A.L. Phosphorylcholine-based polymers and their use in the prevention of biofouling. *Colloids Surf. B Biointerfaces* **2000**, *18*, 261–275. [CrossRef]
29. West, S.L.; Salvage, J.P.; Lobb, E.J.; Armes, S.P.; Billingham, N.C.; Lewis, A.L.; Hanlon, G.W.; Lloyd, A.W. The biocompatibility of crosslinkable copolymer coatings containing sulfobetaines and phosphobetaines. *Biomaterials* **2004**, *25*, 1195–1204. [CrossRef]
30. Luca, C.; Mihailescu, S.; Popa, M. Polymers containing quaternary ammonium groups based on poly(N-vinylimidazole). *Eur. Polym. J.* **2002**, *38*, 1501–1507. [CrossRef]
31. Racovita, S.; Vasiliu, S.; Neagu, V. Solution properties of three polyzwitterions based on poly (N-vinylimidazole). *Iran. Polym. J.* **2010**, *19*, 333–341.
32. Sabaa, M.W.; Mohamed, N.A.; Mohamed, R.R.; Khalil, N.M.; El Latif, S.M.A. Synthesis, characterization and antimicrobial activity of poly (N-vinyl imidazole) grafted carboxymethyl chitosan. *Carbohydr. Polym.* **2010**, *79*, 998–1005. [CrossRef]
33. Bamford, C.; Schofield, E. Non-classical free-radical polymerization: Degradative addition to monomer in the polymerization of 1-vinylimidazole. *Polymer* **1981**, *22*, 1227–1235. [CrossRef]
34. Behari, K.; Pandey, P.; Kumar, R.; Taunk, K. Graft copolymerization of acrylamide onto xanthan gum. *Carbohydr. Polym.* **2001**, *46*, 185–189. [CrossRef]
35. Badwaik, H.; Sakure, K.; Alexander, A.; Ajazuddin Dhongade, H.; Tripathi, D.K. Synthesis and characterisation of poly(acryalamide) grafted carboxymethyl xanthan gum copolymer. *Int. J. Biol. Macromol.* **2016**, *85*, 361–369. [CrossRef]
36. Bhattacharyya, S.; Maldas, D. Graft copolymerization onto cellulosics. *Prog. Polym. Sci.* **1984**, *10*, 171–270. [CrossRef]
37. Taghizadeh, M.T.; Khosravy, M. Kinetics and mechanism of graft copolymerization of vinyl monomers (acrylamide, acrylic acid and methacrylate) onto starch by potassium dichromate as redox initiator. *Iran. Polym. J.* **2003**, *12*, 497–505.
38. Sirirat, T.; Vatanatham, T.; Hansupalak, N.; Rempel, G.L.; Arayapranee, W. Kinetics and modeling of methyl methacrylate graft copolymerization in the presence of natural rubber latex. *Korean J. Chem. Eng.* **2015**, *32*, 980–992. [CrossRef]
39. Mostafa, K.; Samarkandy, A.R.; El-Sanabary, A. Grafting onto carbohydrate polymer using novel potassium persulfate/tetramethylethylene diamine redox system for initiating grafting. *Adv. Polym. Technol.* **2011**, *30*, 138–149. [CrossRef]
40. Neagu, V.; Vasiliu, S.; Racovita, S. Adsorption studies of some inorganic and organic salts on new zwitterionic ion exchangers with carboxybetaine moieties. *Chem. Eng. J.* **2010**, *162*, 965–973. [CrossRef]
41. Talu, M.; Demiroğlu, E.U.; Yurdakul, Ş.; Badoğlu, S. FTIR, Raman and NMR spectroscopic and DFT theoretical studies on poly(*N*-vinylimidazole). *Spectrochim. Acta Part. A Mol. Biomol. Spectrosc.* **2015**, *134*, 267–275. [CrossRef]
42. Coutinho, D.F.; Sant, S.V.; Shin, H.; Oliveira, J.T.; Gomes, M.E.; Neves, N.M.; Khademhosseini, A.; Reis, R.L. Modified Gellan Gum hydrogels with tunable physical and mechanical properties. *Biomaterials* **2010**, *31*, 7494–7502. [CrossRef]
43. Lu, Y.; Zhao, X.; Fang, S. Characterization, Antimicrobial Properties and Coatings Application of Gellan Gum Oxidized with Hydrogen Peroxide. *Foods* **2019**, *8*, 31. [CrossRef]
44. Vasiliu, S.; Bunia, I.; Racovita, S.; Neagu, V. Adsorption of cefotaxime sodium salt on polymer coated ion exchange resin microparticles: Kinetics, equilibrium and thermodynamic studies. *Carbohydr. Polym.* **2011**, *85*, 376–387. [CrossRef]
45. Higuchi, W.I. Diffusional Models Useful in Biopharmaceutics. *J. Pharm. Sci.* **1967**, *56*, 315–324. [CrossRef]

46. Korsmeyer, R.W.; Gurny, R.; Doelker, E.; Buri, P.; Peppas, N.A. Mechanisms of solute release from porous hydrophilic polymers. *Int. J. Pharm.* **1983**, *15*, 25–35. [CrossRef]
47. Sepulveda, P.; Binner, J.G.P. Persulfate–Amine Initiation Systems for Gelcasting of Ceramic Foams. *Chem. Mater.* **2001**, *13*, 4065–4070. [CrossRef]
48. Constantin, M.; Mihalcea, I.; Oanea, I.; Harabagiu, V.; Fundueanu, G. Studies on graft copolymerization of 3-acrylamidopropyl trimethylammonium chloride on pullulan. *Carbohydr. Polym.* **2011**, *84*, 926–932. [CrossRef]
49. Srivastava, A.; Behari, K. Studies on Graft Copolymerization of *N*-Vinyl-2-pyrrolidone on to Carboxymethylcellulose (Sodium Salt) and Metal Ion Sorption Behavior. *J. Macromol. Sci. Part A* **2006**, *43*, 1065–1081. [CrossRef]
50. Clark, W.M.; Lubs, H.A. The Colorimetric Determination of Hydrogen ion Concentration and its Applications in Bacteriology PART I1. *J. Bacteriol.* **1917**, *2*, 1–34. [CrossRef]
51. Sorensen, S.P.L. Enzyme studies II. On the measurement and the importance of hydrogen ion concentrationduring enzymatic processes. *Biochem. Z.* **1909**, *21*, 131–200.

Sample Availability: Not available.

molecules

MDPI

Article

Influence of Chitin Source and Polymorphism on Powder Compression and Compaction: Application in Drug Delivery

Linda Al-Hmoud [1], Deeb Abu Fara [1,*], Iyad Rashid [2], Babur Z. Chowdhry [3] and Adnan A. Badwan [2]

1 Chemical Engineering Department, School of Engineering, University of Jordan, Amman 11942, Jordan; l.alhmoud@ju.edu.jo

2 Research and Innovation Centre, The Jordanian Pharmaceutical Manufacturing Company (JPM), P.O. Box 94, Naor 11710, Jordan; irashid@jpm.com.jo (I.R.); adnanbadwan@gmail.com (A.A.B.)

3 School of Science, Faculty of Engineering & Science, University of Greenwich, Medway Campus, Chatham Maritime, Kent ME4 4TB, UK; b.z.chowdhry@greenwich.ac.uk

* Correspondence: abufara@ju.edu.jo; Tel.: +962-799182424

Academic Editor: Leonard Ionut Atanase

Received: 15 October 2020; Accepted: 9 November 2020; Published: 12 November 2020

Abstract: The objective of the research reported herein is to compare the compaction properties of three different chitin extracts from the organisms most used in the seafood industry; namely crabs, shrimps and squids. The foregoing is examined in relation to their polymorphic forms as well as compression and compaction behavior. Chitin extracted from crabs and shrimps exhibits the α-polymorphic form whilst chitin extracted from squid pins displays a β-polymorphic form. These polymorphs were characterized using FTIR, X-ray powder diffraction and scanning electron microscopy. Pore diameter and volume differ between the two polymorphic powder forms. The β form is smaller in pore diameter and volume. Scanning electron microscopy of the two polymorphic forms shows clear variation in the arrangement of chitin layers such that the α form appears more condensed due to the anti-parallel arrangement of the polymer chains. True, bulk and tapped densities of these polymorphs and their mixtures indicated poor flowability. Nevertheless, compression and compaction properties obtained by applying Heckle and Kawakita analyses indicated that both polymorphs are able to be compacted with differences in the extent of compaction. Chitin compacts, regardless of their origin, showed a very high crushing strength with very fast dissolution which makes them suitable for use as fast mouth dissolving tablets. Moreover, when different chitin powders are granulated with two model drugs, i.e., metronidazole and spiramycin they yielded high crushing strength and their dissolution profiles were in accordance with compendial requirements. It is concluded that the source of chitin extraction is as important as the polymorphic form when compression and compaction of chitin powders is carried out.

Keywords: chitin; polymorphs; chitin sources; chitin characterization; disintegration; dissolution; crushing strength; compression analysis

1. Introduction

Chitin (CH; poly-β-(1,4)-*N*-acetyl glucosamine) occurs in the exoskeletons of crustaceans, insects and the cell walls of algae and fungi. CH has a protective and supporting function; it is responsible for the rigidity of the exoskeletons of insects and crustacean, and its presence in the cell walls of fungi and algae facilitates movement [1]. Currently, there are no synthetic methods for preparing CH. As a result, methods to obtain CH are dependent on extraction from crustaceans and algae. Usually, these methods are carried out in the following sequence: deproteinization, demineralization and discoloration [2]. The methods of preparation, as well as physical and chemical characterization of CH, are the subject of an extensive review by Daraghmah et al. [1].

Usually, CH exists in different organisms in its "complexed" forms with proteins, carbohydrates, or lipids. This means that CH, in order to interact with other chemical functional groups in the tissues of organisms, must have free functional binding groups. This is clear evidence that CH must have some deacetylated domains which means it cannot be isolated in pure form. Consequently, purification of CH involves freeing this insoluble material from various debris and other active matter such as deacetylated derivatives; namely, chitosan (CHS) and glucosamine. It seems that there is no clear demarcation between extraction and purification of CH which is therefore obtained with different molecular weights ranging from hundreds to thousands of kDa. Regardless of variations in molecular weight, CH can occur in three distinct existing polymorphic forms: α, β and γ. CH polymer chains in the α polymorph are aligned in an anti-parallel fashion, whilst in β polymorphs the alignment occurs in parallel. In the γ form the CH polymer chains are arranged in a pattern whereby two parallel chains are in one direction whilst the third is in the opposite direction [1]. Whether or not the anatomical variation of CH in a particular organism affects the mechanical properties of extracted CH is still the subject of active research.

CH is a polymer which is insoluble in water and other conventional solvents. It has a structure deprived of active functional groups. This leaves the CH surface without any reactive functional groups with no ability to react with any chemical moieties; thus, it is chemically inert. In addition, its lack of absorption in the human body allows it to be considered as a pharmacologically inert material. This means it is an ideal material to be used as a pharmaceutical excipient from both a chemical and pharmacological perspective [3,4]. Furthermore, CH is classified as grass in the FDA classification for materials safety usage [5]. This unique characteristic keeps the door open for future use as a pharmaceutical excipient particularly in liquid and solid dosage forms. The challenge in utilizing CH in solid pharmaceutical preparations is its powder flow and mechanical characteristics, including compaction and compression [6].

It may be advantageous to highlight the similarity of CH with the most popular excipient in pharmaceutical solid dosage form preparations, i.e., cellulose and its derivatives such as micro crystalline cellulose (MCC) [7]. MCC shares with CH its β-(1,4) glycosidic bond, and its relative chemical and pharmacological inertness. Researchers in the CH field are attempting to emulate and test if the same techniques which have been successfully used to modify MCC can be applied in the field of CH research, particularly in solid pharmaceutical dosage form preparations [8,9].

It is well known that, like MCC, CH and CHS powder flow properties suffer from low bulk density. This causes unsmooth flow which is attributed to the fibrous nature of these natural polymers [6,10]. Attempts to alleviate this drawback have been tested by adding silicon dioxide to CH and CHS powders [11,12]. The addition of silicon dioxide facilitates smooth flow in tableting machines. Furthermore, combining CH/CHS with Avicel PH201, starch 1500, calcium carbonate or gelatin have been tested and were found to dramatically improve the flow behavior of the composite powders [13–15]. The improvement in flow properties has encouraged novel applications of these composites in producing pharmaceutical solid dosage forms excipients, especially as direct compression excipients. It is worth mentioning that various attempts to utilize CH as a novel solid drug delivery system were carried out by Daraghmeh et.al. [16], whereby CH was co-processed with mannitol to produce orodispersible tablets able to disintegrate in the mouth within a few seconds. Furthermore, CH was formulated, by Gana et.al. [17], with cephalosporins and metal silicates in order to obtain an insight into the effect of pH on the CH surface and its influence on drug stability. Abu Fara et al. [8] used roller compaction to improve CH powder flow qualities and to make its compression and compaction characteristics compatible for use in industrial pharmaceutical machines. Additionally, there are some on-going trials to extend the use of CH and CHS as excipients in a similar method followed in improving MCC powder flow using spray drying [6,18]. Such studies, show that

CH is a suitable solid pharmaceutical excipient. As a result, it can be concluded that CH powder modification is essential in order to commercialize this novel excipient.

Natural polymers usually contain crystalline and amorphous domains in their structure. The balance between these two components controls the powder flow behavior. It has been reported that during the compaction of MCC, its crystalline domains are responsible for fragmentation while its amorphous domains are responsible for polymer plasticity, which induces bonding [7]. This is confirmed in the case of lactose, whereby amorphous lactose offers better compactibility than the crystalline form [19,20]. Accordingly, it seems that an amorphous powder is more compactable but the bulk density is usually lower than the crystalline material. Consequently, studying the impact of the crystalline-amorphous balance on powder flow is a pre-requisite to its industrial use. However, a review of the scientific literature does not show if CH polymorphs differ in their response towards compression and compaction.

CH has diverse functions in different tissues in an organism; e.g., the CH in the lining of the gastrointestinal tract in some organisms is more flexible to suite its functional role and has more elasticity than exoskeleton CH, which is a harder material. Such functionality raises the question whether CH polymorphs behave in a similar manner when compressed, or does the functional use dictate their mechanical strength, as expressed in different tissues of organism? As sea food remains from industrial packaging contain a mixture of CH with various origins, it would be interesting to compare extracted CH from the most utilized CH sources, namely shrimp, crab and squid. These organisms are known to have α (shrimp and crab) and β (squid) CH polymorphs. Such polymorphs can be differentiated by their powder X-ray diffraction patterns. It would be interesting to test each polymorph individually and discern the differences in their compaction and compression behavior.

The present work focuses on pharmaceutical powder compression and compaction properties of CH extracted from the most widely used sea food organisms namely crabs, shrimps, and squids. The influence of variation in α and β CH polymorphic forms due to CH extraction source on compression and compaction is explored. Hence it may be possible to ascertain whether CH sources must be processed separately or collectively prior to their extraction.

2. Results

2.1. Bulk, Tapped Density and True Density

The bulk, tapped, and true densities of CH from different sources, as well as of their mixtures (Sh9Sq1, Sh7Sq3, and Sh5Sq5) are presented in Table 1. Crab (CH-Cr) and shrimp (CH-Sh) CHs have much higher bulk and tapped densities than squid CH (CH-Sq).

Table 1. True, bulk and tapped densities, HR, and CI of CHs from different sources and their mixtures.

Sample *	Bulk Density (kg/m^3)	Tapped Density (kg/m^3)	True Density (kg/m^3)	HR	CI
CH-Cr	332 ± 9	531 ± 9	1400 ± 30	1.60 ± 0.04	60 ± 4
CH-Sh	280 ± 6	401 ± 6	1700 ± 40	1.43 ± 0.02	43 ± 2
CH-Sq	190 ± 4	338 ± 4	1200 ± 20	1.78 ± 0.03	78 ± 3
CH-Sh9Sq1	255 ± 6	402 ± 6	1650 ± 35	1.58 ± 0.03	58 ± 3
CH-Sh7Sq3	252 ± 6	403 ± 6	1550 ± 30	1.60 ± 0.03	60 ± 3
CH-Sh5Sq5	242 ± 5	403 ± 5	1450 ± 25	1.67 ± 0.02	67 ± 2

* CH-Cr = crab CH, CH-Sh = shrimp CH, CH-Sq = squid CH, CH-Sh9Sq1 = mixture of 90 wt% shrimp CH and 10 wt% squidCH, CH-Sh7Sq3 = mixture of 70 wt% shrimpCH and 30 wt% squid CH, CH-Sh9Sq1 = mixture of 50 wt% shrimp CH and 50 wt% squid CH.

To study the effect of mixing CH from different sources, CH-Sh was chosen as representative of α-CH, and mixed with CH-Sq (β-CH) in different compositions, as shown in Table 1. With respect to bulk and tapped densities, such mixtures manifested intermediate values between the low (CH-Sq) and high (CH-Sh) bulk and tapped densities.

In comparing the powder flowability of CH from different sources, the Carr index (CI) and Hausner ratio (HR) were determined [21]. Flowability results, as illustrated in Table 1, indicate that CH from different sources, as well as their mixtures, exhibit extremely poor flowability. Flow interpretation criteria based on HR and CI is given in the literature [21]. As for densities, the mixtures manifested intermediate HR and CI values between the low and high values presented by CH-Sh and CH-Sq, respectively.

2.2. *Water Content*

The water content of CH from different sources are presented in Table 2. The water content of CH-Cr and CH-Sh is almost half that of CH-Sq.

Table 2. Water content of CH extracted from different sources.

Sample	CH-Cr	CH-Sh	CH-Sq
Water content (wt %)	5.44 ± 0.06	4.84 ± 0.05	8.65 ± 0.06

2.3. *Porosity of Powders*

The porosity parameters of the three CH powder samples are shown in Table 3. The pore volume of crab and shrimp CHs are of similar values, whereas that of squid CH is ~20% less. On the other hand, crab CH has the largest pore diameter and squid CH has the smallest.

Table 3. Porosity parameters for CH-Cr, CH-Sh, and CH-Sq.

Sample	Pore Volume (cm^3/g)	Average Pore Diameter (nm)
CH-Cr	$(7.308 \pm 0.003) \times 10^{-3}$	5.68 ± 0.01
CH-Sh	$(6.977 \pm 0.003) \times 10^{-3}$	4.32 ± 0.01
CH-Sq	$(5.353 \pm 0.002) \times 10^{-3}$	3.70 ± 0.01

2.4. *FTIR*

The Fourier-transform infrared (FT-IR) spectra of α-CH (CH-Cr and CH-Sh) and β-CH (CH-Sq) are shown in Figure 1. For α-CH, the amide I band is split at about 1650 and 1620 cm^{-1} (Figure 1a,b), whereas it is a single sharp band at about 1657 cm^{-1} for β-CH (Figure 1c). The amide II band appears at about 1555 and 1559 cm^{-1} for α- and β-CH, respectively. Both polymorphs show strong absorption bands in the 3100–3285 cm^{-1} region which correspond to the N–H group. Bands in the 2840–2960 cm^{-1} region are due to CH, CH_2, and CH_3 in both CH polymorphs [1].

2.5. *X-ray Powder Diffraction (XRPD) Analysis*

The XRPD profiles of CH from different sources are presented in Figure 2. The XRPD patterns of crab (CH-Cr) and shrimp CH (CH-Sh) show four sharp crystalline reflections indicative of α-CH at $2\theta = 9.3, 19.1, 20.6,$ and 23.2°.

The XRPD pattern of squid CH (CH-Sq) shows two broad crystalline peaks at 8.7 and 19.8° which indicates that squid CH is of the β-CH type. The crystal size of CH-Cr and CH-Sh is almost the same. Squid CH has a smaller crystal size than crab and shrimp as its pattern peaks at 9° and 19° are much wider for CH-Sq than the corresponding ones for CH-Cr and CH-Sh. Table 4 shows the crystalline index (I_{CR}) of the three CH samples. These results indicate that α-CH has a more crystalline structure because of its inter-sheet and intra-sheet structure [22,23]. Crystalline index (I_{CR}) is used to estimate the degree of deacetylation (DDA) of CH using a method reported by Zhang et al. [24]. DDA results presented in Table 4 shows that α-CHs (CH-Cr and CH-Sh) is less deacetylated than β-CH (CH-Sq), which conforms with findings reported in the literature [1,2,22].

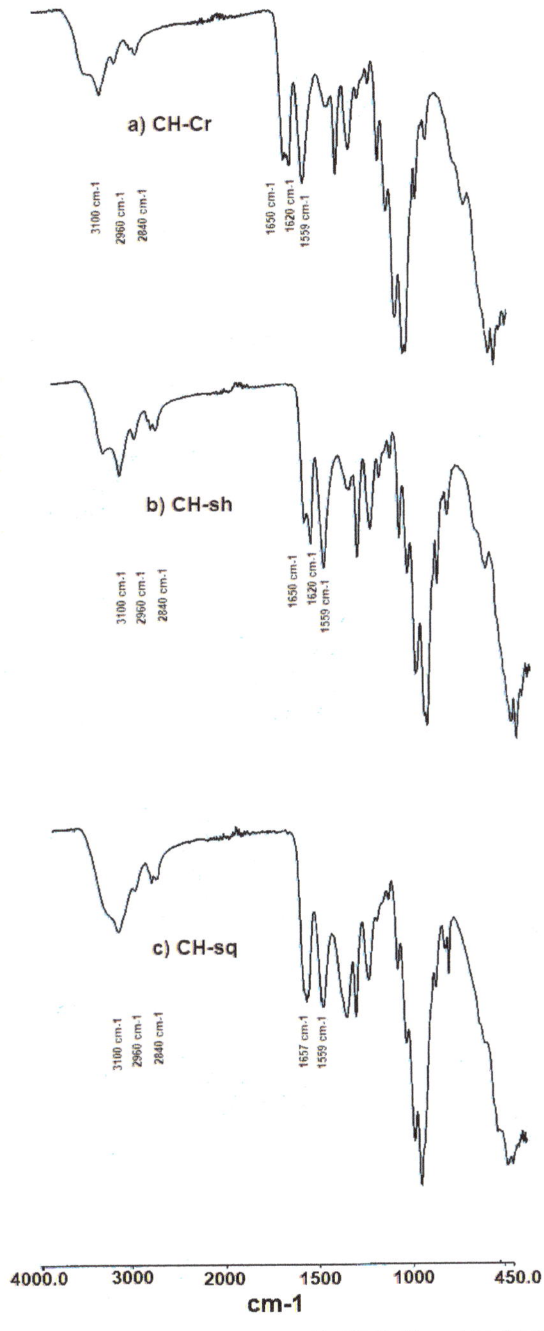

Figure 1. FTIR spectra of (**a**) CH-Cr, (**b**) CH-Sh (**b**), and (**c**) CH-Sq.

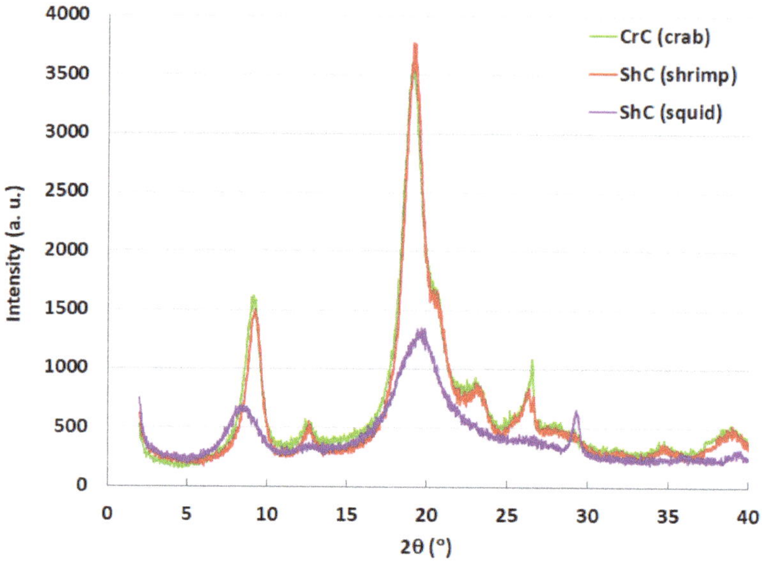

Figure 2. X-ray powder diffraction (XRPD) patterns of CH-Cr, CH-Sh, and CH-Sq CH.

Table 4. Crystalline index (I_{CR}) and degree of deacetylation (DDA) of CH extracted from different sources.

Sample	CH-Cr	CH-Sh	CH-Sq
I_{CR} (%)	84.3 ± 0.1	85.2 ± 0.1	72.1 ± 0.2
DDA (%)	26.1 ± 0.6	24.9 ± 0.5	42.3 ± 0.7

2.6. Scanning Electron Microscope (SEM)

SEM was used to highlight the contribution of micro-irregularities to powder physical properties by visualizing the surface of CH particles from different sources. Figures 3–5 present SEM images of crab (CH-Cr), shrimp (CH-Sh), and squid (CH-Sq) CHs at a magnification of 8000×, 16,000×, and 30,000×, respectively. Figure 6 presents SEM images with measurement of fibers thicknesses of shrimp (CH-Sh) CH at 16,000× and 60,000× magnifications, and of squid (CH-Sq) CH at 5000× magnification.

At 8000× magnification (Figure 3) it is difficult to see any significant difference amongst the different CH samples. At 16,000× magnification (Figure 4) it can be noticed that squid chitin fibers are more organized than crab and shrimp CH samples. At 30,000× magnification (Figure 5) shrimp CH showed two types of fibers; one forming the layers, and the other connecting the layers. Squid CH layers have no connecting fibers.

The data in Figure 6 shows that the fibers forming shrimp CH layers have a thickness of ~200 nm (Figure 6a), where the layers are about 1 μm apart, and the fibers connecting them have a thickness of 35–45 nm (Figure 6b). On the other hand, squid chitin layers are around 60 nm apart and no connecting fibers can be observed at ×50,000 magnification (Figure 6c).

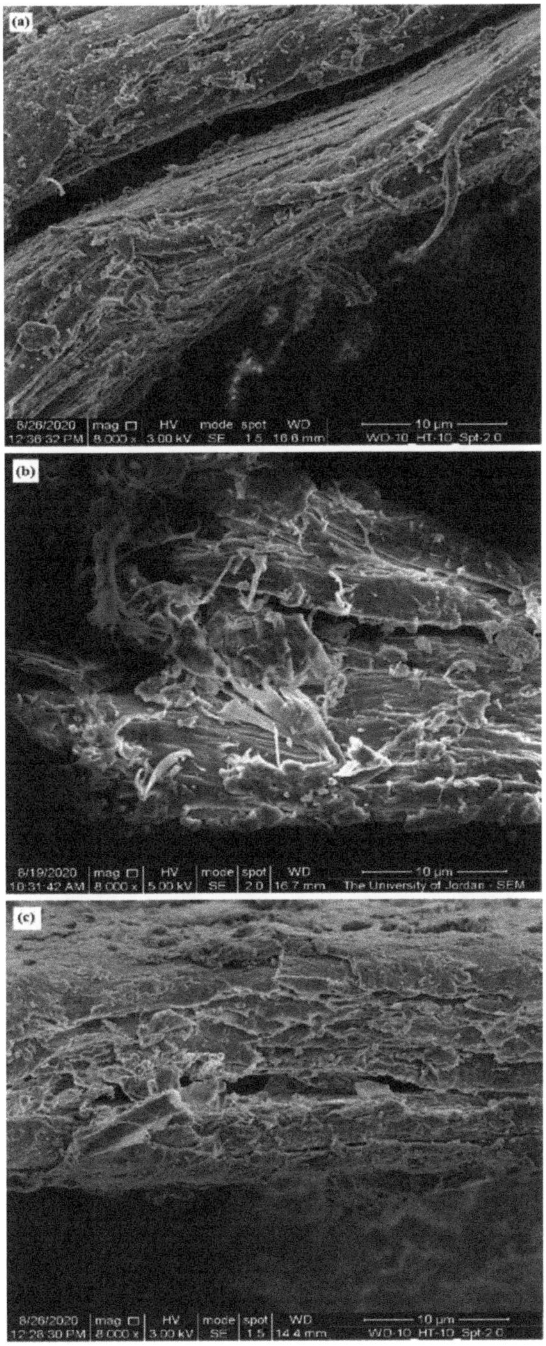

Figure 3. SEM image of (**a**) crab (CH-Cr), (**b**) shrimp (CH-Sh), and (**c**) squid (CH-Sq) CHs at 8000× magnification.

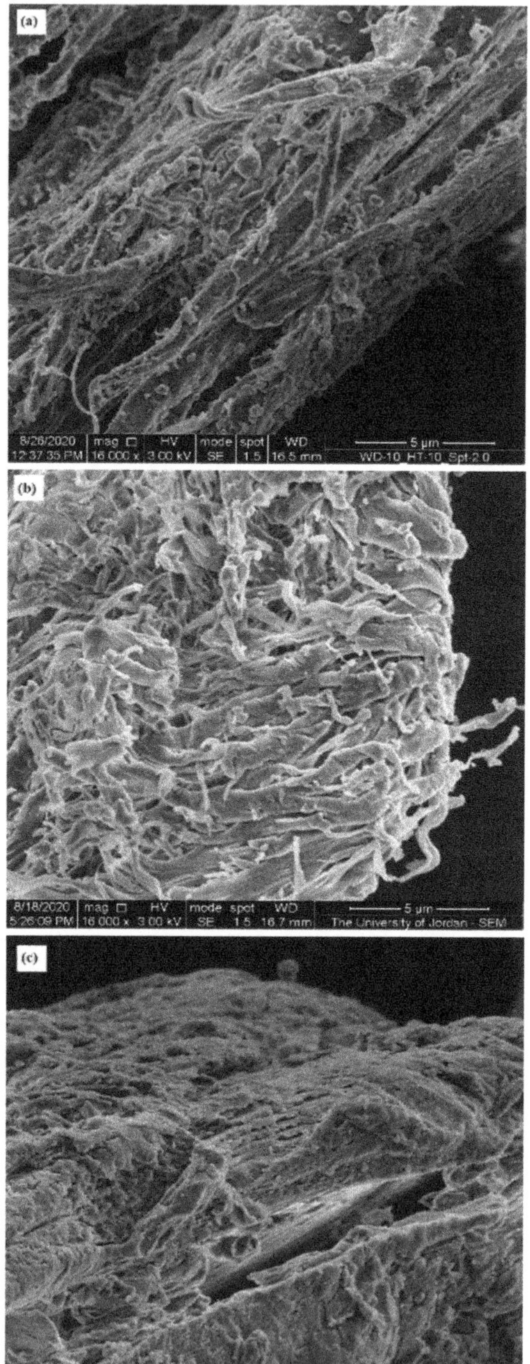

Figure 4. SEM image of (**a**) crab (CH-Cr), (**b**) shrimp (CH-Sh), and (**c**) squid (CH-Sq) CHs at 16,000× magnification.

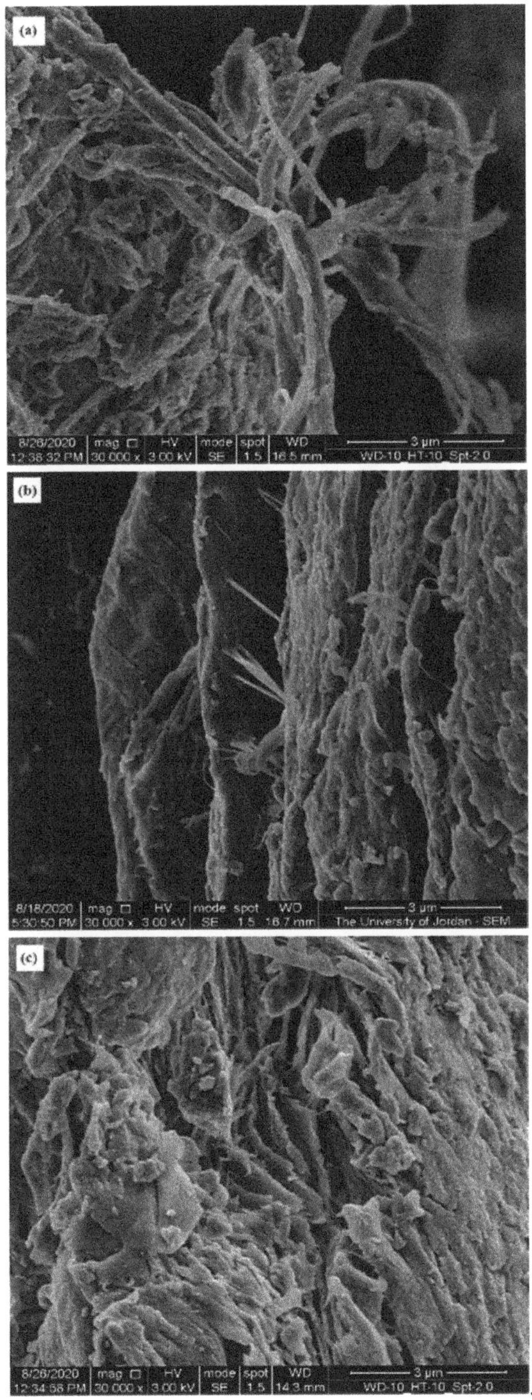

Figure 5. SEM image of (**a**) crab (CH-Cr), (**b**) shrimp (CH-Sh), and (**c**) squid (CH-Sq) CHs at 30,000× magnification.

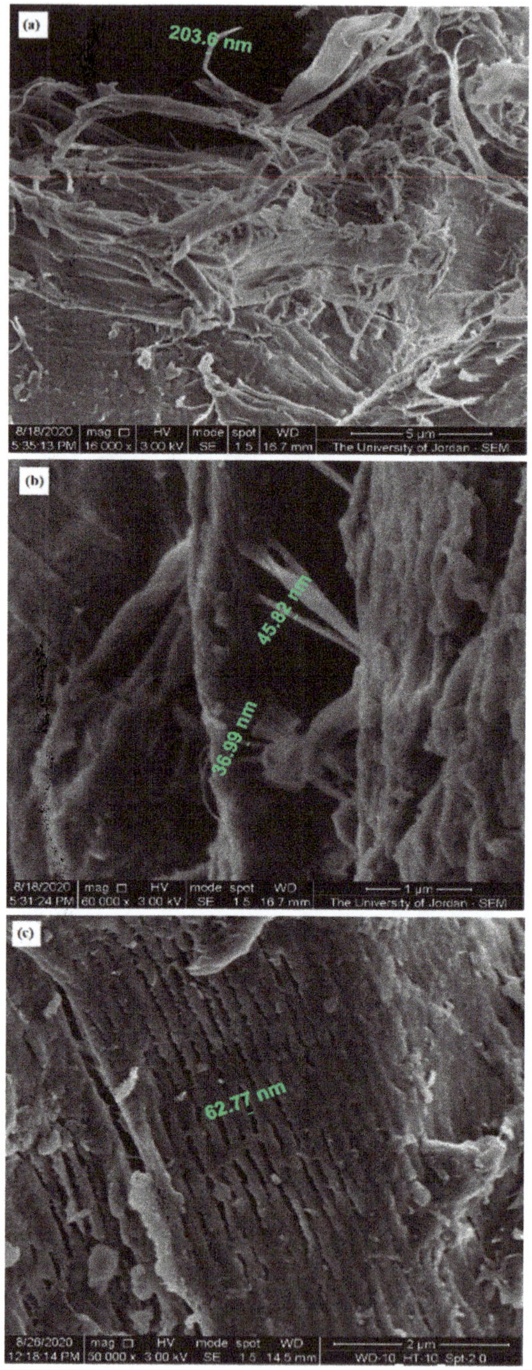

Figure 6. SEM image with measurement of fiber thickness of (**a**) shrimp (CH-Sh) CH at 16,000× magnification, (**b**) shrimp (CH-Sh) CH at 60,000× magnification, and (**c**) squid (CH-Sq) CH at 50,000× magnification.

2.7. Compression Analysis

The three main parameters (a, P_k and ab) obtained via Kawakita analysis (as illustrated in Section 4.2.7. Compression Analysis) were analyzed in an attempt to interpret the compression behavior of the three samples of CH and their mixtures. The data in Figure 7 shows the Kawakita plots of CH-Cr, CH-Sh, and CH-Sq CHs, and Figure 8 shows the Kawakita plots of shrimp/squid CH mixtures; CH-Sh9Sq1, CH-Sh7Sq3, and CH-Sh5Sq5. Table 5 presents the values of Kawakita parameters for the different chitin types and their mixtures.

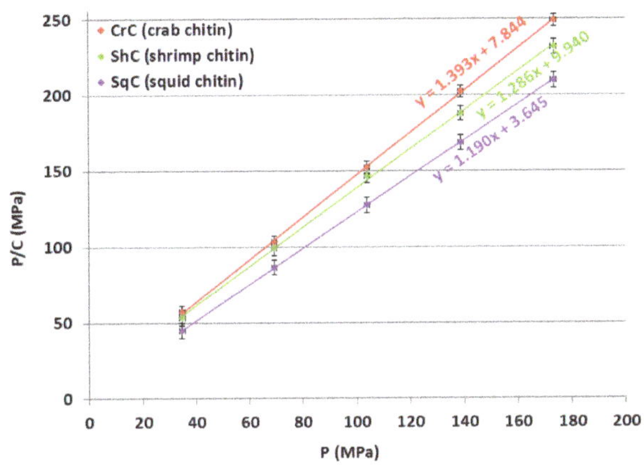

Figure 7. Kawakita plots of crab (CH-Cr), shrimp (CH-Sh), and squid (CH-Sq) CHs.

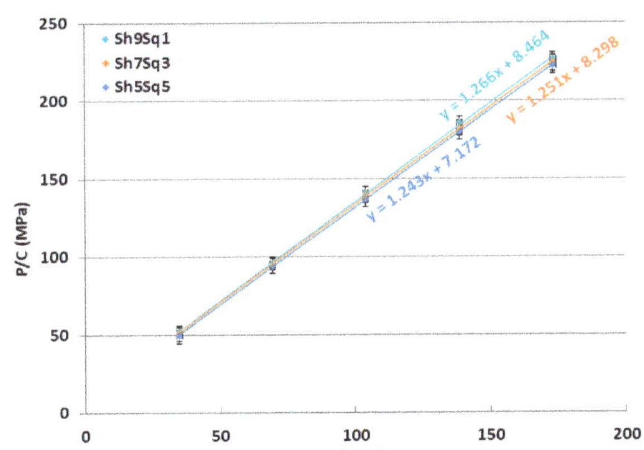

Figure 8. Kawakita plots of CH-Sh9Sq1, CH-Sh7Sq3, and CH-Sh5Sq5 CH mixtures.

Table 5. Kawakita parameters of CH-Cr, CH-Sh, and CH-Sq and their mixtures.

Sample	Slope	Intercept (MPa)	a = 1/Slope	b = Slope/Intercept (1/MPa)	ab (1/MPa)	P_K = 1/b (MPa)
CH-Cr	1.393 ± 0.066	7.84 ± 0.36	0.7179	0.178	0.127	5.63
CH-Sh	1.286 ± 0.061	9.94 ± 0.46	0.7778	0.129	0.101	7.73
CH-Sq	1.190 ± 0.056	3.65 ± 0.17	0.8406	0.326	0.274	3.06
CH-Sh9Sq1	1.266 ± 0.060	8.46 ± 0.39	0.7899	0.150	0.118	6.68
CH-Sh7Sq3	1.251 ± 0.059	8.30 ± 0.38	0.7997	0.151	0.121	6.64
CH-Sh5Sq5	1.243 ± 0.059	7.17 ± 0.33	0.8046	0.173	0.139	5.77

The maximum volume reduction that can be attained (a) illustrates that CH-Cr has the lowest a value with CH-Sq having the highest volume reduction when a compression force is applied.

In addition, the results of the three mixtures show that the volume reduction value increases with increasing percentage of squid CH in the mixture.

P_K, which represents the pressure needed to reduce the value of (*a*) to half its initial value, is the most important Kawakita parameter to be tested. This is because it represents how hard the CH granules are, and therefore their ability to be used in direct compression applications. Table 5 shows that the P_K value of shrimp CH (CH-Sh) is the highest among the three types of CH. P_K value of mixture containing shrimp and squid CHs at different fractional content; CH-Sh9Sq1, CH-Sh7Sq3, and CH-Sh5Sq5, decreases with increasing squid CH fraction.

The last Kawakita parameter used in this work to describe the compression behavior is *ab*. This parameter gives an indication of the degree of rearrangement of powder particles. As Table 6 shows, squid CH (CH-Sq) has the highest the extent of particle rearrangement (*ab*) upon compression. The value of *ab* is the lowest for shrimp CH (CH-Sh). The data in Table 6 also indicates that mixtures containing shrimp and squid CHs in different fractions; CH-Sh9Sq1, CH-Sh7Sq3, and CH-Sh5Sq5, show *ab* values that increase with increasing squid fraction of CH.

Table 6. Heckel parameters of CH-Cr, CH-Sh, and CH-Sq and their mixtures.

Sample	Slope (K)	Intercept (A)	P_Y(MPa)=1/K
CH-Cr	6.71×10^{-3}	0.7336	149
CH-Sh	2.69×10^{-3}	0.5739	372
CH-Sq	1.08×10^{-2}	0.8627	93
CH-Sh9Sq1	3.20×10^{-3}	0.5782	306
CH-Sh7Sq3	4.40×10^{-3}	0.6051	227
CH-Sh5Sq5	4.94×10^{-3}	0.6868	202

Compression analysis was further examined using the empirical Heckel model of compression analysis. In this model, the yield pressure (P_Y) represents a critical outcome whereby it reflects the type and extent of deformation, i.e., plastic/elastic or brittle-fracture. The data in Table 6 and Figure 9 illustrates the yield pressure values for all tested powders. Results show that P_Y ranking of pure CHs follows the order: shrimp chitin > crab chitin > squid CH. Adding squid CH to shrimp CH reduced the P_Y value, and the more squid CH content in the mixture, the lower its P_Y value: CH-Sh9Sq1 > CH-Sh7Sq3 > CH-Sh5Sq5.

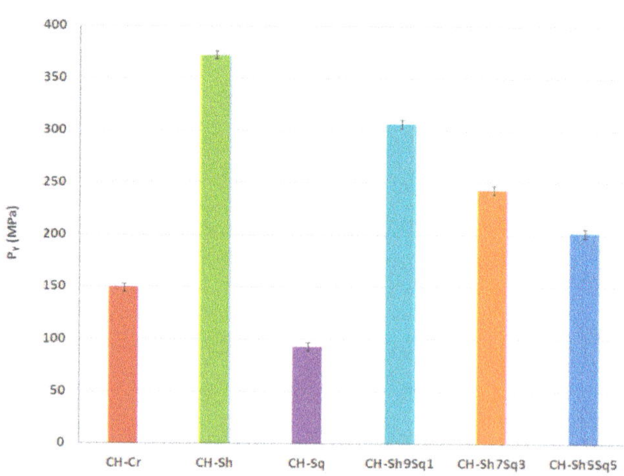

Figure 9. Heckel parameter (P_Y) for CH-Cr, CH-Sh, and CH-Sq and their mixtures.

2.8. Work of Compression

When powders of different types are compressed using the GTP, the instrument displays the force displacement curve during the descending/compression and decompression of the powders. The work of compression (W_C), which is the area under the compression curves of the of the F-D profiles, is shown in Figures 10 and 11. W_C values were calculated at each compression force used. The results shown in Figure 10 indicate that at all loads, the W_C of shrimp CH (CH-Sh) > W_C of crab CH (CH-Cr) > W_C of squid CH (CH-Sq).

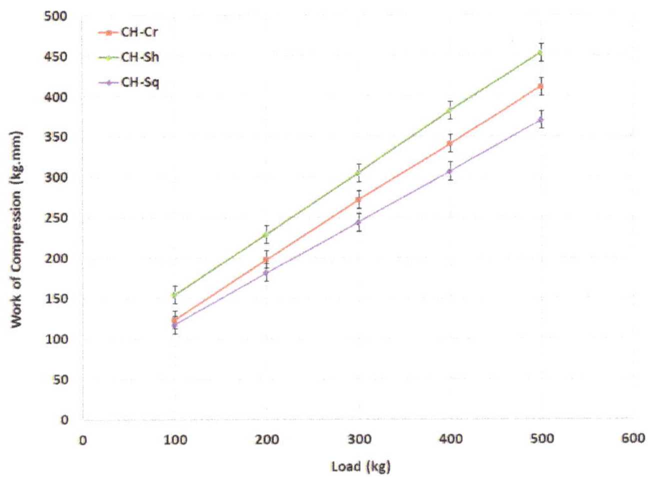

Figure 10. Compression work for CH-Cr, CH-Sh, and CH-Sq CHs.

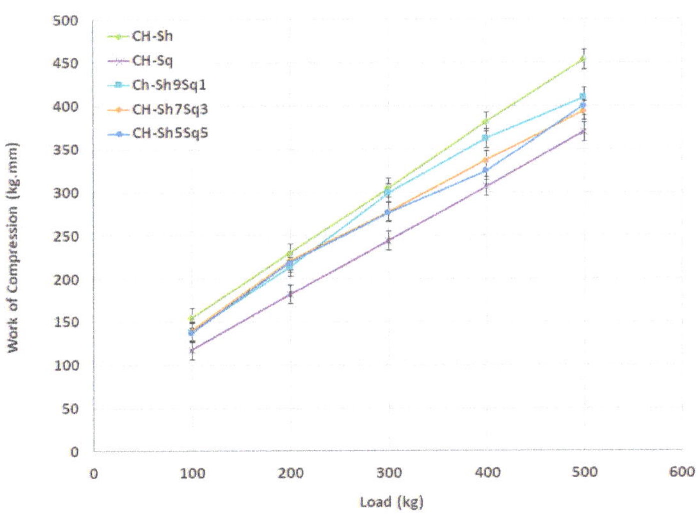

Figure 11. Compression work for CH-Sh9Sq1, CH-Sh7Sq3, and CH-Sh5Sq5 CH mixtures in comparison with pure CH-Sh and CH-Sq.

The data in Figure 11 demonstrates the effect of adding β-CH (CH-Sq) to α-CH (CH-Sh) in different fractional contents. Mixtures containing different amounts of β-CH have W_C values close to each other at low compression loads (100 and 200 kg), while at higher loads W_C decreases with increasing the amount of added CH-Sq.

2.9. Tablet Crushing Strength

Tablet crushing strength for CH from different sources are presented in Table 7. CH-Cr and CH-Sh tablet crushing strength is almost half that of CH-Sq tablet crushing strength. The crushing strength of the mixture is closer in value to that of CH-Sq.

Table 7. Tablet crushing strength.

Sample	CH-Cr	CH-Sh	CH-Sq	CH-Sh9Sq1	CH-Sh7Sq3	CH-Sh5Sq5
Crushing Strength (*N*)	54 ± 2	115 ± 3	310 ± 5	121 ± 3	155 ± 4	195 ± 4

2.10. Tablet Disintegration

All tablets have recorded disintegration times of less than 10 s irrespective of CS source or mixtures comprised thereof.

2.11. Characterization of Metronidazole/Spiramycin Tablets Comprising Drug/CH

Tablet dissolution data are illustrated in Table 8 and in Figures 12 and 13. Figure 12 demonstrates that full drug release of metronidazole/α-CH (crab and shrimp) matrix is achieved within 15 min of dissolution time. This is faster than tablets comprising metronidazole/β-CH (squid) matrix which achieved full drug release within 30 min of dissolution time. Tablets comprising metronidazole/CH mixture (50 wt% shrimp and 50 wt% squid) as well as Rodogyl® tablets also achieved full drug release within 30 min of dissolution time.

Table 8. Time for full drug release for metronidazole® (125 mg) and spiramycin® (0.75 M UI) tablets comprising compacted pure CHs from different sources, and CH mixture (CH-Sh5Sq5) compared to Rodogyl® tablets.

Excipient Used in Tablet Preparation	Metronidazole® Complete Dissolution Time (min)	Spiramycin® Complete Dissolution Time (min)
CH-Cr	15	45
CH-Sh	15	30
CH-Sq	30	45
CH-Sh5Sq5	30	45
Rodogyl®	30	60

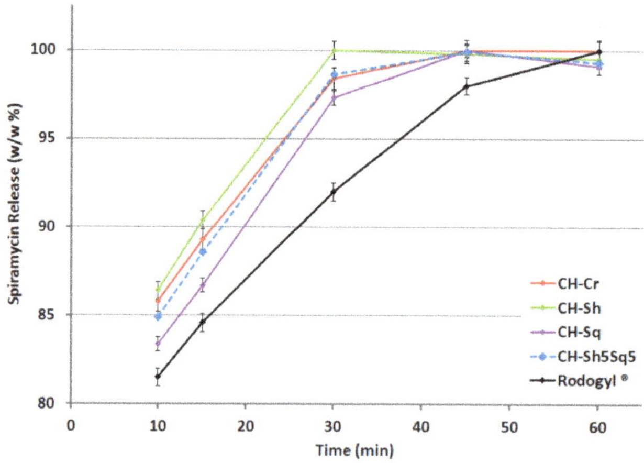

Figure 12. Dissolution profile for metronidazole (200 mg) tablets comprising compacted CHs or CH mixture (CH-Sh5Sq5) compared to Rodogyl® tablets.

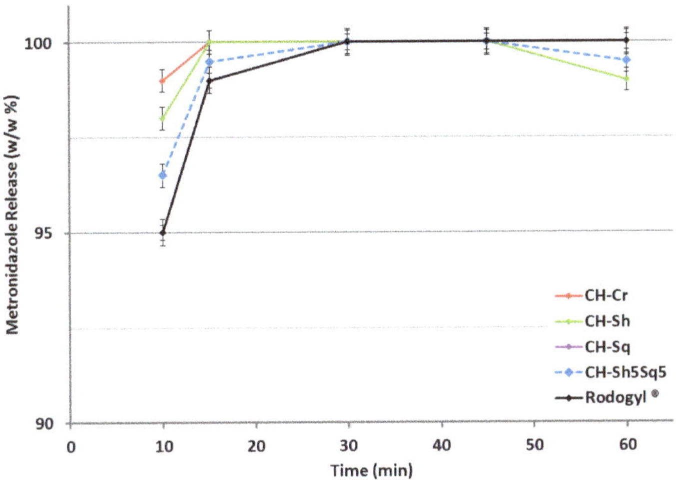

Figure 13. Dissolution profile of spiramycin (0.75 M UI) tablets comprising compacted CHs or CH mixture (CH-Sh5Sq5) compared to Rodogyl® tablets.

Results presented in Figure 13 demonstrates that full drug release of spiramycin/shrimp CH matrix was achieved within 30 min of dissolution time, and within 45 min of dissolution time for the other types of CH, as well as for CH mixtures. This was faster than Rodogyl® tablets which achieved full drug release within 60 min of dissolution time.

3. Discussion

The prime objective of the present investigation was to test CHs extracted from different sources to assess their similarity and suitability for use as a future excipient in pharmaceutical solid dosage form preparations. The final goal is to reach a conclusion whether remains from the sea food packaging industry need to be classified according to their source prior to CH extraction process. This concern becomes evident when one screens the different scientific literature on CH, where there are indications that CH from different sources are chemically similar, but their mechanical properties vary [25].

The strategy followed in this work was to compare CH from crabs and shrimps, both of which belong to α polymorph, in order to find out if the variation in CH source has any influence on the extracted polymer compression and compaction properties. Furthermore, a comparison between α and β-CH polymorphs extracted from squid pins, regarding their compression and compaction behavior, was also conducted. Mixtures of the two polymorphs were also prepared, and their physical and mechanical properties were studied. Whether or not these polymorphic forms could facilitate compaction of two model drugs, namely metronidazole and spiramycin, which are difficult to compress on their own, was tested.

It is well known that shrimps, crabs and squids are the most used organisms in the sea food packaging industry [26]. These organisms differ in their species and their extracted CH is confined to the three well characterized and known polymorphs; α, β and γ. It is essential to first confirm this primary information. The Fourier-transform infrared (FT-IR) spectra of CH-Cr, CH-Sh, and CH-Sq are shown in Figure 1. Distinct vibrational bands appear at a wavelength of 1620–1660 with a distinct split in the band for α-chitin (Figure 1a,b), while there is a single sharp band for β-CH (Figure 1c). This confirms different previous reports on CH of α and β polymorphs [23,27]. Further confirmation of the polymorphic type was carried out using X-ray powder diffraction, Figure 2. The γ polymorph is scarce due to its limited occurrence in the lining of the digestive system of different organisms such as squid. Consequently, the present investigation was limited to a comparison of the compression and compaction behavior of the most commercially available two polymorphs, α and β. Such a comparison helps to clarify if the source of CH extraction causes variation in their related mechanical properties when compacted.

In addition to identifying polymorphic forms of molecules, XRPD has been used as a tool to shed light on the crystallinity and arrangement of CH in different polymer fibers. Figure 2 confirms α polymorph in CH-Cr and CH-Sh, and β polymorph in CH-Sq. The crystalline index, Table 5, shows that the α polymorph is more crystalline than the β polymorph; 84–85% in CH-Cr and CH-Sh, and 73% in CH-Sq. SEM images, Figures 3–6 show the parallel arrangements of β-CH sheets with clear spacing. This is indicative of their ability to hydrate and gel much more than α polymorph [23,27]. Interestingly, the β polymorph shows a smaller pore size and pore volume which may indicate that the value of the parameter a (the capillary water effect) may contribute to its functional strength and flexibility in squid pins. Indeed, the water content determination of CH-Sq shows almost double the value of Ch-Cr and CH-Sh (Table 3). This dramatic variation in water content is an important difference which can be beneficial in the use of CH as excipients for pharmaceutical solid dosage forms. Water content results are consistent with the fact that β-CH (CH-Sq) is more susceptible to intra-crystalline hydration than α-CH (CH-Cr and CH-Sh) due to its lack of intra-sheet hydrogen bonds [23].

Bulk densities of the three types of CH (Table 2) are different although shrimp and crab are similar in polymorphic form. The difference between shrimp and crab CH measured bulk densities may be due to a difference in material arrangements in nature in those living organisms depending on their functional use. Such variation has been previously reported, where extracted CHs from male and female grasshoppers show some differences in CH content and surface morphology [27]. Accordingly, one would expect to observe some variations in CH extracted even from the same species. The higher crab and shrimp chitin bulk densities indicate that these powder samples comprise denser aggregated particles. Such aggregation could be a result of particle-particle adherence which may be attributed to the highly fibrous structure of CH-Cr and CH-Sh, whereas CH-Sq displays a structure which is much less fibrous (Figures 5 and 6). The plain surface of CH-Sq with no fibrous extensions is responsible for the absence of an interlocking structural network, which consequently leads to the formation of aggregated particles in CH-Cr and CH-Sh.

As expected, all types of CH did not provide good flowability as indicated by their HR and CI values (Table 2), which is attributed to their low bulk densities. The foregoing has led, various workers, to suggestion that these powders be treated by roller compaction or slugging before using them as excipients [8]. Thus CH-Sh would be more favorable for tablet compression than CH-Cr and CH-Sq since it has the lowest HR and CI values.

Parameters from Heckle and Kawakita equations indicate that shrimp powder is more resistance to compression than the other two sources. This is reflected in the force required for compression (P_K) which is significantly higher in value than for the other two CH samples (Table 6). Such variation makes the work required for compression higher, as emphasized in the data in Figure 10. Indeed, this is a real indication that differences in CH properties, such as bulk and tapped densities, are reflected in the compression forces required to convert the powders into compacts. The mechanical properties of CH are similar to cellulose derivatives e.g., micro crystalline cellulose (MCC), where reports have shown that various cellulose polymorphs behave differently in response to compaction forces. MCC requires a disintegration agent when used in tablet manufacturing whilst CH powders are self-disintegrating. This is advantageous in terms of the use of CH for pharmaceutical dosage forms which need to be dissolved in the buccal cavity.

The three parameters extracted from the Kawakita equation, namely a, ab, and P_K, are interesting. CH from different sources, which show differences in particle morphology and crystallography, has concurrently shown differences in powder compression behavior. Among the three types, CH-Sq (β-CH) displays the highest volume reduction (a), particle rearrangement (ab), and the lowest compression force (P_K) to reduce the powder bed volume (Table 6). In other words, in order to produce a compact, CH-Sq requires the smallest force to yield the highest volume reduction in comparison with CH-Cr and CH-Sh (α-CHs). This may be attributed to the anti-parallel and parallel arrangement of α-

and β-polymorphs, respectively. As a matter of fact, CH-Cr and CH-Sh responses to compression vary although they belong to the same polymorph. This may be due to variation in the mechanical strength of the extracted materials. Concurrently, examining the results from Heckel analysis (Table 7) CH-Sq is the highest plastically deforming material, it can be stated that compression of the less fibrous and less aggregated material, i.e., CH-Sq, is more mechanically engaged in the deformation of the highly plastic material than the more fibrous one, i.e., CH-Cr and CH-Sh. Such a high extent of deformation shown by CH-Sq contributes to the high reduction of its powder bed volume upon compression compared to CH-Cr and CH-Sh. Moreover, the high extent of deformation of CH-Sq enables the appearance of fresh new surfaces for surface-to-surface contact and bridging [28,29]. Thus, compacts made of CH-Sq have a higher crushing strength than compacts made of CH-Cr and CH-Sh (Table 8).

Although the α-polymorph is more resistant to reduction in size compared to the β-polymorph, such behavior does not hinder both polymorphs from possessing an excellent crushing strength and very short disintegration time. This means direct compression of these individual powders can be utilized in preparing tablets or films which can dissolve in the mouth within seconds liberating the active content. This property can be highly advantageous and utilized in oral dissolving tablets, which are required to dissolve fast but at the same time be composed of a hard compact. The extracted CH polymorphs fit very well for such a function and future research potentially bodes well in this direction. However, due to its fast dissolving characteristics, CH can be used as a disintegrating agent; indeed, this has been previously reported [1,11]. Both polymorphs can be used as direct compression excipients in pharmaceutical formulations, either as whole or partially, and their need for a driving force for compaction is low, allowing ease of use in tableting machines.

Bearing in mind that the work/energy of compression is the product of force and displacement, CH-Sq manifested the lowest energy needed for compression (Figure 10), since it manifested the lowest compression force (or lowest P_K) needed to produce the hardest compacts (Tables 6 and 8). Irrespective of the fact that hard compacts can be made using CH-Sq, a disintegration time of less than one minute, like other CH sources, makes the super-disintegration power of CH independent of tablet hardness.

The dissolution performance of formulations containing CH from different sources was studied and compared with Rodogyl® formulation, which contains a superdisintegrant and a filler with disintegrant action [29] in addition to two drugs: spiramycin and metronidazole. The drug release profile for the two drugs formulated with CH-Cr, CH-Sh, and CH-Sq and the mixture CH-Sh5Sq5 showed that 8.3% CH was sufficient to attain faster and complete drug release compared to Rodogyl®, for both spiramycin and metronidazole (Figures 12 and 13). However, the release of the two drugs in formulations containing CH-Sq was slightly slower than that in formulations containing CH-Cr and CH-Sh. Nevertheless, the mixture containing CH-Sh5Sq5 speeded up the drug release from CH-Sq formulations for both drugs. This makes it necessary to use more than one source of CH in the drug formulations. Thus, a combination of CH sources is recommended to attain optimized tablet physical properties and optimum drug release profile.

Mixtures of CH polymorphs demonstrate an optimized action when compacted. This results in a satisfactory mixture suitable to be compressed and compacted. At first glance, this may suggest to gather remains from sea food packaging industry without classification, but such process shall yield CH with different characteristics depending on the percentage of squid remains involved in extraction process. Such operational procedure will result in lack of consistency in the extracted CH in each produced batch, which would in turn influence the manufactured dosage forms. Consequently, separating the organisms before CH extraction would yield a homogeneous CH powder, enabling pharmaceutical formulator to take advantage when formulating dosage forms by adding the required polymorph in the formulation stage depending on required properties. Alpha polymorph CH is reasonably less expensive than beta polymorph which reserve this polymorph for certain required formulation functions. Indeed, this polymorph can be used to form tablets or films with excellent hardness capable of withstanding mechanical vibrations encountered throughout manufacturing, packaging and transferring steps. The main advantage of CH as an excipient is its ability to form hard compacts with a very short disintegration time. This gives this excipient an advantage compared to excipients based on cellulose derivatives, without any need to add a disintegrating agent to formulations containing CH. Thus, CH is a suitable future excipient. Differences in sources or polymorphic form extend its usage particularly as a solid dosage form excipient suitable for drugs required to be released in a very-short time. CH must be extracted using a single species process, whereby the extracted CH has a well-defined function.

As a general remark, these raw excipients must be exposed to a process of compaction; for example, by using roller compaction, as has been previously reported by our group [8]. The variation in compaction properties between different sources of CH and their mixtures suggest that separation of sources yielding different polymorphs has an advantage in producing excipients with functional properties e.g., as a disintegrant or as filler. CH is chemically and pharmacologically inert and can be considered as a future pharmaceutical excipient.

4. Materials and Methods

4.1. Materials

CH powder from shrimp (CH-Sh) was obtained from G.T.C. Bio Corporation (Qingdao, China). The powder was sieved and the portion of particle size less than 90 μm was used for further analysis, characterization and testing. CH samples from crab (CH-Cr) and squid (CH-Sq) were also obtained from G.T.C. Bio Corporation in the form of flakes and fibers, respectively. They were ground using a ring mill, the powder was sieved and the portion of particle size less than 90 μm was used for further analysis, characterization, and testing. To study the effect of mixing chitin from different sources, CH-Sh was chosen as representative of α-chitin, and mixed with CH-Sq (β-chitin) using the compositions shown in Table 9; these mixtures were characterized and analyzed.

Table 9. Composition of CH mixtures (weight %).

Mixture	CH-Sh	CH-Sq
CH-Sh9Sq1	90	10
CH-Sh7Sq3	70	30
CH-Sh5Sq5	50	50

4.2. Methods

4.2.1. True, Bulk, and Tapped Density Measurement, and Flow Determination

The bulk density of chitin powder samples (shrimp, crab and squid) in g/mL was measured by pouring the powder into a 25 mL volumetric cylinder. The bulk density of all samples was calculated as the ratio of the mass over the volume it occupies. Tapped density measurements were carried out by physical tapping of the cylinder for 100 mechanical taps then dividing the mass over the tapped volume. The cylinder was tapped again for 200 mechanical taps. If the decrease in volume (V_{100}–V_{200}) was less than 2 mL then the V_{200} was considered. If the difference was greater than 2 mL, the increments are repeated, such as the 200 taps, until the difference between succeeding measurements was less than or equal to 2 mL.

The reduction in powder bulk volume due to tapping is considered to be an indication of powder flowability which was evaluated by the Hausner ratio (HR) and Carr Index (CI). As HR and CI increase in value, the flowability is reduced.

HR is calculated using Equation (1), and CI is calculated using Equation (2):

$$\mathrm{HR} = \frac{\rho_{tapped}}{\rho_{bulk}} \tag{1}$$

$$CI = 100 \times \left(\frac{\rho_{tapped} - \rho_{bulk}}{\rho_{bulk}} \right) \quad (2)$$

Flow interpretation criteria based on HR and CI is given in the literature [21]. True density was determined by fitting compaction data, i.e., compaction pressure versus tablet density, according to the method reported by Sun [30].

4.2.2. Water Content Determination

CH samples were analyzed for water content by placing the samples in a porcelain crucible and drying them in a conventional oven at 105 °C to constant weight.

4.2.3. Determination of the Porosity of Powders

Brunauer-Emmett-Teller (BET) pore volume, and pore diameter were determined by physical adsorption of nitrogen gas using a Nova 2200 multi-speed high gas sorption analyzer (version 6.11, Quantachrome Co., Syosset, NY, USA). Samples were subjected to nitrogen gas for adsorption under isothermal conditions at 77 K. The samples were initially placed in a vacuum oven at 60 °C for 24 h. An empty reference cell (Sartorius, analytic, A120s, Göttingen, Germany) and the sample (~500 mg) were placed in the chambers.

4.2.4. FTIR Spectrophotometry

IR spectrophotometry was carried out using Perkin Elmer Spectrum Two UATR FTIR spectrometer (Akron, OH, USA) with a resolution of 4 cm^{-1}, data interval of 2 cm^{-1} and a scan speed of 0.2 cm/s operating in the range of 450–4000 cm^{-1}. The ATR sample base plate was equipped with a Diamond ZnSe crystal; an infrared background was collected for all FTIR measurements. Samples (2–5 mg) were placed on the ATR crystal and a pressure was applied to compress the sample in order to obtain the spectra. The IR spectra of CH samples (raw, ball milled, and roller compacted) were examined.

4.2.5. X-ray Powder Diffraction (XRPD), Crystalline Index (ICR) and Degree of Deacetalyion (DDA)

XRPD test was carried out using an X-ray powder diffractometer (Bruker, Karlsruhe, Germany) in 2-theta range of 2–40° 2θ in reflection mode. The X-ray compartment is a D2 Phaser comprising a copper tube, using Kα X-rays of 300 watts of power at 1.54184 Å wavelength. DIFFRAC.SUITE™ computer software was used to analyze the data obtained.

The crystalline index (I_{CR}) was calculated from the normalized diffractograms. The intensities of the peaks at 110 lattices (I_{110}, at $2\theta \cong 20°$ corresponding to maximum intensity) and at $2\theta \cong 16°$ (amorphous diffraction) were used to calculate I_{CR} using Equation (3) according to Al Sagheer et al. [22]:

$$I_{CR} = 100 \times \left(\frac{I_{110} - I_{am}}{I_{110}} \right) \quad (3)$$

Chitin degree of deacetylation (DDA) is estimated using a method reported Zhang et al. [24], which reported the linear correlation between the crystalline index (I_{CR}) and the degree of chitin deacetylation (DDA) shown in Equation (4):

$$I_{CR}(\%) = 103.97 - 0.7529\, DDA(\%) \quad (4)$$

4.2.6. Scanning Electron Microscopy (SEM)

The morphology of samples was determined using a Inspect F50 SEM (FEI Company, Eindhoven, The Netherlands), operated at an accelerating voltage from 1–30 kV. Samples (≈0.5 mg) were mounted on graphite tape to an aluminum stub. The powder was then sputter-coated with platinum (Emitech K550X, Qourum Technology, Lewes, UK).

4.2.7. Compression Analysis

CH powder samples (crab, shrimp, squid, and their mixtures) were compressed into tablets using an instrumental single punch bench top tablet press (GTP-1, Gamlen Tablet Press Ltd., Nottingham, UK). Compression was carried out at a punch speed of 60 mm/min by applying five different loads: 100, 200, 300, 400, 500 kg. The samples poured into the die of the GTP had a common weight of 100 ± 1 mg. The diameter of the die was 6 mm. The machine was run by software to display the force-displacement (F-D) curve. Kawakita and Heckel models (Equations (5) and (7), respectively) were utilized to describe the compression analysis of the powders [31,32].

Kawakita analysis describes a linear relationship between the ratio P/C and P, Equation (5), where P is the pressure in MPa, and C is the volume reduction. From the slope and intercept of this relationship, constants '*a*' and '*b*' can be deduced; '*a*' represents the maximum volume reduction that can be attained by the powder. '1/*b*' or 'P_K' is another parameter that represents the force required to reduce the powder bed volume to half its maximum value:

$$\frac{P}{C} = \frac{P}{a} + \frac{1}{ab} \tag{5}$$

Volume reduction (*C*) is calculated using Equation (6):

$$C = 1 - \frac{\rho_b}{\rho_c} \tag{6}$$

where ρ_b and ρ_c are bulk and compact densities (kg/m^3), respectively.

Heckel analysis describes a relationship between the logarithm of the inverse of compact porosity (ε) and the pressure applied, as shown in Equation (7):

$$\ln\frac{1}{\varepsilon} = KP + A \tag{7}$$

Porosity is calculated using the Equation (8):

$$\varepsilon = 1 - \rho_r \tag{8}$$

where ρ_r is the relative density of the compact, and is calculated using the Equation (9):

$$\rho_r = \frac{\rho_c}{\rho_T} \tag{9}$$

where ρ_c and ρ_T are compact and true densities (kg/m^3), respectively.

The inverse of the Heckel equation slope or '1/*K*' is an important parameter which assigns Sh5Sq5the type of deformation of materials; whether plastic/elastic or brittle-fracture. This parameter is called the yield pressure and is signified by the symbol 'P_Y'.

4.2.8. Application of Chitin Excipient Using Metronidazole/Spiramycin as Model Drugs

For dissolution analysis, a sample (100 g) comprising 56.3 g of spiramycin, 30.1 g of metronidazole, 1.9 g of sorbitol powder (solubilizer), and 8.7 g of CH (either CH-Cr, CH-Sh, CH-Sq, or the mixture CH-Sh5Sq5) were granulated with 100 mL of 3% *w*/*w* povidone K30 aqueous solution. The granules were dried at 60 °C for 3 h, then sieved at mesh #18 (particle size = 1 mm). A sample of 415 mg was compressed using a Manesty single punch tablet machine (Manesty F3 single stroke tablet press; West Pharma Services Ltd., Dorset, UK) at an applied force of 35 kN using a 10 mm circular shallow biconvex punch. Rodogyl® tablets were used as a reference.

Apparatus II (USP) dissolution tests were performed using an Erweka DT6 system (Langen, Germany) with paddles running at 50 rpm. 900 mL of 0.1 N HCl was used as the dissolution medium. The amount of drug released was analyzed by measuring the absorbance using a UV spectrophotometer (LABINDIA UV/VIS, UV 3000, Maharashtra, India) at a wavelength of 320 nm for metronidazole, and of 230 nm for spiramycin.

For crushing force and disintegration tests, 415 mg of chitin powder (either CH-Cr, CH-Sh, CH-Sq, or the mixtures CH-Sh9Sq1, CH-Sh7Sq3, and CH-Sh5Sq5) was compressed using the Manesty single punch tablets machine at an applied force of 35 kN using a 10 mm circular shallow biconvex punch. The average crushing force and disintegration time of 10 chitin tablets produced were measured using crushing force tester (Pharma Test PTB 311E. Hainburg, Germany) and disintegration tester (CALEVA, Dorest, UK), respectively.

5. Conclusions

Chitin is chemically and pharmacologically inert and can be considered as a future pharmaceutical excipient especially for immediate release tablets, as it can form hard compacts with a very short disintegration time. Depending on the source, CH varies marginally in compression properties. Consequently, CH must be extracted using a single species process. Variations in polymorphic form would not hinder the use of chitin as a pharmaceutical excipient. In order for an excipient comprising chitin to perform ideally in compression and compaction properties, α- and β- polymorphic types would be mixed. In this perspective, α-CH will provide the high bulk density, good flowability, and low water absorption aspects, whereas the β-polymorph will provide the mechanical strength. These aspects were attributed to the difference in crystallinity, fibrous nature, pore size and pore volume of α- and β-CHs.

Author Contributions: Conceptualization, D.A.F. and A.A.B.; data curation, L.A.-H. and I.R.; formal analysis, L.A.-H. and D.A.F.; investigation, L.A.-H., D.A.F. and A.A.B.; methodology, L.A.-H. and D.A.F.; project administration, A.A.B.; software, L.A.-H.; supervision, D.A.F.; writing—original draft, L.A.-H., D.A.F., I.R. and A.A.B.; writing—review & editing, B.Z.C. and A.A.B. All authors have read and agreed to the published version of the manuscript.

Funding: This research received no external funding.

Acknowledgments: The authors would like to thank The University of Jordan and University of Greenwich for their ongoing support. The authors also wish to thank the Jordanian Pharmaceutical Manufacturing Co. (JPM) for providing materials, laboratory and testing facilities. The authors would like also to thank Hatim Al-Khatib from the School of Pharmacy, University of Jordan for providing lab facilities for some sample analysis.

Conflicts of Interest: The authors declare no conflict of interest.

References

1. Daraghmeh, N.H.; Chowdhry, B.Z.; Leharne, S.A.; Al Omari, M.M.; Badwan, A.A. Chitin. *Profiles Drug Subst. Excip. Relat. Methodol.* **2011**, *36*, 35–102. [CrossRef]
2. Roberts, G.A.F. *Chitin Chemistry*; Palgrave Macmillan Press Ltd.: London, UK, 1992; pp. 54–115.
3. Budavari, S. (Ed.) *The Merck Index*, 13th ed.; Merck and Co.: Kenilworth, NJ, USA, 2001.
4. Uragami, T.; Tokura, S. *Material Science of Chitin and Chitosan*; Springer: Berlin, Germany; Kodansha: Tokyo, Japan, 2006.
5. Bellich, B.; D'Agostino, I.; Semeraro, S.; Gamini, A.; Cesàro, A. "The Good, the Bad and the Ugly" of Chitosans. *Mar. Drugs* **2016**, *14*, 99. [CrossRef]
6. Badwan, A.A.; Rashid, I.; Al Omari, M.M.; Darras, F.H. Chitin and Chitosan as Direct Compression Excipients in Pharmaceutical Applications. *Mar. Drugs* **2015**, *13*, 1519–1547. [CrossRef]
7. Abu Fara, D.; Rashid, I.; Al-Hmoud, L.; Chowdhry, B.Z.; Badwan, A.A. A New Perspective of Multiple Roller Compaction of Microcrystalline Cellulose for Overcoming Re-Compression Drawbacks in Tableting Processing. *Appl. Sci.* **2020**, *10*, 4787. [CrossRef]
8. Abu Fara, D.; Al-Hmoud, L.; Rashid, I.; Chowdhry, B.Z.; Badwan, A.A. Understanding the Performance of a Novel Direct Compression Excipient Comprising Roller Compacted Chitin. *Mar. Drugs* **2020**, *18*, 115. [CrossRef] [PubMed]

9. Patole, R.K.; Sav, A.K.; Amin, P.D. A Novel Directly Compressible Co-Processed Excipient for Sustained Release Formulation. *J. Appl. Pharm. Sci.* **2013**, *3*, 89–97.
10. El-hefian, E.A.; Yahaya, A.H. Rheological study of chitosan and its blends: An overview. *Maejo Int. J. Sci. Technol.* **2010**, *4*, 210–220.
11. Rashid, I.; Eftaiha, A.F.; Badwan, A.; Al-Remawi, M. Chitin–Silicon dioxide coprecipitate as a novel superdisintegrant. *J. Pharm. Sci.* **2008**, *97*, 4955–4969. [CrossRef] [PubMed]
12. Alonso, B.; Belamie, E. Chitin–Silica Nanocomposites by Self-Assembly. *Angew. Chem. Int. Ed.* **2010**, *49*, 8201–8204. [CrossRef]
13. Rojas, J.; Buckner, I.; Kumar, V. Co-proccessed excipients with enhanced direct compression functionality for improved tableting performance. *Drug Dev. Ind. Pharm.* **2012**, *38*, 1159–1170. [CrossRef]
14. Chaheen, M.; Sanchez-Ballester, N.M.; Bataille, B.; Yassine, A.; Belamie, E.; Sharkawi, T. Development of co-processed chitin-calcium carbonate as multifunctional tablet excipient for direct compression. *J. Pharm. Sci.* **2018**, *107*, 2152–2159. [CrossRef] [PubMed]
15. Chaheen, M.; Bataille, B.; Yassine, A.; Belamie, E. Development of Co-processed Chitin-Calcium Carbonate as Multifunctional Tablet Excipient for Direct Compression, Part 2: Tableting Properties. *J. Pharm. Sci.* **2019**, *108*, 3319–3328. [CrossRef] [PubMed]
16. Daraghmeh, N.; Chowdhry, B.Z.; Leharne, S.A.; Al Omari, M.M.H.; Badwan, A.A. Co-Processed Chitin-Mannitol as a New Excipient for Oro-Dispersible Tablets. *Mar. Drugs* **2015**, *13*, 1739–1764. [CrossRef]
17. Gana, F.Z.; Rashid, I.; Badwan, A.; Alkhamis, K.A. Determination of Solid-State Acidity of Chitin–Metal Silicates and their Effect on the Degradation of Cephalosporin Antibiotics. *J. Pharm. Sci.* **2012**, *101*, 2398–2407. [CrossRef] [PubMed]
18. Wei, Y.; Huang, Y.H.; Cheng, K.; Song, C. Investigations of the Infuences of Processing Conditions on the Properties of Spray Dried Chitosan Tripolyphosphate Particles loaded with Theophylline. *Sci. Rep.* **2020**, *10*, 1155. [CrossRef]
19. Abu Fara, D.; Rashid, I.; Alkhamis, K.; Al-Omari, M.; Chowdhry, B.Z.; Badwan, A.A. Modification of a-lactose monohydrate as a direct compression excipient using roller compaction. *Drug Dev. Ind. Pharm.* **2018**, *44*, 2038–2047. [CrossRef]
20. Abu Fara, D. A comparison of industrial lactose obtained by roller compaction with spray dried lactose and α-lactose monohydrate using compression analysis techniques. *J. Excip. Food Chem.* **2019**, *10*, 41–54.
21. The United States Pharmacopeia. Powder Flow, Page 618 of PF 28(2) [March–April 2002]. 9D. Available online: https://www.usp.org/sites/default/files/usp/ducument/harmonization/gen-chapter/g05_pf_30_6_2004.pdf (accessed on 15 October 2020).
22. Al Sagheer, F.; Al-Sughayer, M.; Muslim, S.; Elsabee, M. Extraction and characterization of chitin and chitosan from marine sources in Arabian Gulf. *Carbohydr. Polym.* **2009**, *77*, 410–419. [CrossRef]
23. Jang, M.K.; Kong, B.G.; Jeong, Y.I.; Lee, C.H.; Nah, J.W. Physicochemical characterization of α-chitin, β-chitin, and γ-chitin separated from natural resources. *J. Polym. Sci. A Polym. Chem.* **2004**, *42*, 3423–3432. [CrossRef]
24. Zhang, Y.; Xue, C.; Xue, Y.; Gao, R.; Zhang, X. Determination of the degree of deacetylation of chitin and chitosan by X-ray powder diffraction. *Carbohydr. Res.* **2005**, *340*, 1914–1917. [CrossRef]
25. Hepburn, H.R.; Chandler, H.D. Tensile Mechanical Properties and Transconformational Changes of Chitins. In Proceedings of the First International Conference on Chitin/Chitosan, MIT Sea Grant Program, Cambridge, MA, USA, 11–13 April 1977; pp. 124–143.
26. National Marine Fisheries Service. Fisheries of the United States, 2018. U.S. Department of Commerce, NOAA Current Fishery Statistics No. 2018; 2020. Available online: https://www.fisheries.noaa.gov/national/commercial-fishing/fisheries-united-states-2018 (accessed on 15 October 2020).
27. Kaya, M.; Lelešius, E.; Nagrockaitė, R.; Sargin, I.; Arslan, G.; Mol, A.; Baran, T.; Can, E.; Bitim, B. Differentiations of Chitin Content and Surface Morphologies of Chitins Extracted from Male and Female Grasshopper Species. *PLoS ONE* **2015**, *10*, e0115531. [CrossRef] [PubMed]
28. Alakayleh, F.; Rashid, I.; Al Omari, M.M.; Al-Sou'Od, K.; Chowdhry, B.Z.; Badwan, A.A. Compression profiles of different molecular weight chitosans. *Powder Technol.* **2016**, *299*, 107–118. [CrossRef]
29. Abrégée, F. Spiramycine/Metronidazole Biogaran 1,5 MUI/250 Mg Cp Pellic. Available online: https://www.vidal.fr/Medicament/spiramycine_metronidazole_biogaran-68905.htm (accessed on 15 October 2020).
30. Sun, C.C. A Novel Method for Deriving True Density of Pharmaceutical Solids Including Hydrates and Water-Containing Powders. *J. Pharm. Sci.* **2004**, *93*, 646–653. [CrossRef] [PubMed]

31. Ilkka, J.; Paronen, P. Prediction of the compression behaviour of powder mixtures by the Heckel equation. *Int. J. Pharm.* **1993**, *94*, 181–187. [CrossRef]
32. Nordström, J.; Klevan, I.; Alderborn, G. A Particle Rearrangement Index Based on the Kawakita Powder Compression Equation. *J. Pharm. Sci.* **2009**, *98*, 1053–1063. [CrossRef]

Sample Availability: Samples of the compounds are not available from the authors.

MDPI
St. Alban-Anlage 66
4052 Basel
Switzerland
Tel. +41 61 683 77 34
Fax +41 61 302 89 18
www.mdpi.com

Molecules Editorial Office
E-mail: molecules@mdpi.com
www.mdpi.com/journal/molecules

www.ingramcontent.com/pod-product-compliance
Lightning Source LLC
Chambersburg PA
CBHW061615210926
51850CB00009B/2331
* 9 7 8 3 0 3 6 5 1 6 7 6 9 *